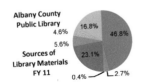

Albany County
Public Library

Sources of
Library Materials
FY 11

16.8%
46.8%
4.6%
5.6%
23.1%
0.4%
2.7%

- Donated Items
- Individual Cash Gifts
- Replacement Fees
- ACPLF
- City Sales Tax
- County Sales Tax
- Friends of the Library

EARTH

THE OPERATORS' MANUAL

ALSO BY RICHARD B. ALLEY

The Two-Mile Time Machine: Ice Cores, Abrupt Climate Change,
and Our Future

EARTH

THE OPERATORS' MANUAL

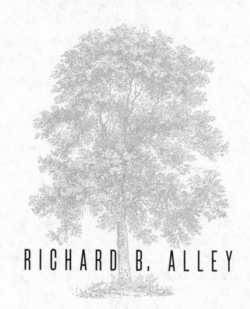

RICHARD B. ALLEY

W. W. NORTON & COMPANY

NEW YORK · LONDON

For information about permission to reproduce selections from this book,
write to Permissions, W. W. Norton & Company, Inc.,
500 Fifth Avenue, New York, NY 10110

For information about special discounts for bulk purchases, please contact
W. W. Norton Special Sales at specialsales@wwnorton.com or 800-233-4830

Manufacturing by RR Donnelley, Harrisonburg, VA
Book design by Chris Welch
Production manager: Anna Oler

Library of Congress Cataloging-in-Publication Data

Alley, Richard B.
Earth : the operators' manual / Richard B. Alley. — 1st ed.
p. cm.
Includes bibliographical references and index.
ISBN 978-0-393-08109-1 (hardcover)
1. Energy development—Environmental aspects—History.
2. Renewable energy sources. 3. Global warming. I. Title.
TJ163.2.A43 2011
621.04209—dc22

2010054016

W. W. Norton & Company, Inc.
500 Fifth Avenue, New York, N.Y. 10110
www.wwnorton.com

W. W. Norton & Company Ltd.
Castle House, 75/76 Wells Street, London W1T 3QT

1 2 3 4 5 6 7 8 9 0

To my wife, Cindy, and daughters, Janet and Karen,
with thanks and hope.

CONTENTS

PART III. THE ROAD TO TEN BILLION SMILING PEOPLE

PREFACE

Today, almost seven billion of us live side by side with whales and woodlands because we get almost all of the energy we use from oil, coal, and natural gas. If we suddenly quit using these fossil fuels and returned to burning whale oil and wood, we wouldn't come close to powering enough of our tractors, trucks, and irrigation pumps to feed us all.

The good we get from fossil fuels is a mixed blessing, though. If we keep using them, and accept the risks of oil-well blowouts and mountaintop removal, for long enough, the fossil fuels will run out—we are burning them about a million times faster than nature saved them for us. We thus must decide whether to burn most of the fossil fuels and then look for replacements, or to learn while we burn and while we still have a fossil-fuel safety net in the ground.

Our decision must be made under the shadow of global warming from the CO_2 released by burning fossil fuels. Sensors on heat-seeking missiles are affected by the interaction between CO_2 and energy transfer through the air, and so is the climate. The atmosphere really doesn't care whether we study it for warring or warming, and military as well as civilian physics research shows that CO_2 matters. We thus have high scientific confidence that continuing to burn fossil fuels will cause large climate changes, which will make life more difficult for poor people living in hot places now, and for most people in future generations.

But if CO_2 emissions especially harm poor people, are we wiser to reduce emissions, or to reduce poverty by helping people become wealthier? Not surprisingly, economics answers "Yes"—do some of each. How much of each may depend on national security, jobs, ethics, and insurance, as well as economics.

You will hear a lot of shouting from the wings that partially drowns out discussion of these important issues, but this shouting is nothing new. President Abraham Lincoln's administration found ways to get good scientific advice through the noise, and following his example still works. I will try to stick closely to Lincoln's example in giving you the best scientific insights. There are things that science doesn't know, more things that I don't know, and I am far from infallible, but much of the science really is solid, and I will tell you when it isn't.

Lincoln also showed us part of the solution for powering the planet. Earth offers vast, sustainable energy resources with the potential to improve the economy and generate a lot of fortunes, including the wind that so intrigued Lincoln on the Illinois prairie long before his presidency.

Being a truly honest broker on such complex topics may be impossible, but I will do my best for you. This could be easier for me than for most people: I enjoyed working for an oil company and benefited from its largesse, my political registration is right of center, and I have won scientific awards for helping show just how bizarrely Earth's climate can behave without any interference from us. But I also helped the U.S. National Academy of Sciences and played a small role in the Nobel Prize–winning effort of the United Nations Intergovernmental Panel on Climate Change (IPCC), where my knowledge of Earth's history and behavior contributed to the confident realization that the CO_2 from our fossil-fuel burning is highly likely to change the world in fundamental ways that will increasingly make life harder for future generations. Our two lovely daughters give me a personal as well as a professional stake in the search for a stable, sustainable world. Onward!

PART I.

THE BURNING QUESTION

1

Prepare to Come About

Synopsis. We humans have always used energy and always will. We now rely greatly on fossil fuels, which promise to make our lives much harder before they run out. But there are plenty of ways to get rich and save the world by remaking our energy system.

May you live in interesting times. —Old curse[1]

POWER ON

More and more of us are living better than ever before. In most of the world, an expectant mother can be reasonably confident that she will deliver a healthy baby, who will parent the next generation and live long enough to help educate the generation after that. Wars continue, but the all-out disaster of World War II is ancient history for many of us, and a fading memory for the rest. We have used our accumulated knowledge and wisdom, and the things we've built, to convert a world that might support a few million hunter-gatherers[2] into home for over six billion of us, heading for nine or ten billion. When problems arise, we usually invent and cooperate to solve them. We have never fully agreed on the purpose of our existence (and we are not likely to agree in the near future), but if we approve of any form of "The greatest good for the greatest number," then this really may be the best time in history.

And yet, we can easily believe we are cursed to live in these interesting times, with disasters waiting at every turn. Perhaps a billion or more people—one in six of us—exist in such poverty or violence that they cannot reasonably expect their children to live long and prosper. Various accounting methods suggest that we are using, and often using up, nearly half of everything that the planet makes available to us and to all other species, with rising population and expectations pushing us rapidly toward using 100 percent.

We have removed perhaps 90 percent of the large fish from the ocean; in fact, we have no idea what a natural ocean ecosystem looks like, because we fished out so many species before scientists learned to see what is going on.[3] Roughly one-third of the land surface not covered by ice sheets is now used for cropland or grazing, with logging extending our impact.[4] Water is essential to us, but in many places a large fraction of the water we use is not being replaced, as we pump it out of old deposits in the ground or melt it from old deposits in glaciers much faster than new rainfall or snowfall supply more.[5] The soil that grows our crops is being washed away far faster than nature can produce more,[6] so farming is becoming more difficult, especially as we use up the "easy" deposits of phosphate for fertilizer.

With human population expected to increase, many of us not getting even the minimum that most civilized people believe is needed for a proper life, and almost everyone hoping to improve their lot, it takes an optimist to believe that the demands on the planet will "only" double. If our use is already approaching half of everything supplied by Earth, where will the rest come from? And what does our growing use mean for the other species that share the planet with us, and their ecosystems on which we rely?

A pessimist can easily look past our successes in advancing knowledge and skills and infrastructure and healthy people, and see the history of disasters, wars, environmental refugees, starvation, and failure.[7] Humanity has had more than enough "wins" to show that success is possible, but more than enough failures to show that success is far from guaranteed.

If water runs out, we can desalinate and pump. If soil runs out, we can grow crops hydroponically without soil, or we might dig the dirt out of the reservoirs behind dams and spread it back on the fields while adding key nutrients, much as a home gardener builds raised beds. If phosphate becomes scarce, we can mine lower-grade ores and use our knowledge of chemistry to enhance them.

But desalination uses energy, and lots of it. So does building a hydroponic system, and mining a low-grade ore. So do plowing and shipping, heating and cooling, flying and driving, and so many other things we do. We already rely heavily on energy use to solve our problems: powering pumps to pull water out of the ground to grow our crops, fueling huge shovels to dig phosphate and ships to send it—often great distances— to the farm fields, where fuel-filled tractors spread it, and much more. And our energy use is arguably the most unsustainable part of our lives.[8] Roughly 85 percent[9] of the world's primary-energy production today is from fossil fuels—oil, coal, and natural gas—with only 15 percent from nuclear, hydropower, wind, or other sources. We are using the fossil fuels approximately a million times faster than nature saved them for us, and they will run out (see chapter 4).

We apply cheap energy to almost all of our problems, a "silver bullet" to slay the dragons that trouble us. But if we continue on our present course, we will run out of silver bullets. Our current energy system cannot last. Worse, the by-products of that energy system threaten to change the planet in ways that will make our lives much harder—if we burn all of the fossil fuels before we learn how to use new energy sources, we will have greatly increased the difficulty of our education.

Fortunately, we have a golden bullet in our pocket—our collective cleverness. The amount of energy that Earth makes available, sustainably, dwarfs the amount that we now use, and dwarfs demand for the foreseeable future. Sunshine from just the desert floors of Arizona would power the whole United States, and from the Sahara could power the rest of the world's people, with huge amounts left over. The technologies required are not science fiction—in fact, they already exist or soon will, and some of them are decades or centuries old.

But big projects take a long time to complete. In 1971, the then U.S. president Richard Nixon declared war on cancer. He proclaimed, "The

Figure 1.1 Karen Alley sailing: "Prepare to come about."

time has come in America when the same kind of concentrated effort
that split the atom and took man to the moon should be turned toward
conquering this dread disease. Let us make a total national commitment to achieve this goal."[10] Almost forty years later, although the goal
remains elusive, huge progress has been made. But the first decade or
two of this "war" did not produce the heady victories we hoped for.
Very simply, cancer proved to be a hard problem.

I believe that our energy problem will prove easier to conquer than
cancer, as I discuss in part III of this book. But the effort required is
easy to underestimate. Our oil and coal companies are so good at what
they do that we easily forget the sheer immensity of their achievements.
We occasionally are reminded of the near impossibility of quickly
containing the oil from even one spill when an oil well blows out or
a supertanker runs aground in an ecologically sensitive area, but drill
rigs and tankers are at work continuously. The efficiency with which
some coal-mining companies change the very face of Earth—remov-

ing the tops of mountains, filling valleys with the non-burnable rocks, and moving on—to extract the energy stored beneath is mind-boggling.

Changing the direction of an oil tanker, or any large ship, is a slow process. By the time the pilot sees an obstacle ahead, mutters, "Oh, (very bad word)," and tries to make a correction, a collision may be unavoidable. The inquiry into the sinking of the *Titanic* revealed that when the crew sighted the iceberg, it was already too late to steer to safety: "We had the order, 'Hard-a-starboard,' and she just swung about two points when she struck."[11]

A remarkably wide range of thinkers, scientists, and engineers now see the ship of our energy use on a collision course that will seriously harm our future, unless a correction is begun soon. The change is still possible—we are not the *Titanic*, doomed to hit and sink—but the longer we wait, the harder the change will be, and the more damage we will do as we sideswipe the unpleasant reality. As my younger daughter Karen says to her crew when sailing her Sunfish in much more pleasant times, "Prepare to come about."

THE OPERATORS' MANUAL

When all else fails, read the operators' manual.[12]

I routinely board very large airplanes without having the vaguest idea of how they work or how to fly them—I'm a passenger, and happy to ride along. I take buses, and taxis, and ride shotgun when others drive, without sweating it. But put me behind the wheel of the car, and I suddenly need to know a whole lot more. Manual or automatic? Mirrors adjusted right? Where is the knob for the headlights? Emergency brake, just in case? When I buy a new car, I actually do skim over the manual. And when the brakes went out on the geology van coming down the mountainside during my student days, knowing what to do was a distinct comfort in a tight spot.

I'm educated as a geologist, with climate and ice and water and a

Figure 1.2 Final assembly of the drill and protective dome for deep ice coring at GISP2 in central Greenland. Records from ice cores tell us the history of the atmosphere, including natural and human-caused changes.

bunch of engineering thrown in. I have been an academic most of my professional life, but I worked for an oil company for a bit and enjoyed both the money and the smart people doing interesting things there. My experience was similar to that of many geologists, who for more than a century have been getting good jobs to help people find valuable things in the Earth (oil, coal, diamonds, gold). Geologists also get jobs to help people avoid hazards (volcano! landslide!), and to be entertaining (dinosaurs!). Recently, however, we have been asked to take on another job.

I often have taught geomorphology, the science of why Earth's surface looks the way it does, and the task has been getting harder. More and more, the processes that made Earth's landscape in the past are not the processes that students observe today, because the dominant processes today are "us." We now move more rocks and dirt than nature does—all of the natural landsliding of hillsides and mud washing down rivers and dust blowing through the air are small compared to the work of our bulldozers and steam shovels.[13] Many home gardeners in suburbia are convinced that they have poor soil, and most of them are right—builders often dig a hole for a foundation and perhaps a basement, spread the rock and clay to smooth the lot, throw a bit of "topsoil" on just thick enough to grow grass, and call it a job well done. Digging a hole for a tomato plant then means tapping into a mess of whatever came out to make room for the foundation or basement, plus some nails and shingle pieces and other construction debris. A geomorphology student wanting to learn about nearby natural soils may have difficulty, because most of the easy-to-visit soils have been so greatly disturbed by humans.

A reporter called recently and asked how long it would take Earth to "forget" humanity if we suddenly disappeared. In some sense, we are now unforgettable—the human-caused plant and animal extinctions have emptied biological "jobs" that will be filled over many millions of years by creatures who will owe their existence to us wiping out the competition. We have pumped oil and gas out of the ground that had been there for hundreds of millions of years, through holes that may not be eroded away for additional hundreds of millions of years. The human

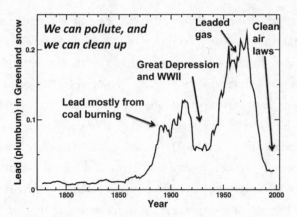

Figure 1.3 Recent history of concentration of lead pollution in Greenland snow.

"layer" of plastic and aluminum foil and heavy metals may be recogniz-able in sedimentary piles hundreds of millions of years from now.

In Greenland, I helped collect ice cores to learn the history of the atmosphere. The folks who study the trace chemicals in the ice can see the clear signal of mining of lead—*plumbum*—used to supply the plumb-ing of the Roman Empire. The post-Roman drop in lead level is fol-lowed by a rise beginning with the Industrial Revolution, a drop for the Great Depression, a huge rise with the use of leaded gasoline and paint after World War II, and then a great drop when we became concerned about lead poisoning and got serious about cleaning up.[14] The lead will be in the ice for a long time if we don't melt it out, and our lead will per-sist in the muds of lakes and the sea floor even if we do melt the ice.

With the amount of stuff we use, and the amount of the world we occupy, we are no longer passengers napping in the back seat of the car. We are everywhere, and changing everything. Hence, many environ-mental scientists are now involved in figuring out what we are doing, how to operate a remarkably complex and involved Earth system, and how to make the ride as enjoyable as possible. This operators' manual is not finished yet—not even close—although we know an amazing amount more than we did even a few years ago, with knowledge com-ing in rapidly. I am proud to have played a small part in this effort. But

I'm also concerned that a lot of people, including some of those who are making laws, still think that they are sitting in the back of the car, looking out the window and enjoying the ride.

I can't possibly cover all of what we know about how the Earth system operates in this one book. Instead, I will focus on a few big ideas: 1) we humans have always used energy and always will, 2) the road we're on now will get us into trouble, and 3) there are plenty of ways to get rich and save the world by remaking our energy system. Along the way, we'll look in on some amazing moments in Earth's history, visit a few of our ancestors, and check in on some fortunes in the making.

2

Burning to Learn

Synopsis. Humans burn things, and burning was probably required to make us human. Despite many problems, burning has helped us, and the more we have burned, the better off we have been in the past.

The mind is not a vessel to be filled but a fire to be kindled. —Plutarch[1]

BRAIN POWER

Have you figured out how to get someone or something else to do most of your work for you?

If you answered yes, congratulations! You're human!

Many things set us apart from most of the other creatures on our planet. Text-messaging and tool-making are a lot easier with an opposable thumb. Baldness is not just about looking distinguished (said the rapidly balding man . . .)—lack of hair over large expanses of our bodies lets us run marathons without overheating, a useful ability for a hunter-gatherer chasing a meal or trying not to become dinner for a pursuing jungle cat. Upright posture lets us wear ties to business meetings, but also lets us see what we're chasing or what is chasing us, and lets us throw rocks or swing clubs at predators or carry our tools and tots to the next meal. Our big brains allow us to develop language and agriculture better than any other creatures on the planet.

But how do we afford those big brains? When we are resting, 25 percent of our energy use goes to the brain, compared to just 8 percent averaged across the apes.[2] A resting newborn baby's brain accounts for a whopping 60 percent of energy use. Our brains, like older desktop computers, use a lot of energy even when idling.

How did our ancestors find the energy for their on-board computer power? One way is that, evolutionarily, they skimped on stomachs.

Hay for dinner, anyone? A cow can do it, but it takes four stomachs and a lot of chewing the cud in the evening. Much of the energy the cow gets from the grass goes to keeping those stomachs alive and working, and keeping the cow alive and chewing to aid in the digestion.

Our closer relatives, such as chimpanzees, eat a variety of things, and have relatively longer guts than we do. A chimp can successfully digest foods such as leaves that would pass almost unaltered through me. But some of the energy the chimp gets from digesting that food is used to

Figure 2.1 Cows can be curious creatures, as demonstrated by these Penn State Holsteins investigating the filming of Earth—The Operators' Manual. *But cows are not the brightest lights in the intellectual firmament, with more stomach than brains, at least in part because of the difficulty of digesting their diet.*

keep the chimp's gut alive to do the digesting, leaving less to power the chimp's brain or other parts of the body. Adding a longer gut would allow more digestion, but with diminishing returns as that longer gut used more energy itself.

Therefore, someone with a shorter gut, with fewer gut cells to keep alive, can divert the energy those cells would have used to power their brain. But a gut that is too short would allow undigested food to be dumped out the rear end, wasting energy, so a shorter gut saves energy only if it is supplied with a more-digestible dinner.

Meat-eating is one solution. Let some other creature do the hard work of collecting and chewing and digesting vegetables and reassembling their parts into something more useful to an animal, and then eat that other creature. Our ancestors were clearly using tools and eating meat more than 2 million years ago, and their brain sizes increased between about 1.9 and 1.6 million years ago, when there is increasing evidence of increasing meat-eating.[3]

A fascinating new hypothesis links our very humanity to another way of improving digestibility: cooking. If our ancestors used fire to start digesting food outside their bodies, a shorter gut inside could finish up the job, allowing them to be well fed and smart at the same time.[4]

We know surprisingly little about the advantages of cooking, considering how much cooking we do, but recent research points to an important role for cooking in improving digestibility as well as in killing diseases.[5] In one experiment, mice that were fed cooked meat gained almost one-third more weight than those eating raw meat. In another study, switching pythons from a raw-hamburger diet to a cooked-hamburger diet reduced the energy they used to digest their dinners by almost one-fourth. Furthermore, people eating primarily raw foods tend to lose weight. Some paleoanthropologists have speculated that cooking was a prerequisite for our spurt of brain growth beginning 1.9 million years ago, although evidence of organized, consistent fire use at that time remains sparse at best; perhaps more likely, fire contributed to the next big spurt of brain growth, within the last half-million years, leading to the large size of our brains today.

Figure 2.2 Native American fire circle, Wind River Mountains, Wyoming. A skilled cook could use this to prepare food efficiently, conserving firewood, but the energy used in cooking was similar to the energy available to people in the cooked food.

Cooking thus looks like a great deal for people. Burn some trees, use the heat released to make the food more digestible, and we get lots of energy from the food with a small investment in supporting our guts. The extra energy powers our minds, and we're human.

But what does this mean for the trees? Perhaps 10 percent or less of the heat from an open cooking fire actually warms the food.[6] When I was on a field expedition to study the ice-age deposits of glaciers of the Wind River Range in Wyoming, the archaeologist Dave Putnam showed us neat, small, efficient stone circles that native Americans had used, probably about 10,000 years ago. Carefully tended, fires in such fire rings might have achieved 20 percent efficiency or more, much closer to the efficiency of a modern stove. If you cooked grain on such an efficient fire ring, the energy you obtained from eating dinner might be slightly more than the energy used in cooking dinner.[7] For a sloppy fire, you might use twice as much energy in cooking as you obtain from eating the cooked food. (Don't even think about how inefficient cooking a marshmallow over a campground fire is.) And most of the energy from our food goes to the body, with only a little bit of the extra energy making our big brains possible. So, a lot of trees had to give their lives, and a lot of termites were deprived of their dead-wood dinner, to make us smart.

Some of the science discussed here is new, and research is ongoing. But it appears that being human means relying on energy from outside—we learn because we burn. And our reliance on the energy of others, and the difficulty of supplying that energy from our traditional sources for billions of people, are daunting indeed.

WHAT IS ENERGY, ANYWAY?

We can usefully divide the universe into two categories: stuff, and the ability to make the stuff do something. You might call these "matter" and "energy." Albert Einstein is famous in part because he showed that matter, or mass, and energy are really the same thing—wind up an old-style alarm clock and you have made the spring a tiny, tiny, tiny bit more massive, and the extra mass will be turned to energy to move the clock's hands and make the ticktocks over the next day.[8] Blow up an atomic bomb, and although you would have difficulty measuring the amount of mass lost, you will easily see the energetic effects of that loss. For most ordinary purposes, though, it is sufficient to treat mass and energy separately.

If you put the right chemicals together in the right pattern, you can make a diamond that looks pretty and scratches things really well, or a hammer that can break things. Use the hammer to break the diamond into tiny bits, toss the dust out the window, and the diamond is no longer useful to you. The stuff is still there—mass is conserved, as the physicists say—but you have to gather the mass in the right way to make it valuable.

Energy is much the same. If you store a lot of energy where you want it, you can use it to make the stuff do what you want. Put a lot of heat energy in a small place, and you can cook a turkey. Or you can let the heat drive the expansion of gases to drive a piston to drive a car to drive you across the country to visit grandma for a turkey dinner. When you are done, the energy is not gone, but it is spread out—you heated the room while cooking the turkey, and then the heat from the warmer room spread out into the outdoors and eventually was radiated to space. Once the heat energy is too spread out, you can't use it. Just like the dust from your broken diamond, the energy exists but it isn't valuable any more.

MEASURING THE BURN

Comparing the energy used in a cooking fire to the energy contained in the cooked foods is easier if we can put numbers on the energy amounts. But how do we measure energy? One way is to use our dining habits.

The U.S. Food and Drug Administration (FDA) suggests appropriate foods and nutrients that we should eat. These FDA recommendations are often given assuming a diet of 2000 calories per day. "Calorie" is a measure of the energy available, and "calories per day" is the rate at which you use this stored energy, converting it to spread-out heat. How rapidly you use stored energy is also called "power"—your personal power output is 2000 calories per day.[9]

Most other branches of modern science, and nonscientists in most other countries, don't measure energy storage in calories, but in joules (named after an English physicist who started out as a brewer, and who experimented with electricity in part by shocking his brother and letting his brother shock him.)[10] The scientists who use joules for energy storage also use watts for power (abbreviated W, and named after James Watt, a Scottish inventor of steam-engine fame, who according to his epitaph "enlarged the resources of his Country, increased the power of Man, and rose to an eminent place among the most illustrious followers of science and the real benefactors of the World").[11]

If you follow the FDA guidelines and burn 2000 calories per day inside of you, your personal power output averages 100 W. Not that long ago, when almost all of us lit our houses with heat-until-they-glow incandescent lightbulbs, everyone understood what watts are—a big turned-on lightbulb was using 100 W. (With modern compact fluorescents, 100 W is four or five bright lightbulbs.)[12] If you ever touched a 100 W incandescent lightbulb that had been on for a while, you know that it was hot! So you shouldn't be surprised that you and the lightbulb are using energy at about the same rate. You may use 150 or even 200 W during part of the day, then idle down while sleeping, but your average will be in the neighborhood of 100 W. A Tour de France bicycle racer may eat 10,000 calories per day, to average a power output of 500 W. A whole lot of that energy goes into waste heat, but at the very highest level for a human, an elite bicycle racer may put close to 500 W into moving the bicycle up a mountain stage. The rest of us can only dream about that, so when we consider amounts of energy, just remember that the energy each of us generates is 100 W.[13]

Recently, total energy usage in the United States has been running notably higher than the personal burning inside of us. In fact, the total energy use in the United States in a day, divided by the number of people, gives approximately 240,000 calories per person per day, or somewhat more than 10,000 W for each person.[14] This is the equivalent of each of us having more than 100 people doing our bidding—100 energy "servants" apiece. We are really better off than that, though. If each of us actually had 100 people to do our bidding, we wouldn't get this much work out of them, because most of their time and most of their energy would be used to keep themselves fed and clothed and toileted and otherwise alive.

Some people become discouraged when they think about the huge energy and environmental challenges facing us. Giving up is wrong, though. If you start to get discouraged, just remember this: Abraham Lincoln ran his mind and body on the same amount of energy as a single heat-until-it-glows lightbulb. You, and Einstein, and Beethoven, and Michelangelo, taken together, use or used less energy than a single chandelier. The careful mathematics of science yield the same answer that we teachers learn from our students—we have more than enough brainpower to figure out how to power our brains!

BURNING TO EARN

Richer countries, and richer people, use more energy.[15] And using the energy helps the people. (I will come back to the problems from the energy use later, and there really are problems, but the good from energy use still outweighs the bad.)

Across the countries of the world, those that use more energy per person generate more economic activity per person. There are notable variations, with countries such as Russia, Saudi Arabia, and Canada using more energy to make a dollar (or a euro or a ruble) of economic activity than countries such as Japan, Italy, and the United Kingdom. These differences matter, but they are not huge, with the less efficient countries usually using no more than about twice as much energy per dollar as the more efficient countries. We all know that the correlation between energy use and economic activity does not necessarily tell us that using energy helps the economy. The world economy has grown over the same time that my hair has fallen out, but the extra sunscreen I buy to cover my expanding forehead during soccer games is not economically significant globally, so my spreading baldness is not responsible for the rising wealth of nations. But overall, we know that the wealthy use more energy, and serious thinkers typically find that the good things obtained from that energy help create the wealth.

Suppose that I keel over from a heart attack while sitting at my computer. My neighbor will use a phone powered by electricity to call a hospital, which will dispatch an ambulance powered by gasoline. A machine powered by electricity might keep me breathing and keep my heart beating until a doctor can fix the problem. The knowledge and dedication of my neighbors are essential to save my life, but so are the machines and the energy they use. My recovery would be aided by the plentiful food in the climate-controlled hospital, with the air-conditioning or heating,

the food refrigeration and cooking and trucking, the plowing and planting and fertilizing, all relying on energy sources that go far beyond the
2000 calories per day that I actually eat.

In an economic sense, the wealthy are not using more energy because
they are wasteful.[16] In general, the wealthier you are, the more dollars of
economic activity you get from a given amount of energy. As an example, in the United States from 1973 to 2008, economic activity (the gross
domestic product, with the effects of inflation removed) more than doubled, while energy use increased by only one-third. Thus, the people of
the United States in 1973 used slightly more than twice as much energy
to earn a dollar as the people in 2008. The effects of inflation have been
removed, so energy efficiency really went up. But energy use still rose,
because the economy grew even faster than energy efficiency grew.

There are lots of ways to "spin" these sorts of data in public dis-
course, and you should have no difficulty finding people spinning this
discussion many different ways. Because the wealthy generate more
dollars of economic activity per barrel of oil, you could argue that the
wealthy are more efficient than the poor. Because the wealthy use much
more energy, and much more stuff per person than the poor, you could
argue that the wealthy are much less efficient. Both are correct in some
sense. The best interpretation of the data probably is: 1) energy use helps
increase wealth, which increases energy use; 2) the wealthy energy users
find ways to improve efficiency; but 3) the effects of item 2 are not large
enough to offset the effects of item 1.

I will come back to facts and figures later. But first, to explore the
role of energy in our lives, the ways we have gotten it, the costs and
benefits, let's first look at some of the history of our energy use. I will
tell you a few stories that interest me, without in any way subjecting you
to a comprehensive treatise.

3

Peak Trees and Peak Whale Oil

Synopsis. Our ancestors moved to new energy sources in part because the old sources were running out. Much of the natural world we enjoy now owes its survival to our use of fossil fuels instead. We cannot go back to our old ways as the fossil fuels are exhausted.

A king's head is solemnly oiled at his coronation, even as a head of salad. . . . Think of that, ye loyal Britons! We whalemen supply your kings and queens with coronation stuff! —Herman Melville[1]

THE SINKING ROCK

One of the many attractions of the Cape Cod National Seashore, in Eastham, Massachusetts, near the old Coast Guard station, is a giant boulder that was delivered long ago by ice-age glaciers (see chapter 11). Climbing to the top of Doane Rock is a rite of passage for youngsters—if you can scramble all the way up and back down safely by yourself, you're a big person. My wife's family has been visiting this region for over a century, and those good people have been kind enough to include me over the last few decades. My father-in-law, Niel Richardson, likes to joke that Doane Rock has been sinking. When he was a lad during the 1930s, he would sit atop the rock and watch the waves breaking on the beach, half a mile away. Now, the pitch pine and scrub oak are much taller than the rock, and a kid perched on top can barely see Doane Road, 100 feet (30 m) away.

Of course, Niel knows that sinking has nothing to do with it—

Figure 3.1 Doane Rock, Cape Cod National Seashore, was deposited by a glacier during the ice age and is the largest such "erratic" boulder on the Cape.

instead, the trees grew. The Cape Cod that the Pilgrims found in November of 1620 was "so goodly a Land, and wooded to the brinke of the sea."[2] The Pilgrims, and those who followed them, then set about deforesting the Cape. Building ships and homes consumed many of the best trees. The 1700s-era Doane House, where we sometimes stay, has irregular-width oak boards in the floor, including some of such remarkable size that nothing similar could be made from Cape trees today.

Early New Englanders didn't need ocean-going boats for whaling; they could launch small boats from the shore, especially to hunt right whales (possibly so-named because they are the "right whales" to hunt, since they swim close to shore where they can be killed, and they float after being killed). Shore whaling may have begun as early as the 1620s off Cape Cod, and was economically important during the latter 1600s and into the 1700s. Boiling the blubber in the "try works" to extract the

Figure 3.2 The view from Doane Rock. In the early twentieth century, this photo would have showed a treeless plain sloping down to the ocean half a mile away.

valuable oil was started by burning wood for heat, although the cooked-out whale pieces could later be added to fuel the fire, giving a reportedly strong and very unpleasant odor.

The howling winter winds on Cape Cod can be bone-cutting cold, so the fireplace in the Doane House, and similar fireplaces in houses up and down the Cape, provided a welcome relief for the people. But the fireplace burned wood—perhaps an acre of trees (0.4 hectare) per year for a house[3]—which came from the rapidly dwindling forest. In 1690, the town of Eastham, home of Doane Rock, enacted a rule to prevent logging on the common lands except for export to make money. This was extended to all lands in 1694, and even export logging was banned in 1695, but the trees seem to have been gone by 1700.[4] The Cape was almost completely deforested, and then kept bare into the early 1900s, when fossil-fuel burning became important and the demand for trees dropped. My father-in-law grew up with the returning forest of the Cape, a forest that really owes its existence to fossil fuel.[5]

Very simply, the land of the Cape was capable of growing lots of trees that were big enough to use for construction as well as fuel, but the human demand between the 1600s and about 1900 far outstripped the ability of the land to grow those trees; rising use of fossil fuels eventually reduced the demand for Cape Cod trees so greatly that they are now growing back, blocking Niel's view of the ocean from the great rock but providing a beautiful cover to the Cape, which is allowing wildlife to return and eroded soils to re-form. Before the Civil War, scarcely more than 30,000 people kept the Cape nearly treeless; 150 years later, 200,000 more people live on a tree-covered Cape.[6] The difference? Fossil fuels.

THE GREAT PENNSYLVANIA DESERT

I live in Pennsylvania—Penn's Woods. I am fortunate to teach at a great university, Penn State. After class, our students walk off campus into a town that is younger than the university. The university was founded in 1855 far from any city, up the hill from an iron furnace. That furnace still guards the main road out of town. Of the thousands of people who pass each day, I suspect few even notice the furnace and fewer realize that they are there because of the furnace.

Much of the infrastructure of the U.S. East Coast in the early 1800s was built of Juniata iron, which was mined, smelted, and forged in and near the valley of the Juniata River in central Pennsylvania. Beginning in the late 1700s, entrepreneurs converted the rusty soils of the region into pig iron, which was then forged into other useful items. Place names such as Pennsylvania Furnace, Lucy Furnace, Harmony Forge, and the older and more easterly Valley Forge of Revolutionary War fame testify to the influence of the iron industry. Iron-making grew in central Pennsylvania because the region offered the iron-rich deposits, limestone, water power, and timber needed for iron manufacture in those days.

Later, use of coal allowed the iron industry to focus in a few places such as Pittsburgh, but the iron-making of the Juniata Valley was done in widely scattered furnaces fueled by charcoal. The making of the charcoal is a fascinating chapter in history, oddly romantic now but decidedly not romantic to the people who did it.

A single iron furnace needed about six hundred bushels of charcoal per day to operate.[7] As many as one hundred men and boys worked, mainly in the autumn and winter, to cut the wood for this charcoal. Wood was cut into four-foot lengths, and anything thicker than about six inches had to be split to make the conversion to charcoal work better. The four-foot-long logs (just over one meter) were stacked four feet high in eight-foot-long cords; a good cutter could supply three of these cords in a single day.

Figure 3.3 Preparing for charcoaling, June 1942, Wayne National Forest, Ohio. C. A. Masie (left) and David Daniels putting wet leaves and then dirt over the wood to restrict oxygen so that burning will produce charcoal. An immense amount of wood was used to make charcoal.

After a winter of cutting, the colliers took over, doing the "coaling" to make charcoal during dry times in spring and summer. An experienced collier and assistants would set up a camp on a flat, dry spot near the cut wood, and live there for several weeks while tending the coals. Thirty or more cords of wood would be stacked carefully in a "pile" fifteen feet high and twice that wide. This pile was then covered with dirt and sod, but leaving strategically placed holes and a central vent. Then the pile was burned beneath the dirt for up to two weeks. The dirt cover slowed the supply of oxygen, while the holes let in just enough air so that the fire smoldered rather than going out. Such a smoldering fire heated the wood sufficiently to drive off or burn up its water, oxygen, and other volatile materials. The pile shrank as it lost these materials, so workers were forced to walk around on top and pack the soil cover back down, lest the holes get big enough to let in enough oxygen to burn up the valuable charcoal. These workers surely walked carefully, knowing they were one misstep from plunging into the glowing hot pit. Eventually, after enough of the original wood was lost, the holes in the soil cover were blocked completely, the fire slowly went out, and the charcoal—now almost all carbon—was separated, cooled, and hauled by mule cart to the furnace.

A lot of other activities were going on around the furnace as well, including mining the iron ore and the limestone, "flatting" the charcoal so that there were no big chunks to clog the furnace, charging the furnace, blasting air in to raise the temperature high enough to melt out the iron and allow the impurities to combine with the limestone flux, and eventually producing the valuable "pigs" of iron and a vaguely attractive blue or green glass slag. Anyone walking near one of the old furnaces today, or kayaking a river downstream of a furnace, is still likely to find pieces of the slag. If you're a history buff, this is a fascinating corner of our heritage.

If you're not a history buff, why should you care? With the forest yielding roughly 30 cords per acre, and the furnace needing 10,000 to 12,000 cords per year, one furnace required clearing 300 to 400 acres (approximately 150 hectares) per year. The pig iron was then taken to a

Figure 3.4 Charcoal-making today in some parts of the world still resembles the old ways, and always requires much wood, although the vast eucalyptus plantations that feed this charcoal operation near Carbonita, Brazil, are now managed for sustainability.

forge, where conversion to useful wrought iron required almost as much charcoal as went into making the pig iron. Thus, the forged iron production from a furnace accounted for clear-cutting a square mile—640 acres, or 250 hectares—of forest every year. This did not include the wood people used to heat their houses or cook their meals. A well-insulated house in Pennsylvania with an efficient stove might be heated for the winter with three or four cords of wood, so ten houses would require another acre of wood (0.4 hectare) per year for heating; an inefficient open fireplace would consume wood ten times faster.[8] Even without considering construction, the rate of tree use was immense.

The remarkable rate of deforestation was recognized even before iron-making was established in the Juniata region. Pennsylvania's most famous multitasker, the great scientist/diplomat/publisher/writer/

etcetera-er Benjamin Franklin, invented a highly efficient stove, and then set out to make a little money by promoting his invention. He was quick to identify the waste involved in not using his stove.

Franklin discussed the fact that the Swedes, Danes, and Russians conserved wood by burning in stoves rather than

> consum'd . . . as we do in great Quantities, by open Fires. By the Help of this saving Invention our Wood may grow as fast as we consume it, and our Posterity may warm themselves at a moderate Rate, without being oblig'd to fetch their Fuel over the Atlantick; as, if Pit-Coal should not be discovered (which is an Uncertainty) they must necessarily do.
>
> We leave it to the Political Arithmetician to compute how much Money will be sav'd to a Country, by its spending two thirds less of Fuel; how much Labour saved in Cutting and Carriage of it; how much more Land may be clear'd for Cultivation; how great the Profit by the additional Quantity of Work done, in those Trades particularly that do not exercise the Body so much, but that the Workfolks are oblig'd to run frequently to the Fire to warm themselves: And to physicians to say, how much healthier thick-built Towns and Cities will be, now half suffocated with sulphury Smoke, when so much less of that Smoke shall be made, and the Air breath'd by the Inhabitants be consequently so much purer.[9]

His words, translated into modern English, are being spoken and written routinely today. His concerns about efficiency, saving money, improving the balance of trade, avoiding air pollution, and worrying about the availability of fossil fuels sound remarkably like the concerns of our green-energy entrepreneurs, and his focus on making money for Benjamin Franklin has much in common with the modern mood. We will surely revisit these issues that concerned Dr. Franklin!

Despite his optimistic view, however, the wood did not grow as fast as it was consumed, because the people who were his "posterity" did not consume the wood at a moderate rate—iron-making, heating and cooking for a rapidly growing population, and logging for building material chewed through the forest. John Bartram, friend of Franklin and often called the father of American botany, in his 1743 travels through Pennsylvania and southern New York, noted,

> We observed the tops of the trees to be so close to one another for many miles together that there is no seeing which way the clouds drive, nor which way the wind sets; and it seems almost as if the sun had never shone on the ground, since the creation.[10]

A bit over a century later, Joseph Rothrock, who served at my university (then called the Agricultural College of Pennsylvania) as professor of botany and of human anatomy and physiology before becoming Pennsylvania's first Commissioner of Forestry, referred to our "forests" as the "Pennsylvania desert"—across vast tracts of our commonwealth, the trees were simply gone.[11] With the trees went the vast majority of the "charismatic macrofauna": the elk we have now were imported from the Rocky Mountains to replace the native stock that was lost, the fishers are imports from New Hampshire and New York, the last known Pennsylvania mountain lion was killed in 1856, the native wolves and bison are gone.[12]

Don't for a minute believe that this is somehow a unique issue for Pennsylvanians. The Sung dynasty of China of a thousand years ago produced iron in prodigious quantities, with similar effects on the forest,[13] and Europe was a bit ahead of the Americas in doing this as well. In the Sherlock Holmes story "The Adventure of Black Peter," we read,

> Alighting at the small wayside station, we drove for some miles through the remains of widespread woods, which were once part of that great forest which for so long held the Saxon invaders at bay—the impenetrable "weald," for sixty years the bulwark of Britain. Vast sections of it have been cleared, for this is the seat of the first iron-works of the country, and the trees have been felled to smelt the ore. Now the richer fields of the North have absorbed

the trade, and nothing save these ravaged groves and great scars
in the earth show the work of the past.[14]

Pennsylvania, like many other places, now has lots of bears and tur-
keys and deer and a wealth of other wildlife because of the efforts of
a lot of people, good laws, and game management. But this was made
possible by the simple fact that we quit cutting the trees faster than they
grew back. Benjamin Franklin in 1744 was concerned about his posterity. The first U.S. Census in 1790 found 484,000 of his posterity, who
were busily cutting the trees. Immense deforestation had occurred by
the time my university was founded in 1855, when the state population
stood at 2.6 million, and Joseph Rothrock was working for 6.3 million
people in 1900 as Commissioner of Forestry to heal the Pennsylvania
desert. Yet the U.S. census estimated in 2007 that 12.4 million people
lived in the largely forested state. Much of the difference is that we
found other ways to generate most of our energy and much of our building material.

PEAK WHALE OIL

We were camped in Colorado's Pike National Forest, west of Pikes Peak,
and in the middle of what might politely be termed an "adventure."
The campground was well populated with cars and tents and campers
and coolers containing calories, and the two hungry black bears were
determined to get their share. At the campsite next to ours, an old cooler
with some metal in its walls and a very sturdy latch resisted the bears'
efforts until the larger one finally batted the cooler away—the way it
bent around the tree it hit was rather sobering testimony to the strength
of a hungry bruin. A few coolers in other campsites yielded more easily,
but a bear contemplating a long, cold winter isn't happy with the food
from just a few coolers.

Our group of a dozen more-or-less hardy souls had locked our food
in the van, but the bears seemed convinced that they needed to inspect
everything, including our tents, just to make sure. So while some of our

group slept, the rest of us sat around a small fire and watched for the bears. When two or more eyes gleamed at the edge of the firelight, we tossed a half cup of white gas onto the fire. After a little calibration, we found the right volume to make a fireball big enough to scare the bears and small enough to avoid torching the trees overhead (or us, if we got it wrong). Thereafter, we succeeded in keeping the bears out of our camp, as they focused on cleaning out the campsites of people hiding in their solid vehicles.

We stoked our fire with Coleman fuel thrown from Dixie cups, somewhat higher-tech weapons than were available to our ancestors of half a million years ago. But the issues are virtually unchanged—we, or the foods we collected, are of value to big hungry beasts, but our mastery of fire's light and heat gives us a serious weapon in our effort to avoid supplying someone else's dinner, one way or another.

Burning trees is great for scaring lions, tigers, and bears, and it provides enough light for a little social interaction or food preparation in a dark winter, but the red and flickering light is hardly appropriate for a pioneer quilter trying to sew twelve even stitches per inch. So our ancestors put a lot of effort into inventing and using better ways to see at night. By the 1800s in the United States, a homeowner could choose from many light sources. Candles were common, made from various animal fats (tallow), from beeswax, and even from bayberries on Cape Cod and surroundings. However, candlelight was fairly dim, candles were time-consuming to make, and the smell from tallow candles was often quite unpleasant or downright stinky.[15]

Camphene lamps provided good light from a relatively inexpensive mixture of alcohol and turpentine, but they had a tendency to blow up. For example, an 1853 *New York Times* article detailed the death of a thirteen-year-old Brooklyn girl after a camphene lamp she was carrying exploded. An 1856 front-page article in the *Times* described how alert policemen saved a house by rushing inside and tossing into the street a camphene lamp that had exploded, and the stove that the panicked residents had knocked over while running outside to escape the original explosion. In 1854, the *Times* reprinted a story from the Frankfurt, Kentucky, *Commonwealth* describing the horrible death of three daughters of a Methodist preacher in a camphene-lamp explosion.[16]

Various oils have long been burned in lamps, including oils extracted from plants, animals, and seeping petroleum. Olive oil is mentioned in the Hebrew Bible in Exodus and figures prominently thereafter; a long-burning drop of olive oil in the temple led to the celebration of Hanukkah. Olive oil also appears in the Quran. Olives and oil are so tightly wedded in history that our words "oil" and "olive" share a common origin—"oil" originally came from olives, and only later was the word broadened to include other things such as petroleum.

For northern Europeans, North Americans, and others far from the Mediterranean, however, olive oil was not the best solution because olives don't grow well in Pennsylvania and colder places. For our more northerly ancestors, whale oil was often the favored illuminator. Whale oil was expensive, but the best whale oils gave a good, clear light with little smoke, little smell, and little danger of explosion.

People were willing to pay extra for whale oil, so other people went to work to get it. Whales were used to supply a host of products, from whalebone stays in corsets to combs, decorative scrimshaw teeth, meat, and more. (The February 22, 1922, issue of the *New York Times* reported the death of a young Zurich skier who had been stabbed in a fall by one of her whalebone corset stays, and noted that it was the second such accident in Switzerland that winter.)[17] But the oil was the centerpiece for the whalers of the 1800s.

Real money and jobs were involved. According to a summary of relevant documents by the New Bedford Whaling Museum in Massachusetts, there were 10,000 seamen on ships in the New Bedford fleet at the peak in 1857.[18] In addition, many industries operated onshore to supply the whalers and to process the products returned to shore.

A summary of key data from the 1878 book *History of the American Whale Fishery* reveals a remarkable history.[19] From a very low level, whale-oil production increased past 2.5 million gallons per year around 1820, quintupled to about 12.5 million gallons per year at about 1850, and fell back to 2.5 million gallons per year by 1875. The initial rise in

Figure 3.5. Dangers of the Whale Fishery. *The great risks of whaling became even larger as the increasing scarcity of "easy" whales in nearby waters forced the whaling boats to the ends of the Earth.*

production was accompanied by a fall in price, from a high of about $15 per gallon (in modern dollars as reported in 2007) to a low of about half that, a slow rise, and then a jump to almost $25 per gallon as the first major drop in production occurred. The price then dropped back a bit, but it remained volatile and relatively high to the end of the record. And in the middle of the high-price/falling-production interval, the first modern oil well was drilled, in 1859 in Titusville, Pennsylvania, which is about one hundred miles north of Pittsburgh.

Some people see this history and cheer the working of the free market: as the price rose, alternatives were developed; therefore, the system works, and we have nothing to worry about in our future. Other people see the oceans being swept free of whales, a tragedy of the commons as a few people profited from the natural treasure of the seas and overhunted some whale species nearly to extinction, with the wealthy enjoying what should have been the common good while the poor were burned to death in explosions of the camphene lamps their poverty forced them to use. The reality is somewhat more nuanced than either of these particular morality plays, but both of them have more than a little truth.

The drop in Yankee whaling had multiple causes.[20] The U.S. Civil War didn't help. The South sank many whaling ships during the war. The North sank more, turning old whalers into the "Stone Fleet," sailing them to Charleston harbor, and scuttling them in the main channels in an attempt to block Confederate shipping. Recovery of whaling following the war was damaged by some big disasters in the Arctic: thirty-three ships were crushed in the ice off Alaska in 1871, and 12 more were lost there in 1876. Insuring a vessel that stood a good chance of sinking on the Arctic grounds didn't seem like a smart bet. (The whalers were in the Arctic rather than off Cape Cod because, just as in 2010 the oil-well drillers were in the deep waters of the Gulf of Mexico rather than up on shore, the easy-to-get resource had been exhausted.) Whale-oil prices in the late 1800s were below their peak, and the availability of superior fossil-fueled lighting products was increasing. The U.S. whaling fleet slowly faded away, as U.S. capitalists took their investments elsewhere.

Whaling did not die with the U.S. fleet, however. Other nations, especially Norway and also Japan, expanded their whaling. By adopting faster chase boats, cannons firing exploding harpoons, factory ships for whale processing, new target species such as blue whales that were too fast for the old U.S. technologies, and new markets for whale products including margarine and a resurgence in whalebone corsets, whaling continued. Eventually, major depletion of many whale species, and the threat of extinction of some, led to the 1946 International Convention for the Regulation of Whaling, to "provide for the proper conservation of whale stocks and thus make possible the orderly development of the whaling industry."[21] Under the International Whaling Commission, a

186   VANITY FAIR.   [APRIL 20, 1861.

GRAND BALL GIVEN BY THE WHALES IN HONOR OF THE DISCOVERY OF THE OIL WELLS IN PENNSYLVANIA.

Figure 3.6 "Grand ball given by the whales in honor of the discovery of the oil wells in Pennsylvania."

moratorium was placed on commercial whaling in 1986. Recovery of some of the endangered stocks is underway.

Quite simply, humans burned whales for energy. The ocean didn't produce whales fast enough to meet our demands. We found alternative energy sources that allowed us to leave some whales in the oceans. But those alternatives were and are primarily fossil fuels. At least in part, we have whales because we use fossil fuels, just as we have trees because we use fossil fuels. The publication *Vanity Fair* featured an illustration in 1861 showing the "Grand Ball given by the whales in honor of the discovery of the oil wells in Pennsylvania," with whales in evening dress toasting, "The oil wells of our native land: May they never secede," and noting, "Oils well that ends well." The cartoonist for *Vanity Fair* showed the value of fossil fuels to whales.[22]

The growth in fossil fuels after we switched from whale oil has been phenomenal. The amount of whale oil produced in the United States during the heyday of whaling, the whole nineteenth century, would fit in about four loads of a really big supertanker. The tens of thousands of men, the hundreds of ships, the decades and centuries of sailing to the ends of the Earth to mine whales from the oceans, the thrills and art and tragedy and suffering, all of it produced an amount of oil in one century that would not supply modern U.S. imports for one day. No one who is paying attention could possibly conceive of going back to the old ways as the fossil fuels run out.[23]

FRUITS AND NUTS

Our backyard in Pennsylvania is an edible landscape. From the first strawberries in late May to the last hardy kiwis hanging on the vines in November, we almost always find something ripe and ready. My wife, Cindy, makes wonderful jelly from the raspberries. Our cherry trees

Figure 3.7 A cedar waxwing eating cherries. To a well-fed person, this is an adventure in nature studies. To a hungry person, this would be competition.

are mostly bird feeders. However, the few cherries that the birds miss, and the serviceberries and blueberries, the beach plums and grapes and gooseberries, the apples and pears and more, we mostly eat as they ripen. There is a very reassuring feeling that goes with a bountiful harvest at our very doorstep.

We face a difficulty, however; the neighbor's sycamores threaten to cut off sunlight to many of our plantings. Our plan is to plant more on the other side of the yard, maybe buy more fruit while growing a bit less, and smile about it. We like our neighbors and have no intention of upsetting anyone over a bit of shade and leaf-raking in the fall. But we don't really need the calories that will be lost beneath the spreading, inedible trees. If our survival depended on getting sunlight to our fruits, we probably would have no problem, because the neighbor already would have cut down the trees to allow sunlight to reach his crops. If not, we might be having a rather pointed discussion with the neighbor, and perhaps we would be out there with a chain saw, using fossil fuel to guarantee our food supply.

The Native Americans of Pennsylvania faced this same problem. A forest of sycamore and maple and hemlock is not especially good for humans to eat. And despite the assurances of the late natural-foods guru Euell Gibbons on a televised breakfast-cereal commercial that many parts of a pine tree are edible, there are better plants for dining if you need the calories.

The Native Americans of the eastern United States had developed those better plants, growing beans, squash, and corn (maize) on clearings in a landscape that naturally would have been tree-covered. Cutting down a forest with stone tools is *not* easy; girdling the trees, letting them die, and then burning them is much easier. Fire stands in for a lot of human effort.

The Native Americans also made extensive use of nut and fruit trees. For example, William Bartram, son of the John Bartram we met earlier in the chapter, described one Creek family that had stored up to one hundred bushels of hickory nuts. In years with large crops, a few weeks of gathering may have provided enough highly nutritious acorns to last several years; early European settlers plowing near streams reported uncovering large caches of acorns where they had been buried by the native people to allow the water to leach tannins and improve edibility.[24] Early observers noted that Native American settlements were in regions especially rich in the useful trees, while surrounding areas were dominated by trees less favorable for human food. A growing body of evidence indicates that the native people were not just passively settling in favorable spots; they were actively managing through girdling, burning, and probably planting as well, to create an edible landscape. With neither lawnmower nor chain saw, they were much more efficient than I am at feeding their families from their land.

This burning to manage the environment was not restricted to eastern North America; it was found across broad swaths of human settlement, from Central America to Australia.[25] Our ancestors everywhere were smart, and figured out ways to feed themselves and their families. This always seems to have involved harnessing outside forms of energy, especially fire. It is no accident that so many cultures have an important story or myth about the arrival of fire, from Rabbit stealing the precious gift and delivering it for human use in Creek and some other Native American traditions,[26] to Prometheus doing the job for the Greeks.

The great climate scientist Bill Ruddiman has led a major effort aimed at estimating the impacts of early human burning and agriculture. His provocative hypothesis is that our ancestors generated so much methane and carbon dioxide from inefficient early rice cultivation and from widespread burning that the enhanced greenhouse effect prevented a new ice age.[27] We have had some lively discussions on the topic, and I believe that the hypothesis remains unproved; however, I consider his research impressive and important in showing the extent of preindustrial human activities and potential impacts on the very atmosphere of the planet. And some of the research conducted by colleagues and I also suggests that Native American burning had a measurable effect on the composition of the atmosphere.[28]

ENERGY SERFS

Humans burn things. From killing disease bugs in food and making it more digestible, to heat-treating stones to improve tool-making,[29] to lighting the night and scaring away predators, to clearing land for crops, and to so much more, we have lived and will live by harnessing energy from outside us. Our well-being has risen as we've figured out ways to harness more energy. Even as we've learned to waste less energy, we've burned more. There is absolutely no question that even greater savings are possible, but overall the human demand for energy is virtually guaranteed to rise.

Our history shows that once we get good at using an energy source, we really use it, and we often use it up. The Native Americans didn't run out of trees in Pennsylvania, but the early iron-makers and the European settlers of Cape Cod were running out of trees when they found other sources of energy, and the ocean was well on the way to running out of whales when we agreed to stop commercial whaling.

Today, roughly 85 percent of our energy use is unsustainable—we are burning stores of energy put aside by nature over much of a billion years, stores of energy that will not grow back fast enough to help our children's children's children, even if we stop burning.

Many people today don't worry about this—the free market has found substitutes in the past, and these people believe that when substitutes are needed, the market will work its wonders and produce more. Other people worry a lot—hardships often come with switches in resources, and our reliance on unsustainable energy to keep us alive and patch up the damage we do is so large that these people doubt the market can respond fast enough and well enough this time to avert great hardships.

But this problem is about to get much harder. The science is now very clear: if we meet the energy challenge by burning fossil fuels and discharging the CO_2 into the air, then we have high confidence that the resulting climate changes will make our lives much more difficult, handicapping our search for sustainable solutions while leaving us without the safety net of the fossil fuels if the search proves harder than suspected. The optimal path forward involves starting now to reduce CO_2 emissions and to slow climate change, through some combination of new energy sources, capturing the CO_2 and putting it back into the ground, and increasing efficiency. The large challenge of supplying enough energy becomes much larger when we factor in climate change.

Almost everyone agrees on the need for energy use. Those on the political right may be more enthusiastic than those on the left; some on the left may suspect that many of us are overusing in some places. But the thought of leaving our biggest energy source in the ground, or paying real money to capture its waste and put that waste back into the ground, makes many people on both sides of the aisle very nervous. These nervous people have placed very high demands on the scientists studying this problem, and those scientists have delivered.

The next section of the book leaps into the science of CO_2 and changing climate. We will see how Earth operates, and how this knowledge helps us gain such high scientific confidence that humans are pushing Earth's climate very hard toward bad things. We will meet a fascinating history, a sobering present, and some possibly scary futures. But, if at any point you get discouraged, just remember that internally, you, and Einstein, and Beethoven, and Michelangelo use or used less energy than a single chandelier. The answers are out there, and some of them are waiting in the third part of this book.

PART II.

LEARNING WHILE WE BURN

4

Fossil Fuelish—Some Telling Facts

Synopsis. We are using fossil fuels much more rapidly than nature replenishes them by burying dead plants without oxygen. If we continue with business as usual, we may begin to run out of fossil fuels within this century or not much beyond.

It is with our passions as it is with fire and water,
they are good servants, but bad masters. —Aesop[1]

ENERGIZING SOCIETY

Some health clubs have started offering an entertaining green workout on exercise bicycles that power televisions. And the workout can be a good one—a whole lot of televisions use energy faster than the 100 W average of a typical person, and watching an hour-long show on a big 400 W television would challenge the best athletes in the world.[2] The average internal energy use of ten of us matches a hair dryer, and fifty of us could drive a space heater. To enjoy these modern conveniences, we must rely on "external" energy, from burning wood or coal or oil, or from hydropower, nuclear power, or other sources. Globally, the total use of external energy in the year 2006 averaged about 2600 W per person.

Of that 2600 W, about 200 W came from burning "traditional biomass"—locally and personally collected dung or wood—while 2400 W came from someone else, who sold the energy to make money.[3] And over 85 percent of that 2400 W came from fossil fuels.[4]

That 2400 W of use per person is not distributed very uniformly around the globe. Compare the average use of more than 10,000 W by a person in the United States to perhaps less than 600 W by a person in the poorest places. The United States is near but not quite at the top of the list of external-energy use per person; Canada ranks a bit higher, and Bahrain and Qatar and a few others are also higher. Energy use in Europe typically runs about half the U.S. level.[5]

If instead we look at the energy used per dollar or euro of economic activity, the United States is more nearly comparable to Europe, and a bit lower than the world average; Japan stands out as getting roughly twice as much economic activity out of its fuels.[6] Some care is required when interpreting these statistics, however. A lot of the goods sold in many countries, including the United States, were made using much energy in other countries, such as China. So, total energy use in China would be lower, and in the United States would be higher, if the United States made more of our own goods.[7] But now that China has such large investments in the United States and other countries, maybe a lot of the goods "owned" by people in these other countries really belong to China? Interpretations can be spun in many different ways!

Recall that differences in energy use per dollar of economic activity among the major economic powers tend to be twofold rather than tenfold, anyway. Tenfold differences would have suggested that the big energy users could learn some simple things and drop their use a lot, whereas twofold differences suggest that a lot of the good ideas of the more efficient neighbors have already been adopted by the big users.

Regardless of the accounting by country, or whether we should consider energy use per person or some other measure, humanity uses an immense amount of energy, which comes primarily from an immense amount of fossil fuel. And burning combines all of that fossil fuel with oxygen from the air to make a lot of CO_2.

In the United States, the average use of crude oil in 2006 was just under 1000 gallons per person. If we add in the amount of coal and natural gas used, and figure out how much all that fuel weighs, the average U.S. citizen was using over 17,000 pounds of fossil fuel that year.[8] With

U.S. adults typically weighing in the neighborhood of 175 pounds,[9] each year the average person in the United States is responsible for the burning of about 100 times their own body weight in fossil fuels, putting out more than 250 times their body weight in CO_2, to supply just over 100 times their personal internal-energy output.[10] The approximately 44,000 pounds of CO_2 per U.S. person per year dumped into the air dwarfs the less than 1000 pounds of "trash"—municipal solid waste—that ends up in a landfill each year.[11] At this rate, a resident of the United States will be responsible for CO_2 emissions equal to almost 20,000 times their own weight over an eighty-year lifetime. Imagine part of our legacy to future generations as 20,000 copies of each of us made of CO_2, floating around and altering energy exchange in the air, or diving into the ocean to make it more acidic. (We will explore these topics in the next couple of chapters.)

For the 6.5 billion people in the world in 2006, fossil-fuel burning averaged about 3000 pounds apiece. This produced about 10,000 pounds of CO_2 each.[12]

Let's be very clear—getting this much fossil fuel collected, delivered, and burned is a *huge* job. Getting enough to raise everyone on the planet to the U.S. level of consumption is, to be honest, not very likely to happen, because the fossil fuels will run out.

But, what is this fossil fuel, and why are so many people confident that it will eventually run out? Let's take a look.

"MUC"ING UP THE OCEAN

Veterans of high school biology generally remember photosynthesis, with varying levels of enthusiasm. Photosynthesis is the process by which a plant uses the sun's energy to combine CO_2 and water to make more of the plant, while releasing oxygen.[13]

After plants die, they decompose, which is like photosynthesis run backward, with dead plant material and oxygen combining to release CO_2, water, and energy. If this happens rapidly in a fire, we call the process "burning." If it is carried out in a bacterium or fungus, a bear or a firefly, we call it "respiration." Regardless, the outcome is the same, so we can think of respiration as a metabolic burn. Animals use a little of the plants they eat to grow, but eventually the animal dies, and almost always the dead animal is burned too.

Occasionally, though, a dead plant, or more rarely a dead animal, is buried without oxygen. Think of a green, "scummy" pond, with algae growing wildly and then dying and sinking to the bottom to be burned by bacteria. Because water never holds very much oxygen, the burning of too many dead things in the sediment at the bottom of the pond, or a lake or ocean or in wet soil, will rapidly use up the oxygen there. Tropical-fish enthusiasts need to put "bubblers" in their aquariums because oxygen moves very slowly through water unless it is helped along by currents or bubbles. An aquarium may have sand or gravel at the bottom but not mud, because the oxygen-rich water from the bubbler can move through the large spaces between the sand or gravel grains to reach a catfish that has burrowed in, whereas water moves much more slowly through the smaller spaces of mud.[14] Bubblers are more important in warmer water because less gas can dissolve in it—heating drives the fizz out of a soda, and similarly drives the gases, including oxygen, out of water. So more dead plants, more and finer-grained sediment, warmer water, and weaker currents all favor the burial of unburned dead plants. Less burning of carbon leaves more oxygen in the air, so we can thank the burial of dead plants for the oxygen we breathe.[15]

When the free oxygen is gone in the mud around dead plants, the burning isn't done quite yet because many bacteria can use the nitrate or sulfate, which contains oxygen bonded to nitrogen or sulfur, or certain other chemicals dissolved in water to burn dead things without free oxygen. Some of these can have surprising consequences—using sulfate releases hydrogen sulfide, for example, leading to black stinky mud.[16] If enough organic material is buried in mud, the sulfate and other substitutes for free oxygen run out too, and some of the plant material escapes being burned.[17] This unburned plant material is on the path to becoming fossil fuel.

Let's start with oil, which comes mostly from algae or other "slimy"

Figure 4.1 Rapid burial of plants in warm, still water favors formation of fossil fuels. This bayou in Jean Lafitte National Historical Park and Preserve, Louisiana, is a good starting point.

plants that live in water; we will then switch to coal, which comes from woody plants that live on land. Bacteria begin to break down dead algae while they are sinking to the bottom of the sea or lake, plucking off the more valuable and more easily removed pieces quickly. Favorites include phosphates, proteins with their nitrogen, and oxygen-rich parts such as alcohol groups—a drink with dinner, perhaps? What remains is a carbon-and-hydrogen-rich "molecularly uncharacterized component," or "MUC." (Regardless of what you may have heard, scientists *do* have a sense of humor—possibly a weird one.)[18]

The detailed path from buried dead plant to fossil fuel depends on time, temperature, pressure, chemistry, starting plant material, and more.[19] As muddy sediments pile up, the deeper ones are warmed by Earth's heat and squeezed by the weight above, becoming sedimentary

Figure 4.2 Natural gas, which is primarily methane, is produced by the breakdown of buried organic material without oxygen. The "swamp gas" bubbles being stirred up by the alligator are methane.

rock and perhaps making it all the way to metamorphic rock. At temperatures favorable for living things (below about 150°F or 65°C, found in the top mile or so of sediments), bacteria generate methane—natural gas—from the buried dead-algae MUC. The next couple of miles deeper, temperatures get toasty enough (200–300°F, or approximately 100–150°C) to break down the remaining organic slime to oil. Oil is primarily hydrocarbon molecules—made of carbon and hydrogen—that are smaller than those in the original plants. Oil can flow up into shallower and cooler rocks, but if its path is blocked, it may be carried even deeper, where higher temperatures break down the longer molecules to yield gas, which is just the smallest hydrocarbon molecule, CH_4. At still-hotter metamorphic conditions, any remaining organic material may yield graphite. Any rock heated beyond this can melt to feed a volcano.

Coal has a lot in common with oil. It forms where and when the rapid burial of plants in wet places restricts access of oxygen and slows down metabolic "burning." With time and heat, buried woody material slowly follows a path similar to that used by the charcoal-makers who helped deforest Pennsylvania a century or two ago, as described in the previous chapter. Chemically interesting parts including water and other oxygen-bearing molecules are removed from the wood to leave mostly carbon, which later can be burned to release energy and CO_2. Generations of students have wondered why they were asked to learn that woody plants are converted to peat and then to coal, which starts as lignite, then becomes bituminous, and on to anthracite. Gas is produced with all of these steps. Peat is found with loose sediment, lignite and bituminous, in sedimentary rocks, and anthracite in rocks that would generally be

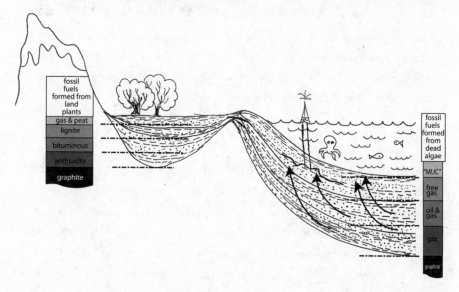

Figure 4.3 Diagram showing the formation of oil and coal. Geologic processes sometimes make basins in which great thicknesses of mud, sand, and other sediments pile up over millions of years. The rising heat and pressure caused by burial contribute to converting buried woody plant material to the various types of coal as shown, and buried "slimy" plant material to oil, with both producing gas. The coal stays put, but at least some of the oil and gas migrate upward, to be trapped on the way or to escape at the surface.

Figure 4.4 Sideling Hill road cut on Interstate 68 in Maryland. The dark beds near the top of the road cut are coal. These rocks first were buried in a sedimentary basin deeply enough to "cook" a layer of woody plant material to make coal, then squeezed up as the Appalachians formed, and ultimately exposed as erosion removed rocks above.

called metamorphic. If anthracite is overheated, it becomes graphite and then a volcano.

After formation, coal is fairly well-behaved—it sits there. However, this does not mean that coal is forever. As mountains are built and eroded, coal may be brought back to the surface, eroded and carried away by streams. While at the surface, the coal may be "burned" by living things, or else buried again in such low concentrations that digging it up would require more energy than burning it would supply. Where a coal seam is exposed at the surface, a forest fire or lightning strike may set it on fire, burning it more rapidly.

Oil and gas are much less well-behaved. In the finest-grained rocks, especially the mud-rocks called shale, oil and gas and water may move so slowly through the spaces that they are trapped even over the long time spans of geology. This keeps the oil and gas in place for us to find later, unless the shale is brought to the surface and erodes away. But we are not very good yet at getting oil out of shale economically, because the oil is held so tightly in the small spaces. We are getting better at economically extracting gas from such shales, though.[20]

Much oil and gas do move through spaces in rock. As heat breaks the old dead algae to make oil and gas, they take up more room within the rock, raising the pressure and often cracking the rock to open a path out. Spaces in rock are typically water-filled, and because oil and gas float on water, these fossil fuels tend to rise to the surface and escape. Natural oil and gas seeps have been known to humans from the earliest times. The animals of the La Brea tar pits, which is today in downtown Los Angeles, were trapped in the sticky residue of leaking oil after the less sticky parts escaped or were burned by bacteria at the surface.

Oil companies typically look for places where organic-rich black shales formed and were buried and heated enough to generate oil and gas, and where these could ooze upward into rocks with lots of spaces (sandstones, or heavily fractured rocks, or old limestones from reefs with big spaces in them, or others), beneath tiny-spaced muddy rocks that sealed in the fossil fuels. When all of these are present, drilling through the fine-grained rocks can tap a lot of valuable oil and gas.

Formation of fossil fuels depends on having the right combinations of rapid plant growth and fine-grained sediment accumulating beneath low-oxygen waters. Most of the land surface is eroding most of the time to feed mud to streams, so plants that grow in most places cannot accumulate to make coal. In many places where sediments do accumulate, they don't get thick enough to bury the dead plants deep enough to make oil—fairly special geological conditions are needed to get the few-mile pile that makes a lot of oil and gas. The organic-rich "juicy" black shales that produced much of our oil often formed as a continent was being torn apart to make a new ocean basin, before the new ocean got big enough to establish a vigorous circulation carrying oxygen to the bottom.

A warm climate over that new ocean basin also favored fossil-fuel formation. As we'll see in the next chapters, adding CO_2 to the air causes warming. But the warmth lowers the oxygen concentration in water, which favors burying the CO_2 in dead plants to form fossil fuels to lower CO_2 while raising oxygen in the air, so fossil-fuel formation may help to stabilize Earth's climate. However, this is a slow process, acting over hundreds of thousands or millions of years. Over mere thousands of years, warming probably raises atmospheric CO_2 a little, as I will discuss soon.

Accidents of evolution may have affected fossil fuels too. The formation of coals from woody plants was not possible before there were woody plants. And soon after they appeared, coals may have formed more easily than they do now because the decomposers had not evolved ways to take the wood apart as well as they do today.[21]

"OH, EXCUSE ME"—GIANT METHANE BELCHES?

Digestive tracts often include low-oxygen environments where bacteria produce methane, the main ingredient in cow belches, swamp gas, and socially unacceptable human emanations that are usually blamed on nearby dogs; methane also occurs in coal-seam gases and associated with oil. By itself methane is odorless to humans, but it often occurs with traces of gases with strong odors, either natural or added by the gas company to help identify potentially dangerous pipeline leaks.

Blowing bubbles underwater requires pushing the water away to make space for the gas. Stir up the mud beneath a Louisiana bayou or your neighbor's pond, and the swamp gas comes bubbling out. Drill into a hot methane pocket deep in Earth, and it may blow your drill rig out of the hole, catch on fire, kill people, and trigger an uncontained gusher of oil and gas, as happened with the Deepwater Horizon oil rig in the Gulf of Mexico in the spring of 2010.

But under sufficiently deep and cold water, something else happens: the methane and water combine to form a sort of ice. This ice is often called "methane hydrate" or "methane clathrate," a water-methane solid that, if brought rapidly to the surface from the seafloor, will burn if touched by a match.[22]

Across broad reaches of the sea floor, buried dead plants feed bacteria, which make methane in the spaces between the mineral grains, forming an icy layer a few hundred feet thick (a couple of hundred meters), sometimes seen right at the sea floor but usually a few hundred feet down. Below this layer is an additional zone of gas bubbles—as more mud accumulates on top, the deeper layers are warmed by Earth's heat until the clathrate breaks down to release methane gas.

Some people have imagined catastrophic stories involving this methane. Small bubbles have been observed coming out of the sea floor in places, and craters in the sea floor show where methane release has removed sediment. Blow too many bubbles in the sea, and a ship can't float on the gassy water and will be lost, so maybe the Bermuda Triangle is just a bubble-blower's bathtub?[23] And if the methane can sink ships, how about changing the whole climate rapidly, giant methane belches causing global warming?[24]

Reality is slightly more boring. We have enough ice-core data for at least the last hundred thousand years or more to know that methane belches big enough to change the climate a lot have not occurred.[25] Before that much gas could accumulate, its pressure would be expected to open a crack through the sea-floor muds to let the gas escape as small bubbles.[26] It may be true that enough methane belching can sink a ship, but it seems very unlikely that there is enough more to change the whole world's climate in a few days or even a few years. In fact, if that were possible, oil-company scientists probably would have already figured out a way to stick a pipe into that methane and sell it. Those scientists are working on ways to extract the methane in and below the clathrates safely and economically, but the distributed nature of the resource means that much more research is required.[27]

Still, there is an immense amount of methane down there—perhaps containing about as much carbon as the entire global stock of economically recoverable fossil fuels.[28] And this methane in the sea floor, and to a lesser extent in the permafrost of the Arctic, is still of great interest, not only because of the commercial potential, but because warming might release notable amounts over centuries to millennia. One really good calculation, although still with large uncertainties, estimated that 85 percent of the modern amount could be released by a sustained warming of 5° to 6°F (3°C).[29]

PEAK OIL?

In 1956, the visionary geologist M. King Hubbert prepared a remarkably modern-looking consideration of the past and future of fossil fuels. In particular, he used some simple curve-fitting combined with trenchant observations to estimate that U.S. oil production would peak within a few decades, probably in the early 1970s, with world oil production peaking sometime not too far from now, and coal production within a century or two.[30] He was right about the U.S. oil production. A lot of articles and books, some scholarly and some apocalyptic, have been written on the coming oil shortage and what it might mean for us.[31] Some people imagine a world in which peak oil is happening right now, and crisis follows soon thereafter.

Clearly, the times of easily retrieving oil are winding down. In considering the 2010 Deepwater Horizon disaster in the Gulf of Mexico, many words were written assigning blame to a particular oil company, or their contractors, or a government agency, or the administration, or a person or group of people, or the whole society. But there can be no doubt that drilling such a deep hole in rocks under such deep water so far from shore contributed to the size and scope of the disaster—a blowout would have been much easier to fix in a shallow well on the plains of West Texas, for example. And oil companies do not invest in immensely expensive operations such as the Deepwater Horizon because of some love of throwing money away, but because that is where they can still find oil and make money.

Furthermore, although environmental regulations may prevent drilling for some "easy" oil, the credible evidence indicates that this does not represent enough oil to notably affect long-term supply projections—

"Drill, baby, drill" is not a sustainable answer.[32] Consider, for example, the oil beneath the federally protected coastal plain of the Arctic National Wildlife Reserve (ANWR). Given the remote location, this oil isn't easy to retrieve, but it is probably easier than some oil being sought now. The U.S. Geological Survey estimated that slightly less than eight billion barrels of oil are technically recoverable, with an uncertainty of about half that much. Two comparisons are instructive: at $70 per barrel, that oil has a market value of over half a trillion dollars, and that oil represents roughly two years of net U.S. imports at the spring 2010 rate.[33] Because exploration, drilling, and recovery of this oil likely would be spread over decades, drilling in ANWR might make a lot of money for a lot of people, but it wouldn't make a big change in the energy independence of the United States.

However, South Africa today makes liquid fuel—oil—from coal at competitive prices. Germany did this during World War II, and scientists at my university have developed ways to make jet fuel from coal. Reasonable arguments can be made that the chemical engineering required to allow one fossil fuel to substitute for another is just not that difficult or expensive, and thus that all fossil fuels should be considered as interchangeable—oil shale and tar sand and coal can be thought of in the same way as oil.[34] So, let's look at how much fossil fuel there is, and how rapidly it was made naturally compared to how rapidly we use it.

Most of the literature on this topic discusses the total amount of carbon in fossil-fuel deposits.[35] Because Earth holds so much carbon, writing about pounds or kilograms requires numbers with too many zeros, and even using tons requires numbers with a lot of zeros, so here I will use gigatons of carbon, or Gt C, with 1 gigaton equal to 1,000,000,000 tons or 1 billion tons. I will use the metric ton, but, to be honest, the estimates of how much coal we have are almost never known within 10 percent, so it really doesn't matter whether we think of 2000-pound English tons or 2200-pound international tons, because they are almost the same in this game.[36]

Estimates of the total reserve of fossil fuels that might be economical for us to dig up and burn are often in the neighborhood of 4000 Gt C,

with some estimates running as high as 6000 Gt C. In the highly speculative limit that the sea-floor methane clathrates are on the high side of
estimates, and we could somehow recover all of them, the 6000 Gt C
might be doubled, although most fossil-fuel estimates don't currently
include the clathrates as technically and economically accessible fuels.

So, the two big questions are probably 1) how rapidly will we use up
this stored fossil fuel? and 2) will nature make more fossil fuel rapidly
enough to help us?

The answer to the second question is no, fossil fuels are not made
naturally at a rate that will help us. At first glance, this might seem
strange. After all, the net amount of plant material that grows each year
globally—called the "net primary production"—is in the neighborhood
of 100 Gt C, so the fossil fuels represent only about sixty years of plant
growth.[37] But virtually everything that grows really does get burned,
quickly and efficiently. Autumn does follow spring in seasonal climates,
there are bacteria and fungi and other hungry organisms everywhere,
and keeping them from burning valuable plants until we can do it is
very difficult.

The rate at which unburned plants are buried naturally to begin the
path to fossil fuels is only about 0.05 Gt C per year.[38] So if there were no
natural loss of fossil fuels, and if all the buried carbon were concentrated
enough to be useful to us, fossil fuels would be accumulating at less than
1 percent of our modern burn rate. However, the rate at which fossil
fuels are lost naturally, by leakage or erosion, is almost the same as the
rate at which they are formed. And almost all of the dead plants that are
buried without burning are too spread out to be useful to us—the total
amount of such dead-plant carbon in the rocks of Earth's crust may be
10,000,000 Gt C, dwarfing the amount that experts expect could be
recovered at even vaguely recognizable prices.[39] No known technology
would allow most of that carbon to be recovered while using less energy
than the carbon contains. Most of our fossil fuels accumulated over the
last 500 million years or so, giving an average rate at which nature saved
them for us of something like 0.00001 Gt C per year, which is so close

to zero in comparison to our recent use of 6 Gt C per year that the production of new fossil fuels can be ignored completely.

This means that we have been given a bank account of fossil fuels, but we should not count on any new deposits—once we burn the fossil fuels, they are gone. So how long are they likely to last?

An energy-supply optimist, taking the high-end estimates, adding the clathrates, and comparing to the current rate of use, might give us two millennia. Even a moderate energy-supply optimist omitting the highly speculative clathrates would give us a millennium. If so, then the fossil fuels will run out if we continue business as usual, but not fast enough to alarm a lot of people purely based on energy-supply issues.

But our use is rising steadily, and we have started by burning the fuels that yield the most energy and that take the least energy to dig up or pump out and get to where we want to burn them. If the world were to burn fossil fuels as rapidly as the United States does, the global use rate would approach 40 Gt C per year, and if the reserve is on the low end, at 4000 Gt C, the fossil fuels would be gone in a century. No one really expects that every bit of fossil fuel can be used before shortages start to occur and stress economies. And to use even a large fraction of the total fossil-fuel reserves will require development and implementation of new technologies, likely including deeper-water drilling and more mountaintop removal, finding ways to use oil shales and tar sands, building huge chemical plants for fossil-fuel conversion, and more.

Realistically, our current pattern of fossil-fuel exploitation must evolve fairly quickly as the easiest and cheapest resources are depleted. And it is well within reasonable possibilities to expect children born today to live long enough to see notable stresses from fossil-fuel depletion. We can fairly say that we will need to switch away from fossil fuels eventually. But eventually may arrive sooner than we expect, as we will see next.

5

Abraham Lincoln or Your Brother-in-Law?

Synopsis. Where science meets politics, public argument is virtually guaranteed regardless of how good or bad the science is. Starting with Abraham Lincoln in the United States, society has set up ways to assess science and provide useful results to people and policymakers. These assessments often are very different from the public debate, but they offer the best guidance for society on what we know and don't know, and what our options are for moving forward.

Man is not the only animal who labors; but he is the only one who improves his workmanship. —Abraham Lincoln[1]

STREETS OF GOLD

The next several chapters present science showing that the wisest path for humanity is to leave at least some of the fossil fuels in the ground, through a measured transition to other energy sources, even though the market value of the fossil fuels, if pumped or mined, is likely to be immense. I can absolutely guarantee that across the range of the blogosphere, or even in the U.S. *Congressional Record*, you will find people who think that this is complete raving lunacy. You will find voices saying that a measured transition is too slow and we must almost instantly stop burning fossil fuels, and others saying, "Drill, baby, drill" and burn it all. You will find accusations of lying, cheating, stealing, bribery, and more. In a world of thousands of bloggers claiming to be experts but expressing widely differing opinions, whom should you believe?

If you have already decided to trust me to tell it to you straight, you can skip to the next chapter. But if you are still a little uneasy, travel with me back to Abraham Lincoln's day, and to a better way to get your scientific information. The next chapters will rely on Lincoln's path heavily.

IRONING OUT OUR DIFFERENCES

The Battle of Hampton Roads during the U.S. Civil War may seem like an odd stop on the way to our energy future, but the two are related. The March 8–9, 1862, battle was the first military engagement between ironclad ships, and it revolutionized naval warfare despite ending in a draw.[2] The battle also helped demonstrate the value of a rather surprising weapon that works well in peace too.

When the Commonwealth of Virginia seceded from the Union, some federal ships stationed along the Virginia coast escaped, but others were burned and sunk in place to prevent their use by the Confederacy. One of those was the steamship USS *Merrimack*. She was subsequently reclaimed and refitted by the Confederate Navy as the CSS *Virginia*, an ironclad.

In an effort to isolate the Confederacy, the Union had quickly launched a naval blockade, including at the important Hampton Roads, Virginia, harbor, and in early March the *Virginia* set out to break the blockade. She was equipped with iron armor and a ram. During the first engagement on March 8, her armor protected her from return fire as she rammed and sank the USS *Cumberland*. The captain of the USS *Congress* recognized the danger and intentionally ran her aground to avoid the *Virginia*, but ultimately the *Congress* was destroyed as well. The *Virginia* was menacing the USS *Minnesota* when darkness and low water intervened.

Overnight, the ironclad USS *Monitor* arrived, and the *Virginia* engaged this worthy foe in the morning. An inconclusive battle led to the withdrawal of both warships.

The value of the ironclads was remarkably clear, however—one rather crude ironclad, hastily built and running on a salvaged steam

Figure 5.1 Detail from The first encounter of Iron-Clads. Terrific engagement between the "Monitor" and "Merrimack."

engine, had taken on and defeated important ships of the Union Navy, but had been neutralized by another ironclad. Although these were not the first ironclads ever built, their success greatly accelerated the conversion from wood to metal for the Civil War combatants and for the other main navies of the world.

As has so often been the case, technological innovation brought new problems. One problem was the rusting of iron armor in salt water. Another was the attraction of the magnetic compasses to the iron plates; protecting your compass and crew with iron armor can leave you uncertain which way to steer on a dark and stormy night.

President Abraham Lincoln's administration and the Union soon met these and other problems with a new weapon. Long before he was elected president, Lincoln was an enthusiast for science and engineering because of the wonderful opportunities they offered to help people.

(We will return to Lincoln later in the book because of his visionary ideas on powering the country by wind.) He also remains the only U.S. president to hold a patent for an invention—his early experience rescuing grounded riverboats led him to design and build a scale model of a device to help a boat avoid grounding by pushing air-filled floats into the water when a river became shallow, reducing the draft of the boat so it would not become stuck.[3]

So Lincoln approved of good scientific advice. In one day, March 3, 1863, the U.S. House of Representatives and Senate passed, and Lincoln signed, a bill establishing the National Academy of Sciences. The act directed that "the Academy shall, whenever called upon by any department of the Government, investigate, examine, experiment, and report upon any subject of science or art." Furthermore, the act directed "the actual expense of such investigations, examinations, experiments, and reports to be paid from appropriations which may be made for the purpose, but the Academy shall receive no compensation whatever for any

Figure 5.2 A Civil War–era compass. The iron armor on the Monitor *and similar ships interfered with the ability of shipboard compasses. The U.S. National Academy of Sciences provided and implemented a scientific solution for this problem.*

services to the Government of the United States."[4] The Academy was not to be about getting wealthy or even about getting paid; it was and is science in service of the public.

The U.S. government quickly asked the Academy for help on many pressing problems, including the issues of rust and compasses on ironclads.[5] The committee on rust was less successful than hoped, learning that no solution proposed at that time solved the problem, and recommending specific new studies to advance knowledge. This is useful advice, by preventing waste of public money on something that doesn't work, but it's not especially satisfying. Decades later, rust remained a difficult problem—very simply, salt water is really good at attacking iron. The Compass Committee succeeded, finding that among the solutions then proposed, one using small magnets to cancel the effects of the iron plates was especially valuable. The committee provided the government

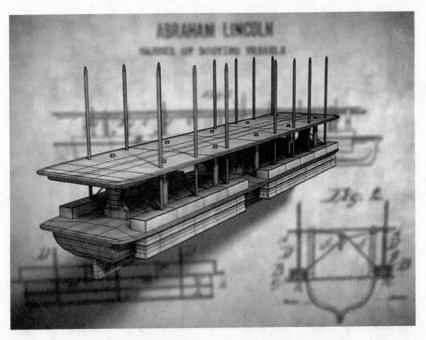

Figure 5.3 Abraham Lincoln's patent No. 6,469, from May 22, 1849, for a "Manner of Bouying Vessels."

Figure 5.4 Representation of the founding of the U.S. National Academy of Sciences, painted by Albert Herter. President Abraham Lincoln is joined on his left by Senator Wilson and founding members (from left) Benjamin Peirce, Alexander Dallas Bache, Joseph Henry, Louis Agassiz, Admiral Charles Henry Davis, and Benjamin Apthorp Gould.

with a seventy-three-page report, and then devoted six months to the conversion of compasses on twenty-seven ships, as well as additional measurements and more actions to enable others to continue this effort. The U.S. Navy sailed through the latter part of the Civil War and beyond secure in the knowledge that their armored compasses knew north, the result of good scientific assessment.

PASSIONATE SCIENCE

But why did Lincoln approve of an Academy for assessing science, following older traditions such as the British Royal Society and the French Academy of Sciences, rather than just calling on a few scientists and asking them for advice, or reading the scientific literature himself? Lincoln probably understood science very well, including how much arguing is involved in science and how difficult that makes the task of the policymaker.

Many of us were asked to memorize a definition of science back in elementary school, and that definition may not have included much discussion on the role of argument in science. So here's a post–elementary school definition, in English.

Science involves recognizing that the world had and has a lot of smart people, so learning from them is wise. But they are people and thus are imperfect, so we should be able to improve on what they did. Scientific improvement means predicting better—what will happen if we perform this experiment, or drill that next core, or build the airplane wing this way? So we look for a new idea that disagrees with a prediction of an old idea, without throwing away things that worked with the old idea, and then make the observations and see which idea "wins." An idea that repeatedly makes predictions that fail is set aside; an idea that succeeds is kept as being useful, and possibly even on the path to truth. Then, we return to the start of the paragraph, and repeat.[6]

Much of my job is teaching bright students, so I am continually reminded that the path to success for those students goes over me, by showing that basically everything I and my generation have ever done is either wrong or incomplete, and replacing what we did with something better. This means students must argue with the older scientists about our science, criticizing it, looking for flaws or inadequacies, and generally trying to break what we have built—the ideas that have been attacked most strongly without breaking are the most reliable ones. When a student demonstrates that my scientific papers were incomplete, or even wrong, that does *not* mean that I am a bad scientist,[7] only that I am a scientist—every scientific paper that matters has been, or will be, shown to be inadequate in some way.

Because science is done by humans, who are never perfect, we scientists may be tempted to make intentional mistakes, or we may get sloppy about keeping out the unintentional mistakes. The competition between my ideas and yours draws on much that is good and bad about people, engaging us in a market of ideas, but at the risk of bringing our egos into the fray. Science is not the result of dispassionate machines spitting out Truth; it involves passionate humans pursuing truth and fame and next week's paycheck, while satisfying curiosity at the same time. Science is powerful because we recognize the passion and humanity, and have built the strongest mechanisms we can to keep us from fooling ourselves.

We use rigorous training, long apprenticeships, shared goals for human betterment, and more, but we know that these will never completely overcome human nature, so we also enforce peer review prior to scientific publication as the real entry point to science. We insist that any published scientific paper must first be evaluated by world-class experts—peers—coordinated by an editor, to ensure that the scientific standards, including honesty, openness, clarity, and acknowledging prior workers and funding sources, are followed. And we continually argue about ways to improve peer review. Peer review is not a magic bullet, but it ensures that scientific papers contribute to science—testing our ideas against the real world, keeping the ones that predict successfully, and setting aside the ones that don't.

The complexity, rapid advance, and obligatory arguing in science mean that, even with peer review, policymakers are not likely to get what they want directly from the scientific literature; they need help. Furthermore, policymakers are wise enough tò know that on certain topics impacting jobs and values, the unavoidable argumentation of science will be amplified by nonscientific issues in ways that make it much more difficult for the policymakers to learn what is going on.

SPINNING SCIENCE

After the great San Francisco earthquake of April 18, 1906, the downed power lines, damaged stoves, and broken gas lines triggered a massive fire, which could not be extinguished easily because the water lines also were broken.[8] For easterners and Europeans thinking of visiting or moving to California, fire was a familiar foe, but earthquakes less so. Very quickly, California boosters, including newspapers and the Real Estate Board of San Francisco, set out to reassure the world that the damage from the earthquake was minor, that fire rather than the earthquake was the culprit in the minor damage, and that there existed little or no danger of future earthquakes. These boosters did not immediately set out to find the best science and communicate it to the world accurately, and they even publicly criticized science and scientists that disagreed with them.

Such a response may be more the rule than the exception. For example, when Rachel Carson published the book *Silent Spring* in 1962 on the dangers of the overuse of pesticides and particularly DDT, a similar story played out.[9] The former Agriculture Secretary reportedly privately suggested to former President Eisenhower that Carson was "probably a communist," and also wondered why "a spinster was so worried about genetics." *Chemical and Engineering News* opined that Carson's book would appeal to "organic gardeners, the antifluoride leaguers, the worshipers of 'natural foods,' those who cling to the philosophy of a vital principal, and pseudo-scientists and faddists." Carson was labeled a "hysterical female" and a "fanatic." The publisher (Houghton Mifflin) was threatened with a lawsuit from a chemical corporation, with counsel suggesting that Carson's work reflected "sinister" influences from the Cold War. Widespread and well-funded advertising campaigns by industrial groups targeted the public to promote the pesticide companies and their work.

Today, pesticides continue to play an important role in agriculture, and while spirited debate remains on how big that role should be, there is little doubt that a sudden moratorium on all pesticides would have a large and immediately bad effect on food production. Yet essentially everyone who is informed today recognizes that overuse of pesticides leads to damages that greatly exceed the benefits, and that Carson discussed a real problem that needed to be fixed. *Chemical and Engineering News* revisited the issue in 2007, with a rather different perspective in citing Carson's great contributions. DDT is in use now to combat malaria, but with a much clearer view of the costs and benefits.[10]

As in the case of DDT, interested groups have fostered public criticism of science they disliked on numerous topics, ranging from the

cancer-causing nature of cigarettes to the ozone-harming effect of chlo-
rofluorocarbons.[11] People tend to fight back—vigorously—when they
believe that their jobs, traditions, or beliefs are threatened, and they
often do so by criticizing science and scientists, or paying scientists or
"scientists" to support their side. Because science requires argumentation, and honor and glory come to those who can overthrow the existing science, you can be virtually guaranteed that scientists can be found
to publicly oppose the currently best understanding of nature, even if
those scientists are not publishing or have not published refereed scientific literature on the topic.

Furthermore, this response to environmental problems is predictable
whether the problem proves to be real or imagined. Although some
environmental "alarms" have proved to be all too accurate, requiring a
response to preserve health and welfare, other alarms have proved to be
largely or completely false. No scientist can wait for absolute certainty
before publishing because there is no such thing, so we can be confident
that occasionally a refereed scientific paper will be based on a statistical
accident or a mistake rather than a scientific reality.

Perhaps the most dramatic environmental "problem" that later
shrank greatly or disappeared entirely was the suggestion that electromagnetic fields from power lines cause leukemia or other health problems.[12] The initial report of a problem came from exploratory science,
of the type that can reveal important facts but is more prone to misleading results than some other research. An investigator drove around
Denver, visited the houses of a few known childhood leukemia cases,
and looked for anything that these houses or neighborhoods might have
in common. The investigator noted that a few were not far from power
lines. This led to a scientific paper. A writer picked up on this and produced a rather alarming series of articles in *The New Yorker*, and then a
book. A few quick scientific studies seemed to support the earlier findings. People became alarmed about power lines, electric blankets, wiring in houses, and more. A 1992 movie was made with Eddie Murphy
crusading against power lines in Congress. Groups sprang up to fight

the supposedly deadly power lines. Lawyers closed in for the kill with lawsuits against power companies. And, just as you would expect, the industry countered with its own experts and lawyers, defending itself, its jobs, and its employees.

In this case, the weight of scientific evidence soon came to support the industry. More studies, and more careful ones, found no significant effect.

ASSESSING SCIENCE

This brings us back to Lincoln and the National Academy of Sciences. Although there are a few people who, without irony, will use a computer—a truly great scientific and engineering triumph—to type a paper claiming that science doesn't work, the great majority of people recognize that the best of human knowledge tested against nature is useful to us. The trick is to find the best way to lift the lid on the ferment that is science, reach in and grab the useful parts for the policymakers and the public, and then slam the lid shut and let the scientists get back to what they do.

Policymakers have put a lot of effort into making these scientific assessments work, and have been doing so for a long while, as embodied in the act that Lincoln signed to establish the National Academy of Sciences. The keys are to harness the expertise of scientists for the public good, in publicly and scientifically scrutinized, publicly and scientifically responsible, not-for-profit ways that bring in the full range of views. Like referees in sporting events, the assessors bring order out of the very serious game that is science. But, unlike referees, the assessors can take enough time to look at *all* the instant replays, reread the rules, and make very sure that they are doing the best possible job.

A single scientist in a public venue can easily make comments that are not representative of the field, but this becomes very difficult in a serious assessment, because too many experts are watching. Assessments tend to be a bit slower than we would like, and are often fairly cautious—if the experts cannot agree, the report says so, and frequently recommends more research to solve the problems and to reduce the uncertainties, which takes time. In some sense, this is self-serving, because the scientists are likely to be the people who do that research.

Notice, however, that history shows that scientists will indeed recommend cutting off their own funding if that is the best path. When the original scare about power lines was racing around the globe, the U.S. government asked the National Academy of Sciences to assess the possible health effects of exposure to electromagnetic fields in the home. In 1996 the committee appointed by the Academy completed a three-year review. As described by physicist Robert Park, roughly half of the members of the Academy panel had been funded to study this topic, and they could have lost research funding if the report found the electromagnetic fields to be safe. Park wrote that the vice chair of the committee had "staked his reputation" on these fields causing cancer, raising special concern that the committee was going to use inappropriate reasons to find the fields to be dangerous. Yet, the committee unanimously stated that "the current body of evidence does not show that exposure to these fields presents a human health hazard."[13]

Assessments are not perfect either—after all, humans are involved. But despite imperfection, the assessments are the best thing we humans have ever come up with to distill science into a useful form for policymakers and others in the public. The various sides have agreed beforehand (in 1863 in the United States) on the mechanism for learning what the science really says. Sometimes, the political process sticks to this prior agreement and moves forward with the assessment results. Often, the "sides" are so entrenched, well funded, and anxious to "win" that they choose to attack the assessors and the assessment process instead.

WARMING TO THE ACADEMY

Not surprisingly, the National Academy of Sciences, and many other assessment bodies worldwide, including the World Meteorological

Organization/United Nations Intergovernmental Panel on Climate Change (IPCC), have been asked many times to assess the effects of the CO_2 released from fossil-fuel burning.

In 1975, the Academy had noted that warming seemed likely, but that natural cycling, such as the progression of ice ages, presented many questions, warranting further research.[14] By 1979, the Academy pointed to the clear evidence for warming from continuing fossil-fuel burning, with essentially the same estimate of how much warming would be caused by rising CO_2 that we still use.[15] This assessed view has not changed, despite many reexaminations by many independent committees.

Consider one prominent study since. When George W. Bush was elected president, the political campaign had included some discussions that were rather skeptical of human-caused climate change. The White House in 2001 requested the Academy's "assistance in identifying the areas in the science of climate change where there are the greatest certainties and uncertainties," and "views on whether there are any substantive differences between the IPCC Reports and the IPCC summaries." The National Research Council (NRC) rapidly assembled a committee of distinguished climate scientists. This committee included the professor who probably is the world leader in combining skepticism about the dangers of human-caused climate change with a history of scientifically important contributions in climate-related fields. The report, while pointing out the notable uncertainties and the opportunities for additional research to narrow those uncertainties, provided a clear and cogent summary:

> Greenhouse gases are accumulating in Earth's atmosphere as a result of human activities, causing surface air temperatures and subsurface ocean temperatures to rise. Temperatures are, in fact, rising. The changes observed over the last several decades are likely mostly due to human activities, but we cannot rule out that some significant part of these changes is also a reflection of natural

variability. . . . The committee generally agrees with the assessment
of human-caused climate change presented in the IPCC Working
Group I (WGI) scientific report (p. 1).[16]

To gain a full appreciation of the report and its nuances, you would
need to read the report—the main text is only twenty-four pages long.
Subsequent reports have generally strengthened its results. Regardless,
the broad range of scientists, when acting for the public in a committee
convened by the Academy, reached a clear consensus, driven by science
and not by preconceived notions, following the model of the earlier
committee on electromagnetic fields and leukemia, and thousands of
earlier panels dating back to 1863 and Lincoln.

The IPCC came to essentially the same answer, but with increasing scientific confidence, in four assessment reports, dated 1990, 1995,
2001, and 2007. The major professional organizations in these fields,
and other national academies, have reached similar results. I know of
no properly and publicly convened authoritative assessment that points
to any other result than the one just quoted.

Just to re-reiterate: the assessors do *not* make policy, pass laws, or
otherwise control the outcome of the political process. The assessors
volunteer their time to tell the policymakers and the people what is
solid, speculative, and silly in the science, what options may exist, and
what they might entail. Then the policymakers and the people decide
what to do about it.

In discussing global warming, I will try to stay true to assessments at
all times. If you are deeply interested in the topic, I urge you to continue
your researches by reading the assessments, starting with the National
Academy of Sciences or the IPCC.

There are people on many sides of many issues who have been assessors and who have signed assessments, but who make rather different-
sounding public pronouncements. There are people who speak on
contentious issues who do not submit themselves to the requirements
of scientific integrity inherent in the refereed scientific literature and in

the assessment process, and some of these people simply mislead the public and policymakers, often about the assessments. The assessment process surely can be improved; there is a never-ending effort and discussion on ways to improve it, and I am among those who invest time in this effort. But if you want the best information available, go to the assessments—Lincoln knew what he was doing!

6

Red, White, and Blue-Green

Synopsis. Military and civilian research in fundamental physics shows that CO_2 is a greenhouse gas, and that raising the atmospheric concentration of CO_2 will have a warming influence on the climate. By itself, a doubling of CO_2 warms Earth's surface approximately 2°F (just over 1°C), a number that is amplified by the processes discussed in chapter 7.

The only faith that wears well and holds its color in all weathers, is that which is woven of conviction and set with the sharp mordant of experience.
—J. R. Lowell and C .E. Norton[1]

ALL TOGETHER NOW

On the plane back from the great Jakobshavn Glacier in Greenland, the U.S. senator's aide asked me, "What is the single most important piece of evidence supporting your global warming theory?"

After hemming and hawing for a moment, I replied, "What is the single strand that is most important in holding a rope together?"

For a scientist, there may be a brief moment when a new idea, a flash of what may be brilliance or stupidity, actually does hang by a single thread. Even then, though, the idea is informed by a lifetime of study and experience, and the wisdom and knowledge of generations of prior scientists. But after admiring a new idea for a few seconds, a scientist's job is to test it, to see if it is consistent with the basic laws of science and if it makes successful predictions about experiments not yet conducted.

By the time a new idea is published in the refereed scientific literature, it is woven in with many other strands. But it still faces a long road before it can be called a "scientific theory"—one of the big, well-accepted, powerful, organizing ideas of science.[2] It must be tested many more times, in many different ways, by different groups, and braided together with many more strands of scientific knowledge to make an interdependent rope. Finding one single strand that can be cut to break the rope is then very unlikely.

Yet, the senator's aide asked a good question, and a better answer may be useful in understanding our climate future.

First, I don't have a theory of global warming, but the scientific community does have a theory of climate, which predicts that increasing CO_2 in the air will have a warming influence on Earth's surface. And while there isn't a single strand supporting this understanding of climate, we can ask whether there might be an especially strong strand or strands in the rope. I believe almost all of the scientists in the field would start with the interaction of atmospheric CO_2 with Earth's heat energy.

Even here, though, there are many strands. The effects of CO_2 on energy transfer are observed in the laboratory and the field with various instruments, and calculated independently from quantum-mechanical principles. Satellites looking down see that Earth's atmosphere is blocking energy in just those wavelengths that the laboratory measurements and the calculations show are blocked by CO_2 and other greenhouse gases, and the satellites have seen increasing blockage over time as the levels of greenhouse gases have risen, while this increasing blockage is beautifully explained by the physical understanding of the greenhouse gases.[3] But to make sense of all this, we need a bit more information about the energy balance of the planet.

GLOWING BRIGHTER—THE GREENHOUSE EFFECT

What is the use of a house if you haven't got a tolerable planet to put it on?
—Henry David Thoreau[4]

Of the planets of the solar system, Mars, Venus, and Earth could be three sisters. Earth and Venus especially are almost exactly the same size and quite similar in composition, while Mars isn't too different. But cold, red Mars or hot, white Venus would have been intolerable for Thoreau's house, with only blue-green Earth being just right. Why?

There is an easy answer: planets closer to the sun are hotter. But Venus is especially bright in our night sky because almost all the sunlight reaching Venus is reflected away without warming the planet. A naive calculation including this reflection would make the surface of Venus almost as cold as Mars and frozen solid, not hot enough to melt lead as documented by space probes.[5] Let's take a short look at the temperature of your dinner, to see why Venus is so warm and why fossil-fuel burning will turn up Earth's thermostat.

Touch the heating element—the "burner"—of an electric stove or toaster that has been turned off for a long time, and you will find that the burner is the same temperature as the rest of the room. Now, turn on the stove. By doing so, you are using energy to push electrons through the burner, and the resistance to their motion generates heat. Initially, that heat primarily goes into warming the burner. But as the burner gets hotter, it begins to dump more of that extra heat into its surroundings, warming them through some combination of conduction, convection, and radiation, as we will see next.

If you were silly enough now to touch the burner, you would very quickly understand how it got its name. The rapidly vibrating atoms of the burner would collide with the atoms of your skin, speeding them up and perhaps knocking them out of your skin altogether as heat is conducted into your finger to make it sizzle and smoke. If you are wiser and keep your finger away, the air right next to the burner will be heated instead, expanding and rising to warm its surroundings by convection.

And as you watch the burner, you will see that it is producing light, first a faint deep red glow, then brighter and yellower as the burner warms. This glow is called "electromagnetic radiation," or just "radiation" for short (not to be confused with radiation from an atomic bomb, which includes both electromagnetic radiation and rapidly moving particles; we will learn more about this kind of radiation later, in the chapter on nuclear power).

Eventually, if you keep supplying electricity at the same rate, the temperature of the burner will stabilize, which occurs when the burner transfers energy to the air or your food at the same rate that electrical energy is supplied. If you then turn the stove to a higher setting, supplying more electric energy, the burner will warm before stabilizing at a new, hotter temperature, glowing more brightly and with the color shifted at least a little way from the red end toward the blue end of the spectrum, although the burner would need to get much hotter to look blue. The burner rather quickly reaches the temperature that is needed to send out energy at the same rate that it is received.

To lose heat by conduction, the burner must touch something, and to lose heat by convection, whatever the burner touches must move away. But the radiation that you see as a red glow doesn't require touching anything. All things that are warmer than absolute zero are always radiating energy, glowing. Warmer things glow more brightly, and have more of their glow at shorter wavelengths with higher energy, moving from beyond the infrared range where we humans can't see, into red, orange, yellow, and up to or past blue into the ultraviolet and beyond.[6]

A planet cannot lose or gain enough heat by conduction or convection to be important, because space is simply too empty—there is virtually no stuff out there to conduct or convect. So the temperature of a planet is controlled by radiation.

To start, we can assume that the incoming energy and outgoing energy for a planet are equal.[7] The incoming energy is virtually all sunlight—other stars just aren't close enough to supply enough energy to matter, and the radioactive decay within the planets supplies too little energy to be important (we will come back to this later, because the planet's heat can help power humanity). Venus gets more incoming energy than Earth, and Mars less, because of their distance from the sun.[8] Some of the sunshine reaching a planet is reflected from clouds or snow or other things and bounces right back to space without warming anything. Venus, with its thick cloud layer, reflects about 80 percent of the sunlight reaching it, so we say it has an albedo of 80 percent ("alb" means "white," as in "albino," so albedo is the whiteness). Earth—with dark oceans balanced out in part by puffy white clouds and gleaming snow— reflects about 30 percent, and the almost cloudless Mars is darker yet, with an albedo of only about 22 percent. The rest of the sunshine is absorbed to warm each planet, and must be radiated back to space by the glow of the planet.[9]

From the basic physics of radiation, doubling the absolute temperature (the number of degrees above absolute zero) of an object causes the rate at which it radiates energy to go up sixteenfold. Stated differently, a 1 percent increase in the absolute temperature causes a 4 percent increase in the energy radiated.[10] Many physicists have commented on the remarkable ways in which the universe is favorable for our existence—tweak the fundamental laws or constants a little bit, and we wouldn't exist. Because of radiation physics, it takes a huge amount of energy to raise the temperature much, so we don't fry when the sun comes up in the morning or freeze when the sun sets in the evening, and as a consequence, we're here.

Anyway, if you figure out how much sunshine is arriving at the top of the atmosphere of the red, white, and blue-green planets, subtract off the reflected part, distribute the rest around the planet, and ask what temperature is needed to radiate that much energy back to space, you get some rather surprising answers: Mars is cold at −78°F (−61°C), as expected, but Earth comes in at 0°F (−18°C), and Venus at a remarkable −63°F (−53°C). So much sunshine is reflected from Earth, and even more from Venus, that they should join Mars in being frozen. But they don't. The surface of Mars, beneath its thin air, is 11°F (6°C) warmer than this simple calculation would indicate. Earth, with its thicker atmosphere, also is warmed more, by 59°F (33°C). And Venus, shrouded in its thick atmosphere, has a surface that could melt lead, 855°F (475°C) warmer than indicated by this simple calculation for radiative equilibrium.

The calculations done here are good and based on good physics. So our very simple model must have omitted something. We know this something as the "greenhouse effect."[11]

Start by looking at Venus with a telescope, and you will see sunlight reflected from the tops of clouds, not the rocky surface. If you could put on infrared goggles and see the longer wavelengths being emitted by Venus, you would still be looking at the tops of clouds—"seeing" the surface of Venus through the thick clouds is very difficult.[12] To send out as much energy as it receives, Venus must radiate at −63°F (−53°C), but that is the temperature at the top of the clouds, not at the surface or in the center of the planet, because radiation cannot escape directly to space from those places.

Any mountain climber on Earth knows that a high peak is colder than the base camp, for fairly simple physical reasons.[13] Venus has a really "thick" atmosphere that weighs about ninety times as much as Earth's atmosphere (the surface pressure is a bit over 1300 pounds per square inch, or about 9 million pascals), and that is primarily made of CO_2. As we will see in a moment, the heat-trapping "greenhouse" effect of that CO_2 is what makes Venus's surface hot enough to melt lead.

On the other end, Mars has a tiny greenhouse effect. Look at Mars through either your regular telescope or your infrared goggles and you will still see the surface, or close to it, except when it is fuzzed out by an occasional dust storm. The radiation you see from Mars is coming from closer to the surface than that from Earth or Venus. Mars's atmosphere is mostly CO_2, as on Venus, but the pressure at the surface is only 0.6 percent of that on Earth—not much! The greenhouse effect on Mars is still 11°F (6°C), so the thin atmosphere does a little, but not a lot.

Now, for Earth. Look down from a spaceship or satellite, and at times your eyes would see beautiful blue oceans, green forests, and dazzling white ice caps, but sometimes you would be looking down on the tops of clouds or on smoke and haze from human-caused or natural fires. The

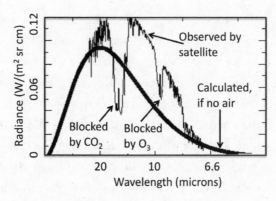

Figure 6.1 NASA satellite data showing the "greenhouse" effect of CO_2 and ozone (O_3) on the energy leaving Earth.

perky television weatherperson sometimes shows you how Earth looks in a particular infrared wavelength that is blocked by water vapor. You see the surface in dry places, but the surface is blocked by water vapor in other places, and you see the little bit of radiation coming from cold places high in the atmosphere above most of the water vapor, highlighting the fascinating swirls of wet storms. At certain other wavelengths, you would see the effects of CO_2, methane, ozone, or other gases, and even see the plume of CO_2 coming off cities.[14]

This greenhouse interaction of the atmosphere with radiation is observed fact. We have had satellites looking down for decades, and we use their views to help forecast weather, monitor pollutants, and in other ways. The atmosphere does absorb most of the energy emitted from Earth in certain wavelengths.[15] Physical understanding based on laboratory measurements and quantum-mechanical calculations does allow very accurate calculation of the radiation measured on Earth and in space. Satellites have seen the drop in radiation leaving Earth in certain wavelengths as greenhouse gases have risen, and that drop is very accurately explained by the physically calculated effects of the measured rise in greenhouse-gas concentrations.[16] Although the blogosphere remains muddied by counterclaims, I know of no credible scientific objection to the warming effect of rising concentrations of greenhouse gases.

How does this radiation blockage work? Start with a molecule of CO_2, the main greenhouse gas on Mars and Venus, and the second most important one (after water vapor) on Earth.[17] A CO_2 molecule consists of a carbon atom with oxygen atoms stuck on opposite sides in a straight line, held together by chemical bonds. Think of the atoms as bowling balls, and the chemical bonds as springs attached to the bowling balls. The molecule can vibrate as the oxygen bowling balls stretch and squeeze the springs, or as they flex sideways on those springs. And as they flex, the molecule can rotate. Getting these rotations and flexings going takes energy. And just as springs tend to vibrate at certain rates but not others, the molecule's vibrations are "quantized"—only certain vibrations are observed, and the energy of a molecule vibrating in one of these ways is always close to a well-known value.

Modern drivers passing potholes may have a useful insight to greenhouse gases (if the drivers can remain calm long enough to recognize the analogy). The giant tire of an earthmover will easily roll over a pothole. If traffic is safely blocked, a child could "drive" a tiny toy car down one side of the pothole, across the bottom, and up the other side without any trouble. But if the tire on your car drops into the pothole, the tire blows, the rim bends, and you're stuck. In somewhat the same way, both long and short wavelengths of light will pass a CO_2 molecule without interacting with it, just as the huge tires of the earthmover and the tiny tires of the toy passed the pothole. But light at appropriate wavelengths is captured by the CO_2 molecule, in the same way that the "just wrong" tire on your car was captured by the pothole.

After capture, the molecule is vibrating—the wave of light is gone, but its energy has excited the CO_2 molecule. Usually, the excited CO_2 molecule loses this energy by passing it along in a collision, generally to a nitrogen or oxygen molecule because they dominate the atmosphere.[18]

Thus, when CO_2 molecules absorb energy from infrared radiation, the surrounding air is warmed and not just the CO_2 molecules. Occasionally, an excited CO_2 molecule will lose its extra energy by emitting infrared radiation, of about the same wavelength that was absorbed, and that emitted radiation goes in a random direction. Because most of the infrared radiation absorbed by CO_2 and other greenhouse gases was going up, and the radiation subsequently emitted by these gases goes about equally in all directions, the net effect is to reduce the upward escape of radiation to space, while warming Earth and the atmosphere.

BOGUS BAND SATURATION

A century ago, misinterpretation of an experiment had suggested that the atmosphere contained enough CO_2 that all of Earth's outgoing radiation was blocked, so adding more CO_2 wouldn't have more effect.[19] That was wrong, and the error probably should have been evident even then. However, some decades passed before the scientific community sorted out the problem, and some people outside the scientific community still have not read about the solution and accepted the correct interpretation.

Much of the fundamental research that allowed proper understanding of the greenhouse effect was done for very different reasons. Especially during World War II, the military came to rely heavily on airborne operations, and this led the military to seek better understanding of the air, first for operations and communications and later for such advances as heat-seeking missiles. The hot exhaust of an enemy bomber is a target, but in certain wavelengths, CO_2 and water vapor make "vision" difficult by blocking radiation, whereas other wavelengths are much clearer. Designing sensors to see the hot target through the "swamp" of greenhouse gases in the air required an understanding of those gases. For this and many other reasons, military researchers became interested in radiation processes in the atmosphere.

In the United States, work at the Air Force Geophysics Laboratory at Hanscom Air Force Base in Massachusetts led to what is known as the AFGL Tape, which has an immense store of carefully measured data relevant to radiation in the atmosphere. These data were *not* collected to promote global-warming science or to foment some United Nations–led world government plot; they were collected to aid in military operations, and later were applied in other ways, such as in the development of heat-seeking missiles, that were highly practical for the military. But, the atmosphere doesn't care whether you study it for warring or warming, because the physical processes are the same.

When researchers from universities, industry, and the military applied the improved understanding of greenhouse gases from the military research, they found that indeed adding more CO_2 to the air will have a warming influence, for two reasons. First, the air thins as you go upward, and there will always be some height at which the CO_2 becomes too scarce to block a lot of the outgoing radiation. CO_2 added to the air is quickly mixed by the winds to and above that height, blocking some radiation there and thus raising the height for radiation escape. As described earlier, the energy balance of the planet fixes the temperature at the height for radiation escape, and the air warms downward,[20] so adding CO_2 to raise that height warms the surface.[21]

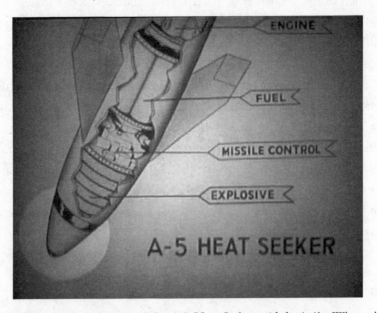

Figure 6.2 Cutaway drawing of the A-5 Heat Seeker guided missile. When others applied the fundamental physical knowledge gained by the Air Force in research for many military applications including heat-seeking missiles, the warming influence of CO_2 became unavoidably clear.

Figure 6.3 U.S. Air Force jet test-fires a heat-seeking Sidewinder missile.

Furthermore, additional CO_2 has a warming effect at all elevations, not just way up high. If you drive over the edge of a pothole, your tire may not fall in. But if it nearly falls into enough potholes, eventually your luck will run out and one will get you. In the same way, CO_2 is really good at intercepting just-right wavelengths, but it stands some chance of intercepting not-quite-right wavelengths. There is always some not-quite-right energy escaping that could be intercepted by additional CO_2 molecules. CO_2 absorbs energy in certain "bands" of the spectrum, but those bands are not "saturated"—adding more CO_2 intercepts more radiation and so has a warming effect.[22]

Any business manager knows that hiring additional people can speed up completion of a job, but that doubling the number of people on the payroll doesn't double the speed because some of the workers will end up duplicating the efforts of others. CO_2 acts in much the same way; in round numbers, if you double the amount of CO_2 in the air, you get some amount of warming, and if you re-double that amount of CO_2 you get that much more warming. Put differently, doubling the warming requires quadrupling the CO_2.[23]

A TINY TERROR?

At first, you may find it odd that the little bit of CO_2 in the atmosphere can have a notable effect on climate. If the entire atmosphere were suddenly turned to liquid at the density of water, you would have a layer of liquid air just over 30 feet thick (about 10 m). Most of that layer would be oxygen and nitrogen. Water would account for about an inch (25 mm) on average, but it would range from almost nothing to a little over 2 inches (50 mm)—all the rain and snow come from that small thickness, because evaporation replaces the water as rapidly as it falls out. Condensing the CO_2 in the air would yield a bit over 0.1 inch (3 mm). Think how differently you would feel working in a humid tropical rain forest with 2 inches of water over you, compared to a desert with little water in the air; you probably can sense shifts in humidity that would translate to a change of 0.1 inch in the water over you.

Before we humans started serious fossil-fuel burning, CO_2 in the atmosphere stood at approximately 280 parts per million (ppm), and we have raised it past 390 ppm.[24] A small number, right? Add another 0.1 inch of nitrogen around the globe to the 24 feet (7 m) already there, and no one would notice. Shrink-wrap the globe in a 0.1-inch-thick black plastic bag, and you would block the sun and kill us all. The U.S. Centers for Disease Control and Prevention recommends that we limit occupational exposure to carbon monoxide, CO, to less than 35 ppm.[25] For hydrogen cyanide gas, the U.S. Department of Labor Occupational Safety and Health Administration reports that just 135 ppm will be fatal in 30 minutes.[26]

I have absolutely no difficulty finding many examples in which a trace of some material is completely insignificant, and other examples in which a trace is highly dangerous. This is why scientists are hired to study these things—we need hard data and real understanding, not vague guesses about how much is too much. And the hard science says that the trace of CO_2 up there is important to our climate, and that changing the CO_2 concentration up there will change the climate.

At first glance, though, the warming from doubling the atmospheric CO_2 level doesn't look hugely scary. There are some complexities because of interactions with clouds and with dust and other particles, but a doubling of CO_2 for conditions similar to those of modern Earth will warm the surface about 2°F (1.1–1.2°C) if everything else is held constant. Despite the occasional frothing of the blogosphere, there is no credible question about this result. There are still plenty of people who seem to question the ability of the military to see the hot exhaust of an enemy bomber, but those who are more interested in results than in arguments have long since moved on. And, as we will see in the next chapter, everything else cannot be held constant, and this is not good news.

7

Canting the Kayak

Synopsis. Warming increases the amount of water vapor in the air, melts reflective snow and ice, and causes other changes that amplify the warming. Over times of years to millennia, these positive feedbacks exceed the negative feedbacks that tend to reduce warming. Doubling CO_2 in the atmosphere is expected to warm Earth about 5°F (3°C), more than twice the direct effect of the CO_2.

> *Sunshine cannot bleach the snow*
> *Nor time unmake what poets know.*
> —Ralph Waldo Emerson, "The Test"[1]

STABILIZING A GOLDEN RETRIEVER

come from a family of bird-watchers. We have found that for water birds, a kayak is almost as essential as binoculars. Whether sneaking up on little "peeps" in a Cape Cod salt marsh, or drifting below eagles perched along a tree-shaded Pennsylvania stream, the boat gives a view that is not available from dry land.

However, a kayak is not quite as stable as dry land. Set a kayak in the water, watch it rocking with the waves, and you might find it hard to believe that anyone sitting in such an unstable contraption could successfully distinguish a semipalmated plover from a piping plover, or even a bald eagle from a box kite.

Fortunately, we humans have a host of built-in stabilizers that allow us to see the birds clearly. When a wave forces you and the boat and the binoculars to wobble, you respond by tightening some muscles and loosening others, tilting you and the binoculars so that the initial wobble is almost neutralized. The response of your muscles is called either a "stabilizing feedback" or a "negative feedback"—when the wave forces your body to wobble, the feedback response of your muscles acts to reduce the initial change.

The first time we put our little kayak in the water is still the only time we unintentionally lost anyone overboard—there have been a couple of really hot days when kayakers intentionally ended up wet, but that's different. When a certain dear person first got in and the boat started to tip, she did the perfectly natural thing for dry land, reaching out to steady herself. However, pushing against water does not work, and the extra weight as she leaned and reached for the water tipped the boat further. Fortunately, she was uninjured, and the water wasn't toooo cold.

This kayaker's response was a positive feedback—the forcing of the leaning boat caused her to reach toward the water, which caused the boat to lean more. In this case, the positive feedback crossed a threshold—a tipping point, if you prefer—and the boat tipped over. Much more commonly, positive feedbacks amplify without tipping. For example, if you go canoeing with an ice chest and a golden retriever, the forcing of the golden retriever leaping to the side to bark at a deer or a bear or an air molecule causes the boat to tip a little, which causes the positive feedback as the ice chest slides to the side to tip the boat more, but the dog and ice chest usually remain dry because the canoe doesn't tip all the way over.

We saw in the previous chapter that if there were no feedbacks, doubling the CO_2 in the atmosphere would cause a warming of about 2°F (just over 1°C). This number comes from the stabilizing effects of radiation—as rising CO_2 traps energy to cause warming, the warmer Earth "glows" more brightly to send more energy to space. But Earth has other feedbacks, as the warming affects water vapor, snow, and more, and these changes affect temperature, so the calculation from the previous chapter is incomplete.

Snow and ice are probably the easiest to understand. Anyone who has walked on snow on a sunny day knows that almost all the sunlight is reflected from the snow. If you are walking toward the sun, the reflected light adds to the direct sunlight, and you should be grateful that modern sunglasses allow you to avoid the rapid onset of snow blindness—essentially sunburn of the eye—and the slow onset of problems such as cataracts.

The British polar explorer Robert Falcon Scott participated in the British National Antarctic Expedition (the Discovery Expedition) of 1902–1904, reaching farther south than anyone had gone before. (He died in 1912 on a subsequent expedition.) While trekking south in 1902, one of his companions, Edward Adrian Wilson, contracted snow blindness. Scott's short description in his 1905 book is sobering:

> December 26. Poor Wilson has had an attack of snow-blindness, in comparison with which our former attacks may be considered as nothing; we were forced to camp early on account of it, and during the whole afternoon he has been writhing in horrible agony. It is distressing enough to see, knowing that one can do nothing to help. Cocaine has only a very temporary effect, and in the end seems to make matters worse. I have never seen an eye so terribly bloodshot and inflamed as that which is causing the trouble, and the inflammation has spread to the eyelid. He describes the worst part as an almost intolerable stabbing and burning of the eye-ball; it is the nearest approach to illness we have had, and one can only hope that it is not going to remain serious.
>
> December 27. Late last night Wilson got some sleep, and this morning he was better; all day he has been pulling alongside the sledges with his eyes completely covered. It is tiresome enough to see our snowy world through the slit of a goggle, but to march blindfolded with an empty stomach for long hours touches a pitch of monotony which I shall be glad to avoid."[2]

The native peoples of the far north understood snow blindness, and how to avoid it, long before modern sunglasses, making goggles from wood or bone with tiny slits to limit the amount of light coming in. Scott and Wilson had slit goggles too, but perhaps not as good or used as well as those of the Inuit.[3]

The sunlight reflected from the snow warms the eyes a little as well as causing damage, but most of the sunlight reflected from Earth's snow and ice goes back out to space without hitting anyone's eyes or otherwise warming the planet. Melt the snow to expose the darker dirt or bushes beneath, and you can take off your sunglasses or slit goggles. You probably can take off your sweater too, because the warmer ground will be warming the air.

Scientifically, this is called the "ice-albedo feedback": warming melts snow and ice to reveal darker materials beneath, allowing more sunlight to be absorbed, causing more warming. Cooling allows snow and ice to be more widespread, reflecting more sunlight, causing more cooling. Calculating the strength of the ice-albedo feedback requires some care.

Figure 7.1 Modern travelers on snow and ice are surely aware of the high albedo, and glacier glasses make life easier than it was for early Antarctic explorers. Here, the author pauses during the filming of Earth—The Operators' Manual, *on the névé of the Franz Josef Glacier, New Zealand.*

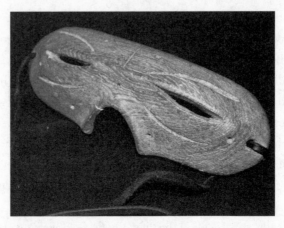

Figure 7.2 Native people of the north lacked dark plastic lenses but found practical—and beautiful—ways to handle the high albedo of snow, such as these Inuit slit goggles.

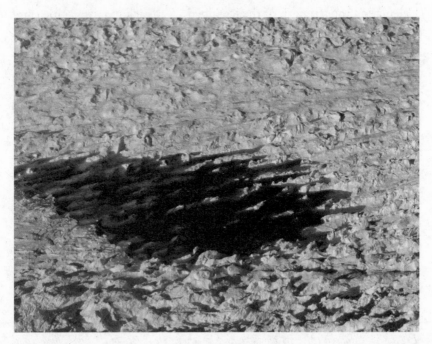

Figure 7.3 When warming begins to melt snow, the albedo drops, allowing the sun to cause more melting. Low places on the surface of the Greenland Ice Sheet often contain dark lakes during the summer, as shown here on Daugaard-Jensen Glacier in East Greenland.

For example, snow in the shadow of a ridge or a pine tree isn't reflecting much sunlight, and some cold places are not snow-covered. But in general, colder places are snowier, cooling favors snow, and snow favors cooling.

Much more important globally is the water-vapor feedback. Many of us experience this in everyday life—hair dryers are built to warm the air they blow because this picks up more water. In the same way, warmer air blowing over the vast ocean or over growing plants picks up more water vapor. For air that is "full" of water (saturated, or in balance with water beneath), warming of 1°F increases water vapor about 4 percent (7 percent per 1°C).[4] Water vapor is a greenhouse gas, and in most places it gives a little more warming than CO_2 does.[5] So warming increases water vapor, and water vapor increases warming.[6]

There has been a lot of effort to find something, anything, that might allow us to escape this simple reasoning and the supporting data—if water vapor didn't amplify warming, we might not need to worry so much about our CO_2 emissions. But despite a lot of smart and hopeful people looking very hard, the data keep pointing the other way, common sense works, and warmer air holds more water vapor to amplify the warming.[7]

This is where things get more interesting, and worrisome. First, suppose you turn up the temperature by 1°F by turning up the sun, or adding CO_2 to the air, or doing something else, while holding everything else constant.[8] Now, let the snow and ice change but nothing else, and you get a bit more warming—for the sake of this example, say that the total warming is 1.1°F.[9]

Next, go back to the original 1°F warming with all else held constant, and let the water vapor change but not the snow and ice. Water vapor is found all around Earth, whereas snow and ice are not, helping water vapor have a greater global effect—say, 1.6°F.

But now, what if you let both water vapor and snow change? The warming from the water-vapor feedback will cause some snow to melt, because snow doesn't care why the temperature rose but only that it did rise. Furthermore, the warming from the ice-albedo feedback will increase water vapor, which will cause more warming. The water-vapor and ice-albedo feedbacks together would take the initial 1°F warming and amplify it to 1.9°F.[10]

Still other feedbacks exist. Snow can bury the short plants of the tundra. If warming allows taller shrubs to invade the tundra (as is observed to be happening in some places already),[11] the shrubs may stick up above the snow and allow more absorption of sunlight. This positive feedback may be opposed a little by a negative feedback occurring a little closer to the equator, if warming causes replacement of dark conifers by deciduous trees that allow the winter sun to hit snow. And if warming causes an expansion of grasslands or deserts into darker forests and allows more reflection of sunlight, a little cooling results.[12]

Clouds are especially interesting for the climate. Cloudy days are usually cooler than clear days, and cloudy nights warmer than clear nights. Low, thick clouds have a net cooling effect, with the daytime cooling exceeding the nighttime warming. High, thin clouds have a net warming effect, letting in most of the sunshine but efficiently blocking Earth's outgoing radiation. The net effect of all clouds taken together is to cool the planet slightly.[13]

With warming, the balance of the evidence indicates that clouds have a net positive feedback, amplifying climate changes—the total cooling effect of the clouds becomes smaller as the temperature goes up. The high, thin clouds that warm the planet seem to increase with warming, as water vapor reaches higher in the atmosphere. In some places, such as polar regions where warming is causing sea ice to break up, low, thick clouds that cool the planet may increase as more water vapor reaches them, fighting the warming and giving a negative feedback. In other places, the low, thick clouds seem to decrease with warming.[14]

Put all of these feedbacks together and more, including the positive and the negative, and the sum of them is notably positive. If you double the CO_2 in the air and let things come into balance, the direct warming from the CO_2 alone is in the neighborhood of 2°F (just over 1°C), but the total warming including the feedbacks is likely to be between 5°F and 6°F (about 3°C; this total warming from doubled CO_2 is usually called the "climate sensitivity"). But even with all of the uncertainties, you are really, really optimistic if you expect the warming from doubled CO_2 to be less than 3°F, and there is some chance of going above 8°F or even 10°F (2–4.5°C or higher). The uncertainties are large, mostly because we don't know as much about clouds as we would like, despite a lot of effort to understand them better.[15] The uncertainties include more "room" on the high side because of the interactions among the various feedbacks—if scientists have overestimated the positive feedbacks, the warming is pushed down toward the CO_2-only value, but if the positive feedbacks are underestimated, the interactions among them can cause a very large warming.

There is enough fossil fuel in Earth to quadruple and perhaps sextuple or octuple the level of CO_2 that persisted in the atmosphere for millennia before the Industrial Revolution—enough for two or even three doublings (see chapter 4)—and it is very optimistic to think that we will hold our influence to a single doubling without a big effort. Each doubling has about the same influence on temperature. So an optimist might make a back-of-the-envelope calculation and argue that we face only 3°F (about 1.7°C) of warming from a single doubling with a low climate sensitivity, and we might not get even that much if the CO_2 level drops before the ocean finishes warming, so perhaps global warming is no big deal.[16] A pessimist might fear that we face a global warming of 25°F (14°C) or more in a burn-it-all world. Somewhere in the 12 to 15°F range (7–8°C) under a burn-it-all business-as-usual scenario—a notably larger change than the 9 to 11°F (5–6°C) of warming that took place from the ice age to the preindustrial world (see chapter 11)—may be a useful starting point for discussions.

That warming from the ice age had huge effects on the planet, and we will see in a few chapters that an even larger warming could have even bigger effects on us. However, just taking the amount of CO_2 and an estimate of the climate sensitivity, as I just did, oversimplifies more than we should; we need to put some more physics back into the mix. And to do that, and to get a better idea of what such a warming might mean, we need a quick consultation with an expert—your favorite accountant.

8

Why Accountants and Physicists Care about the Past

Synopsis. The physical understanding of Earth's climate can be used to make useful estimates of future climate changes in response to increasing CO_2, if appropriate care is taken developing and testing the tools used.

Always live within your income, even if you have to borrow money to do so.
—Artemus Ward, American humorist[1]

To state the facts frankly is not to despair the future nor indict the past. The prudent heir takes careful inventory of his legacies, and gives a faithful accounting to those whom he owes an obligation of trust.
—John F. Kennedy[2]

WHODUNIT?

The congressman was animated. Here were scientists, including me, sitting before him claiming that CO_2 causes warming. In the discussion, we had mentioned ice-age cycles. Beautiful ice-core records from Antarctica show 100,000-year cycles, paced by features of Earth's orbit, bumping along through time. Just glancing at a graph of the data, the history of temperature looks almost identical to the history of CO_2 in the atmosphere.

But very careful analysis shows with moderately high confidence that as the ice ages ended, the CO_2 level started to rise a few centuries after the temperature did.[3] The early temperature rise was small, and the CO_2 level and temperature moved together during almost all of the warming period, but the temperature seems to have been leading the CO_2. The congressman stated that he had been told that this invalidates CO_2 as the cause of global warming; if warming causes CO_2 to rise, then CO_2 rise cannot cause warming because the effect always comes after the cause. However, the congressman had been given an argument based on badly flawed logic; I doubt that the congressman or anyone else, on seeing a chicken lay an egg, would argue that this proved that the egg could not hatch into a chicken.

However, in my reply I avoided poultry and instead tried an explanation that I repeat for you here. At the end, the congressman stated that he didn't understand me. I am doubtful that subsequent communications have helped. This probably reflects some combination of my fail-

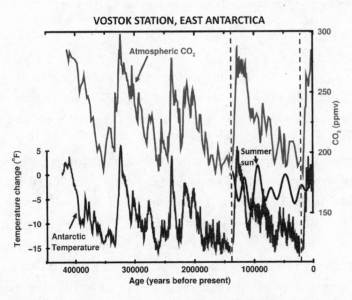

Figure 8.1 *History of temperature and CO_2 concentration from East Antarctica, plus midsummer sunshine for the more recent part of the record. The vertical lines show times of high sun, low temperature, and low CO_2, and also highlight that a tiny bit of warming seems to have occurred before the CO_2 started to rise near these times.*

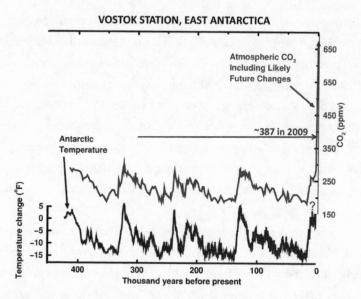

Figure 8.2 The previous figure must be squeezed down to leave room to show just part of the human increase in CO_2 that could happen under business as usual. This curve stops after slightly more than doubled CO_2, but we can raise CO_2 beyond that.

ure to explain clearly, and the near impossibility of having a meaningful exchange of ideas in a short briefing or hearing in a politically charged atmosphere. Anyway, here's another attempt, just in case the congressman or one of his aides is reading this.

Try an analogy first. (This is *not* something that actually happened!) I buy too much stuff with my credit card and go into debt. The credit-card company is gleeful; now it can collect interest payments from me. The interest causes me to go deeper in debt. As I near bankruptcy, I ask an accountant to figure out how I came to be so broke and how my finances might change in the future, to guide me in dealing with the debt.

If my accountant followed the logic reported by the congressman, my interest payments could not have contributed to the debt because they started after the debt, so there are no consequences of ignoring messages from the credit-card company. In my experience, accountants don't make this mistake. The interest payments in this case did not start my problem because they kicked in after the debt arrived, but they are a feedback, and we have seen that feedbacks can be very important.

Still, while my debt is growing, and my interest charges are adding up, I am getting older, and grayer, and balder. My debt correlates with many things, not just with interest payments. "Correlation does not imply causation" is the old saying; if you see a correlation—two things increasing together, or decreasing together, or one increasing while the other decreases—it may be worth investigating more deeply to see if cause and effect are involved. But many things may correlate by accident or because both are being controlled by something else (my debt rose with time, and so did my hairline, but for different reasons!). So how does my accountant tell with high confidence that the interest payments amplified my stupid overspending to get me into really deep doo-doo?

My accountant probably would do several things. First is understanding how the financial system works; anyone with knowledge of money is aware that debt is not generally caused (directly) by receding hairlines, but debt can be deepened by interest charges. Next might be to look for alternative explanations: A careful accountant would surely check whether someone was embezzling money from me, whether the bank was incorrectly applying charges, whether my identity had been stolen and used to buy things, and whether anything else strange was going on. And the accountant would also make sure that the known overspending and interest payments successfully and accurately predict the size of my debt and its history over time.

"Predicting" the past and the present might seem strange, but accountants and others do it all the time. And they do it with models. So if you think that models are the last refuge of scoundrels and fools, read on. Models surely can be abused by scoundrels and fools—witness some of the recent financial shenanigans on Wall Street—but models also are the successful, reliable, day-to-day tools of a lot of very careful people. *Any* tool can be used or misused, and models are no exception.

MODELS AND MODEL VALIDATION

Back to my hypothetical accountant—call her Amy, for convenience—studying my hypothetical debt. I want to know how fast I will go broke if I continue with business as usual, and whether selling my soccer cleats and bicycle and giving all the money to the credit-card company will rescue me. I thus want Amy to give me *projections* of the future—predictions for different "what if" scenarios involving my behavior, changing interest rates, and other factors. With those projections in mind, I presumably can choose a wise path.

To provide these projections, Amy will build a model of my bank balance. She cannot literally be the bank, but she can set up a computerized spreadsheet (the model) and input my expenditures and income, the interest rate and compounding frequency and late fees, the increase in interest rate and the extra late fee that the credit-card company tacked on when it thought no one was watching, and anything else going on with my budget.

Amy starts this model at some time in the past with my bank balance then, and runs the model to the present ("You overspent here, then the late-payment charge was subtracted, then the interest kicked in, then the interest rate went up," and so on). If the model fails to calculate the history of my bank balance correctly, then she or the bank may have made an error, or someone may have stolen my identity and be siphoning off my remaining resources, or something else may have happened. In that case, she will go back to work to figure out what the problem is.

But if the model succeeds in predicting the past by starting further in the past—that is, the right numbers come up at the end—then she has much greater confidence that all of the essential parts of the puzzle are accounted for, and she can attribute my debt to overspending and interest. She then can use the model to project my future finances and advise me on wise behavior ("If you do nothing, then your debt will rise like this, but if you cut up the credit card and sell your stuff and eat nothing but peanut butter and stale bread for the next three years, then . . ."). She does not need to wait for ten or twenty years to test the model against the future before advising me to cut up the credit card and sell the bicycle; she has already used the history of my finances to test the model for accuracy. The model is based on a fundamental understanding of income and outgo and compound interest; the implementation of these fundamentals is tested against history plus whatever small changes occur in my finances while she is working on this; and if the model passes these tests, she uses it with confidence to project the future and provide scenarios that I can use for guidance.

Amy must make sure that she uses a long enough history with enough transactions to provide a rigorous test of the spreadsheet model, especially if she had to make a couple of comparisons with the past to detect errors in the model. She also is likely to use simpler models to check on the main model—performing plain old mental math to check on the spreadsheet, for example ("$80 for soccer cleats, at 25 percent interest for one year, equals $100, bump it up a bit for compounding," and so on).

Climate scientists have done for CO_2 exactly what Amy the hypothetical accountant did for my interest payments. We will see in chapter 11 that the ice ages were caused by sunshine moving around on Earth due to features of its orbit, that the changing ice sheets and temperatures triggered feedbacks that included changes in CO_2, and that the CO_2 in turn amplified the temperature changes. Physical understanding shows that CO_2 changes must affect temperature. Efforts to predict the size and geographic pattern of the temperature changes while ignoring the effects of the CO_2 have failed, while including the effect of the CO_2 successfully predicts the temperature changes. The climate models used to do this may be somewhat less familiar than accounting programs, but they are used and tested in much the same way, as we will see next.

ACCOUNTING FOR CLIMATE MODELS

Climate models range from the very simple—just a vertical column up to space, for example—to truly and wonderfully complex representations of Earth's system.[4] Typically, a model starts with the relevant fundamental physics, including conservation of mass and energy, physically accurate radiative transfer, the "gas laws" you may remember from introductory chemistry, and so on. The model must be told about the fundamental characteristics of Earth, such as how much sunlight hits the top of the atmosphere, the strength of gravity, the distribution of land and water, and the size of the planet and its rotation rate. One test of climate modeling, in fact, is to put in the characteristics of some other planet (Venus or Jupiter, for example) and see how well the model does simulating conditions for those planets.

All models are simpler than the real world—an accountant's model doesn't include the coffee machine at the bank, for example, because it isn't especially important. A climate model realistically cannot represent every tree, or keep track of how the trees grow and when they are cut down. But the wind goes faster near the ground over a grassland than in a forest, so trees must be represented somehow if the model is to get the surface winds right. The solution is to specify a roughness parameter, which is set to a high value in forests and a lower value in smoother spots.[5]

The roughness is constrained by data—we have maps showing which areas are forests, and which are grasslands, and cities, and deserts, and we have measurements of roughness over these different surfaces. But the data always include uncertainties because no measurement can be perfect. So the modelers may adjust the parameters within the range allowed by the data to provide the best fit to "climatology," the average conditions measured over some period.[6] Note that this tuning is *not* done to match the changes in climate over time (that would be cheating), but only to match average conditions for a single time window. And the tuning is *not* free to pick whatever value gives some preconceived answer, but is constrained within rather narrow windows by real measurements.

The power of physics is demonstrated by what happens next. Turn on such a model, and it quickly simulates a climate that looks like Earth's climate. Air rises in the tropics and sinks over the doldrums in the great tropical circulation, storms scream out of the west across the mid-latitudes, ocean currents make great loops with westward intensifications such as the Gulf Stream, oscillations in the coupled ocean-atmosphere system look a lot like El Niño and the North Atlantic Oscillation, temperatures and rainfall and snowfall look a whole lot like the modern climate. If you change Earth's rotation rate, or switch the planet to have no continents, or notably change the sun's output, the model produces very different climates. If you tell the model that CO_2 is not a greenhouse gas, the simulated climate is way too cold and may drop the planet into a completely frozen "snowball" state.[7] But given the right rotation rate and land distribution, solar output and influence of gases on radiation, the model produces a climate that does look very much like Earth's.

Versions of these models are used to forecast weather, and smart companies spend money buying weather forecasts because, although forecasts always are imperfect, they are accurate enough to pay off. In general, a weather-forecasting model is run with a higher spatial resolution and more attention to getting the starting conditions right than is a climate model, but some of the climate models have been run in weather-forecasting mode, and they work just fine.[8] There are surely features that are not perfect—sometimes the model gives two bands of tropical rainfall close together where one would be more accurate, and El Niño may be off a bit in position or timing—but implementing real physics in computer models does produce usefully accurate simulations of the climate. The Gulf Stream and El Niño and other climate features are not put into the model by cheating; they emerge from the physics of the Earth system.

The ability of a climate model to match the modern climate without cheating is a first test of the model, but surely not the last one. Just as my hypothetical accountant checked the model of my bank account against its history and against the changes occurring while she worked, climate modelers check against history and evaluate the accuracy of their projections made years or decades ago.

A lot of information on this testing is available,[9] showing that the climate models pass the tests with flying colors. Earlier climate models projected many climate trends in response to rising CO_2 that are emerging in the data now being collected. These include cooling of the stratosphere, rise in the base of the stratosphere (the tropopause), amplified warming in the Arctic and decline of sea ice there, expansion of the subtropical dry zones, and increase in the fraction of rainfall arriving in short-lived but intense events. These successful predictions certainly strengthen our confidence in the utility of climate models.

The models are still models though, even after they have been tested so many different ways and found to work, and everyone in the climate-change field knows this. Thus, in the assessment process described in chapter 5, other requirements have been added. Before a result is brought forward confidently for policymakers, the assessors ask whether 1) the result can be understood from fundamental science, with no serious questions or disagreements from other fundamental science; 2) the result appears in a range of models, from simple to highly complex, built and run by different research groups on different computers, and with few or no models failing to produce the result; 3) the result appears in a range of paleoclimatic data from different research groups and from different times; and 4) the result is appearing in recent data of various types from various research groups. If not all of these are realized, then the stated confidence is lowered, or the result is not brought forward.

The warming influence of CO_2, for example, is fundamental physics that can be calculated and observed in many ways, is found to occur and to be amplified by positive feedbacks in simple and complex models ranging from one-dimensional columns to full global models, and appears in paleoclimatic and recent instrumental data, as we will see in the next chapters. Thus, we can confidently attribute many past climate changes to the effects of changing CO_2, and we can confidently project that raising CO_2 in the future will have a similar warming effect unless something else bigger intervenes (such as a massive geoengineering project to block the sun, as described in chapter 23).[10] First, however, we need to quickly visit one widely discussed model "test" that really isn't a test at all.

SEEING THROUGH CHAOS

"Your models can't predict weather three weeks ahead, proving that they can't predict climate three decades in the future." I have heard some version of this from friends and policymakers, weather forecasters and scientists in other fields. And while the first part is surely correct, the second part is absolutely wrong. A weather forecast telling you when a particular storm will rain on your picnic is reasonably accurate for no more than a few days and will never be useful beyond a week or two, but accurate climate forecasts over much longer times are possible.

To see why, suppose you roll dice while playing a board game with your family. You will have no idea what numbers will come up next. Roll the dice a huge number of times, and you will have a very good idea what numbers will come up on average. Play with dice that your brother-in-law has cleverly modified by adding a bit of weight on one side and filing off a few corners so that some numbers come up more often than others, and if you play long enough you can have high confidence that someone is cheating. Forecasting the next roll of the dice is like forecasting the weather, forecasting the average of a lot of rolls of the dice is like the work of climate science, and fossil-fuel burning is loading the dice to bring up high numbers more often (and perhaps to add higher numbers).

In this analogy, a weather forecaster would make a prediction after the dice are rolled but before they stop, by watching very carefully and very quickly how the dice are rolling. If the analogy with dice doesn't work for you, consider the popular television game show called *Wheel of Fortune*, in which a contestant spins a large wheel divided into segments; when the wheel stops, the segment beneath the pointer dictates the next action in the game. Before the wheel is spun, no one can accurately predict where it will stop, but as the spinning wheel slows it becomes easier to tell. A forecaster would monitor the already-spinning wheel very carefully, calculating various factors (perhaps speed, friction, and the wind from the cheering crowd) to predict the outcome before it is obvious to the casual observer.

A weather forecaster using a computer model must start the model with the already-spinning weather at some point in time. But if it is late autumn and the temperature in Boston is 43.2°F, is that 43.23°F or 43.19°F? And if it is 43.19°F at Boston's Logan Airport and 43.23°F at Harvard Yard a few miles away, does the temperature vary linearly between the two, or might the temperature rise to 43.4°F as you travel east from Harvard and then drop quickly once you reach the harbor on the way to the airport?

Quibbling over this might seem silly, but it matters a lot in forecasting weather. If you take your forecasting model, feed it some set of temperatures and run it for three weeks, it will give you an answer. If you then go back and tweak those temperatures within the uncertainties (change 43.2°F to 43.19°F at Logan Airport, and make similar tweaks at other observing sites), and run the model again, something very interesting happens. For the first day or two, the tweaks make almost no difference—the forecast you would give to the TV weatherperson would be the same in either case. But two or three weeks out, the model gives a very different answer.

The answers before and after your tweaks are climatologically identical—both model runs still have day and night, and produce weather that is typical of Boston in late autumn. But the individual storms and cold fronts in the two model runs started out almost identical and then diverged, so that three weeks out one run has a sunny day and the other has a storm swirling in wet snow from the northeast. There is simply no way to measure and calculate everything with sufficient precision and accuracy to beat this, so we say that weather is "chaotic," or, perhaps more accurately, "sensitive to initial conditions."[11]

This does *not* mean that "anything goes." The weather produced by the model has temperatures and precipitation and winds that are fully consistent with the time of day, time of year, and location on the planet. The model will not suddenly produce broiling temperatures in the Arctic winter, or freezing temperatures in a tropical summer; the answers are bounded by physical reality. But the weather simulated by the model is no longer useful—a forecast based on the climatological averages for a site is just as good. However, the climate produced by the model *is* useful—to learn about Boston's climate in late autumn, you can average the conditions from either run over a few weeks during autumn, and repeat that for thirty years. Change the CO_2 in the model, repeat the thirty years of runs, and you will have a useful estimate of how CO_2 affects Boston's climate.

So weather is not climate, and the inability of models to forecast weather a few weeks from now provides no information on the ability of models to project climate in a few decades. Our understanding of climate rests on fundamental science, and this fundamental science is implemented in models to help us use that fundamental science to learn about the future. Extensive testing of the models includes verifying the ability of the models to simulate the climate at a time, and to simulate changes over various times in the past. So, let's take a quick tour of the past, to see how well our understanding of climate works there.

9

The Moving Finger Writes

Synopsis. The history of many of Earth's climate changes and their causes is written in the planet's sediments. When combined with our physical understanding, this history shows that many factors, such as drifting continents and changing brightness of the sun, have affected climate, but the changes caused from outer space or from deep in Earth have usually been small or slow.

> *What seest thou else*
> *In the dark backward and abysm of time?*
> —William Shakespeare[1]

WISDOM OF THE AGES

If you were planning to open a new business in a foreign country, or to invade that country, you would probably want to know about the history of the people there. A judge deciding on the sentence for a drunk driver might consider that driver's twenty-three prior driving-under-the-influence convictions. And my hypothetical accountant in the previous chapter, when tasked to help me manage my hypothetical debt, looked at the record of my bank balance to see whether unexplained withdrawals indicated theft from my account. History really does matter—if something happened in the past, it must have been possible, it should be understood, and it can be used to test our understanding.

I often hear from people whose interest in climate history has led them to believe that human-generated CO_2 cannot be guilty of warming the climate now because the climate has changed naturally in the past. These people might not make the best jurors, because the same logic would acquit a modern arsonist if a lightning storm had burned the woods hundreds of years ago. For the modern fire, it would be better to hear all the evidence before deciding whether arson or a thunderstorm was responsible.

So let's look at the history of the climate. Almost surprisingly, the more we scientists have learned about the immense natural changes of the past, the more confident we have become that human CO_2 and other greenhouse gases will warm the climate, with little chance that nature will intervene fast enough to be helpful. On a global scale, the natural changes have almost always been much slower than what we expect in the near future if we continue with business-as-usual burning of fossil fuels, but those natural changes confirm the accuracy of our physical understanding and of the climate models that lead us to expect such rapid business-as-usual changes.

To get to this judgment, though, we need to visit the realm of paleo-climatology, to learn what happened when and why.[2]

READING THE RECORD

The story of prehistoric people is written in the things they left behind, the tracings on rock and the trash heaps behind their houses. The story of the climate is similarly written in the sediment of the Earth system.

We met sediment back in chapter 4, as mud making the oxygen-poor piles rich in dead plants that produced oil, coal, and gas. That sediment contains climate information too. Where coal formed on land, for example, the climate was wet enough to grow plenty of plants and to keep the dead ones under water, certainly not desert conditions. The types of plants give clues to the climate too—tropical trees, or tundra flowers? The sediment deposited in a lake is easily distinguished from a sand dune, and both differ greatly from the unique features produced by glaciers, so the sediment type helps in learning the history of climate.

Many indicators of past climate are more subtle but very informative,

Figure 9.1 A beam used in construction of Long House at Mesa Verde National Park, Colorado. Such tree rings can be used to learn ages because one ring formed per year. In a dry region such as the desert in the southwest United States, the thickness of rings is especially influenced by availability of water, so trees can provide paleo-rain gauges.

often based on chemical or isotopic characteristics of shells or organic matter. To get an idea of the wonderful possibilities of this field, consider one of the many techniques used to learn the history of temperature of the sunlit waters near the ocean surface. Plants need sturdy cell walls to keep their insides in. Some algae construct their cell walls partly of robust molecules called "alkenones," with thirty-seven or thirty-nine carbon atoms in a line. The plant can string the carbon atoms together using either single or double bonds, and changing how many of each helps control how flexible the molecule is. When colder water makes the cell wall more brittle, the plant can adapt by making the molecules more flexible.[3] Because these cell-wall molecules are so sturdy, most of them survive being attacked by bacteria or being eaten and pooped by worms, and so they end up in the pile of mud at the bottom of the sea. Pull out some mud, extract these alkenones, measure the number of double bonds, and you have a temperature estimate for the water in which the algae lived.

Double-bonded cell-wall molecules may not be quite as easy to read as a thermometer on your windowsill, so those of us who study the history of climate like to estimate the temperature in a few different ways. Shells are usually mixed in with the alkenones in the mud. If we observe a change between mostly cold-water types and warm-water types, we have an independent estimate that the temperature changed. Furthermore, both chemical and isotopic ratios in those shells are linked to temperature in the water when the shell was growing. Even pollen washed or blown into the water from nearby land might prove interesting, to see if warm-weather or cold-weather types were common on land at that time. Usually, good agreement is obtained across these and other indicators, giving us high confidence in the estimate of past temperatures.[4]

The age of sediments is also revealed in many ways. In very favorable situations, we can count annual layers, something that I have spent a lot of time doing on ice cores. Summer and winter snow look different, and are chemically and isotopically and physically different. We can check this in many ways, such as chemically identifying little pieces of ash spread around the world by huge historically dated volcanoes. Most ages are estimated from radiometric (or radioactive) techniques, which are tested carefully against historical records and annual-layer counts for the younger samples, and by comparing independent radiometric techniques for older samples.[5]

We need to know not only when the climate changed, and how it changed, but also what caused the climate to change. Some possible causes of climate change come from outside the planet, but many are home-grown. For example, a big, explosive volcanic eruption spreads ash and sulfuric acid around the world to block the sun and cool the climate for a year or two, and the ash then falls into sediments where we can find it and learn that the volcano erupted.[6] The lava flows from less explosive eruptions have little magnetic-mineral "compasses" that

Figure 9.2 When a lava flow cools, such as this one on Kilauea flowing into the sea, the magnetic "compasses" are frozen parallel to the Earth's magnetic field. Later movement as continents drift can be traced using this magnetic record.

"freeze" as the flow cools, and these tell us about the strength of Earth's magnetic field at that time and where the volcano was. So we can find histories of drifting continents and changes in the magnetic field to compare to the climate history.[7]

Some people have suggested that cosmic rays might matter to climate, and we can find a record of them too. Earth's magnetic field interacts with the cosmic rays from deep space, and fewer cosmic rays get through when the magnetic field is stronger. Cosmic rays that do reach Earth break molecules in the air to make odd "cosmogenic" atoms such as beryllium-10 that then fall onto glaciers or into other sediments, so fluctuations in cosmic rays are recorded by variations in the rate at which beryllium-10 was delivered to sediments.[8]

Beryllium-10 also helps us understand changes in the sun. Over the last thirty years or so, we haven't needed "proxy" data such as levels of beryllium-10 because we have had good satellites measuring the sun's total energy output directly.[9] That output has varied just a little (about 0.1 percent from highest to lowest) with the approximately eleven-year sunspot cycle. To learn about the sun before reliable satellites circled the planet, this relation between the sun's brightness and the number of sunspots can be combined with observations of sunspots by astronomers. In addition, the solar wind has been stronger when sunspots were more common, and a more-active solar wind lets fewer cosmic rays reach Earth. Hence, the record of beryllium-10 is influenced by sunspots as well as by the magnetic field. The history of magnetic-field strength from lava flows can be used to learn how much of the variation of beryllium-10 is explained that way, and the remainder of the variation is linked to changes in the sun, providing estimates of the sun's brightness before astronomers started counting sunspots.[10]

If a lot of space dust got between us and the sun, or lots of rocks from space fell on our heads, they would supply material that is chemically and isotopically very different from materials commonly found near Earth's surface.[11] Thus, if we find a widespread layer in ice or sediment containing a lot of these outer-space indicators, we know that more meteorites or space dust were being delivered to Earth, whereas failure to find such layers allows us to eliminate space stuff from our list of possible causes of climate change.

Back on Earth, the level of CO_2 in the air has been recorded continually since 1957 from direct measurements, which were begun and maintained for decades by Charles David Keeling of the Scripps Institution of Oceanography in California.[12] Instrumental data from before 1957 are too scarce or too inaccurate to provide reliable reconstructions. For measurements of CO_2 over the last 800,000 years, we rely on the levels found in air trapped in ice cores. Good agreement is found between measurements of ice-core air trapped since 1957 and Keeling's history. Different Antarctic cores collected from areas with different tempera-

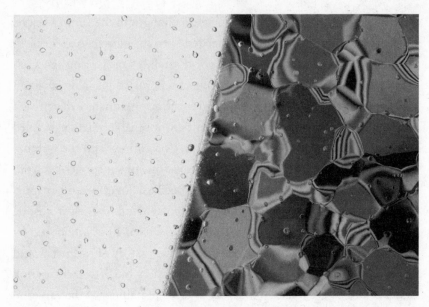

Figure 9.3 Two views of ice from a West Antarctic ice core. The left shows the bubbles that trap old air. The right uses crossed polarizing filters to show the individual ice crystals as well as the bubbles.

tures and snowfall rates and impurity concentrations give the same history of CO_2, and the records pass other tests as well, so we are confident that the history is reliable.[13]

Older than the oldest reliable ice, the history of atmospheric CO_2 is estimated from several other indicators, and agreement or disagreement among these is used to provide insight to uncertainties. For example, almost all land plants "breathe" in CO_2 through little pores called "stomata," but water goes out while CO_2 is coming in. When CO_2 is more abundant, fewer water-losing stomata are needed for plants to get enough CO_2, so the plants grow leaves with fewer stomata. Therefore, counting the stomata on fossil leaves tells us about the CO_2 in the air at the time the plant grew.[14] Other useful techniques to learn past levels of CO_2 have been developed and applied successfully.[15] And we can use "bookkeeping" to estimate the history of CO_2 in the air, by tracking changes in how many volcanoes were belching out CO_2 and how much CO_2 was going to make fossil fuels or shells.[16]

By using all the methods covered in this chapter so far, and all the related science, we can reconstruct how the climate changed over time and what was causing the climate to change. The quality of the estimates ranges from "suggested by the balance of evidence" to "pound-on-the-table near-certainty." A lot of misunderstanding arises in public discussions when the uncertainties get separated from the estimates; I will try to keep them together for you here, and lean on the "pound-on-the-table" parts.

WHAT THE STORY SAYS

When my students and I explore these topics in class, we spend a couple of weeks reviewing how Earth's climate works, and then the techniques for learning what happened when. But the biggest chunk of the semester is devoted to a blow-by-blow history of Earth's climate, starting at the beginning and rolling through to the present. The students' task is to explain this wonderful story in a way that their parents or their senators would find comprehensible and interesting. In this book, we will cover this senatorial version, or maybe the review-for-the-final version, rather than a much more comprehensive textbook treatment.[17]

We will start by looking at the historical perspective on possible causes of climate change, to find out which processes are likely to be important and which ones would be safer to ignore initially. Please note that in some sense this exercise can never end—if I show you that every hypothesized cause of climate change except CO_2 fails to explain the history of climate, you can always generate a new hypothesis. But the odds are fairly good that people have identified the main possibilities, and all of those except CO_2 leave large holes in the explanation of climate change, while adding CO_2 to the mix fills those holes nicely.

Space Dust-Up

Almost surprisingly, climate-change causes from beyond Earth appear much less important than we might imagine—changes in space dust, or in cosmic rays, or even in the sun itself have generally left small or zero impacts on the climate history.

True, there was one really bad day about sixty-five million years ago when a massive 6-mile-wide (10-km) meteorite[18] blasted a 110-mile-wide (180-km) crater in sulfur-rich rocks of the Yucatan, splashing pieces of meteorite and melted rock and sulfuric acid to make a sun-blocking layer high in the atmosphere around Earth, which eventually fell down to leave a layer recognizable in sediments from around the whole planet. Wildfires and great waves from the impact were followed by dark, cold, and acid rain, killing most of the biodiversity on the planet, including the great dinosaurs.[19]

However, this event is looking more and more unique over a very long time, because evidence of similar catastrophes is not showing up in the sedimentary record despite a lot of effort to find some. The initial hypothesis of a dinosaur-killing meteorite was followed by many studies confirming and extending the results, so that now only a small—but sometimes very vocal—band of scientists doubts the power of the meteorite. Even these doubters generally agree that a big meteorite hit and did damage; they just hold out for some terrestrial contribution to the extinction.

In contrast, more recent hypotheses for similar events based on preliminary data typically have been followed by independent studies that did not support the hypotheses.[20] The most recent of these extraterrestrial-impactor hypotheses was the idea that a comet killed the mammoths near the end of the most recent ice age.[21] The debate is still raging as I write this, with well-known scientists and interesting science still supporting the idea, but a wave of independent failures to find confirming evidence is casting a very long shadow.[22]

The further back in time we look, the sketchier the sedimentary record, and especially before half a billion years ago or so, we might have missed big and important events.[23] Way, way back, there clearly were impacts much bigger than the dinosaur-killer, including one that blasted out the material for the moon as Earth was forming.[24] And there have been smaller impacts more recently that might have had important local effects. Still, for understanding climate change, big rocks from space are not very important in our picture now. Equally important, studies searching for smaller, slower changes in the rate at which space dust arrives on Earth find no significant changes, so some sort of sun-blocking space-dust veil also seems very unlikely to have affected the climate.[25]

Finally, for space rocks, note that the idea of a comet or meteorite changing Earth's orbit enough to make a difference to the climate is science fiction. The dinosaur-killer would have changed Earth's distance from the sun by less than an inch, not enough to get excited about, and even the Mars-sized body that blasted out the moon material wasn't big enough to have a huge effect on our orbit.[26]

Solar Stuff

The sun is the big driver of Earth's climate, so anything that the sun does should affect Earth. And, indeed, recent climate history confirms this. The small variations in the sun's output during sunspot cycles have caused small variations in Earth's temperature.[27] The cold of the Little Ice Age of a few centuries ago seems to have involved both a little extra sun-blocking effect from explosive volcanic eruptions, and a small reduction in the sun's output; greater solar output and a reduction in sun-blocking from volcanoes contributed to Medieval warmth and to the warming of the early twentieth century.[28] Events similar to the Medieval warmth and the Little Ice Age occurred over the previous millennia, and many data sets point to a role for solar fluctuations in these as well, although the changes are small enough and the uncertainties large enough that a little doubt remains.[29]

We seem to be fortunate, indeed—if the sun changed a lot quickly, our climate would change a lot, and we would be unhappy. But as far back as we can see clearly, the sun hasn't had large, rapid changes. (We will see that large but very slow changes in the sun haven't mattered much to the climate, with CO_2 as the likely stabilizer.)

There is a small but interesting scientific literature looking for amplifiers that might allow tiny changes in the sun to cause larger changes in climate. The sun might affect features of the shallow currents and overlying winds of the Pacific that are involved in El Niño, for example. And because a switch between El Niño and La Niña in the Pacific can change the global average temperature by a few tenths of a degree, understanding these regional impacts of changing sunshine (and of volcanic eruptions and other causes of climate change) might be important in explaining the size and pattern of the climate response to the sunspot cycle[30] or the Little Ice Age,[31] and in predicting the regional patterns of the next decades of global warming. However, the few tenths of a degree from such influences are very small compared to the possible warming if we burn most of the fossil fuels.[32]

We will come back to the sun in a little while. First, however, we will let the history of the climate show us that the other commonly suggested amplifier of solar changes is very weak at most.

Cosmic Ray-Ban

As I am writing this, cosmic rays are all the rage in the fringe-climate blogosphere, as well as commanding some attention from serious science. The cosmic-ray interest overlaps a lot with earlier work on solar cycles, and in many cases is an outgrowth of the attempt to find some way that tiny solar fluctuations could have a notable climatic impact. There is a history of exploratory science finding some relation between something linked to the sun or cosmic rays and something linked to climate (beyond the obvious direct effect of changes in the sun's brightness). Then, this correlation largely or completely disappears with more data or more careful analysis, but some other possible correlation is proposed in the scientific literature.[33] The history is perhaps reminiscent of the search for a link between leukemia and power lines, which we visited briefly back in chapter 5—the issue is so important that exploratory research is surely warranted, but the press and the blogosphere sometimes get carried away with rather poorly supported hypotheses.

For cosmic rays, there is a hint of a mechanism. Physicists have long used cloud chambers to "see" particles, including cosmic rays. If you set up very careful laboratory conditions, a high-energy particle zipping through the air will ionize it, nucleating small cloud droplets that show the track of the responsible particle. If something similar could happen in the real world,[34] then the effect on clouds might affect climate, with more cosmic rays either cooling (if mostly low, thick clouds are produced) or warming (if mostly high, thin clouds are involved), or having little net effect (if both are produced, with offsetting effects).

Preliminary indications are not encouraging for a cosmic-ray effect, though.[35] In the real world, the tiny ions made by cosmic rays are not the limiting step in producing clouds, so tweaking their numbers by changing cosmic-ray numbers isn't expected to affect cloudiness much.[36]

Nature provided us with a wonderful test of cosmic-ray effects during the ice age, showing that they cannot matter much to the climate. About 40,000 years ago, the magnetic field weakened a lot without switching direction. The sediments and lava flows that formed over about a millennium during this event do not show a strong magnetic alignment—the compasses didn't know which way to point, so they point in all different directions. With the magnetic field weakened, cosmic rays streamed into the Earth system to make beryllium-10 and other isotopes, and a very large spike in the level of these in ice cores and sediment cores marks the event. But the climate record just ignores it—there is no evident correlation between the two.[37]

We cannot exclude some small climatic role for the cosmic rays that made beryllium-10, but we can exclude a large role with high confidence. Similarly, if cosmic rays provided a major control on climate, the history of climate should be closely tied to the history of the magnetic field, with spikes in climate evident at each magnetic reversal. This is not observed. There is still interesting science to be done on cosmic rays and climate, but serious scientists in this field are looking for a fine-tuning knob, not a major control.

Home-Grown Heroes: Volcanoes

As we saw earlier, a big, explosive volcanic eruption blasts ash and gas into the stratosphere, above the rain that cleans the lower part of the atmosphere, so the material can stay up for a year or two and block a little sunshine. Cooling is especially strong if the eruption throws up a lot of sulfur dioxide, which is rapidly changed to sulfuric acid, which forms small sun-blocking droplets. A big eruption often causes globally averaged cooling of a few tenths of a degree.[38]

Like so many other things, if the volcanoes could get organized, they could really change the world. But there is simply no way for a volcano in Antarctica to tell a volcano in Alaska that it is time to erupt. Occasionally, eruptions will occur close together in time purely by coincidence, and such random events seem to have been important, along with the sun, in causing events such as the Little Ice Age.[39] The flexing of Earth's outer shell caused by growth and shrinkage of massive ice sheets, which were made of water that evaporated from and then returned to the ocean, may have organized volcanoes a little as the ice ages came and went, but not enough to make a huge difference.[40]

Home-Grown Heroes: Drifting Continents

Continents drift around the globe like great bumper-cars, tugged along by cold slabs of sea floor sinking back into Earth or sliding off high, warm spreading ridges where volcanoes make new sea floor, while slow convection currents stir the rocks beneath. Collisions close ocean basins, while rifts tear apart continents to make new oceans.

Look at the world tomorrow, or in a year, or a million years, and you would recognize it. But come back in 100 million years, and it would look really different.

A continent that starts at the pole and moves to the equator will surely get warmer. If a continent is torn apart to make an ocean, the climate along the new beaches will be quite different from the conditions before.[41]

Furthermore, these processes may affect the global climate at least a little. Currents would have been different from those observed today when South America was still joined to Antarctica by a land bridge, allowing warm water to reach Antarctica, and when South America was separated from North America by ocean. And because land is often a little more reflective than ocean, putting a continent in the sunny subtropics where it can have a desert like the Sahara may reflect a bit of extra sunshine, cooling the globe a little even if the desert itself is very hot, although this depends somewhat on what clouds do in response to the moving continents.

But our physical understanding and our climate history agree that drifting continents alone do not cause a lot of global temperature change. Move continents around in a climate model, and you find perhaps a few degrees of effect at most. Look at history, and you find both warm and cold times with a lot of land near a pole, or with the land moved away from the poles.[42]

Looking for Help

The history of climate shows that Earth has experienced huge swings. At times ice reached to the equator in "snowball Earth" events, and at other times hothouse climates existed with no ice evident even at the poles.

Yet, looking at a long list of possible causes of climate changes leaves most of the past changes unexplained. The cosmic rays, space dust, and magnetic field are at best fine-tuning knobs, whereas fast changes in the sun have fortunately been small, and the direct effects of drifting continents have been both small and slow.

You won't be surprised, then, when we go back to CO_2 in the next chapter to look for help. We will also revisit the sun, because it does matter in interesting ways.

10

And Having Writ, Moves On

Synopsis. Earth's climate has swung from hot to cold and back, with higher CO_2 levels and warmer times occurring together. Although the causes of climate changes discussed in chapter 9 have mattered, making sense of climate history requires the warming influence of CO_2, as explained in this chapter and the next. A hint of a major difficulty is that climate seems to have changed slightly more as CO_2 levels changed than expected from many models.

True, the billiard-tables were of the Old Silurian Period, and the cues and balls of the Post-Pliocene; but there was refreshment in this, not discomfort; for there are rest and healing in the contemplation of antiquities.
—Mark Twain[1]

THE SAURIAN SAUNA

The dinosaurs of 100 million years ago stomped across a sauna-like world with very hot tropical temperatures—roughly 15°F (8°C) above recent values—and with no permanent ice near sea level, even at the poles.[2] Yet the sun that warmed the dinosaurs is estimated to have been about 1 percent dimmer than recently.[3] The continents were in somewhat different positions from today, accounting for a bit of the warmth but not most of it.[4] CO_2 concentration in the atmosphere was notably elevated, roughly three to four times as high as before the Industrial Revolution, or two to three times recent values.[5] Based on our understanding of the physics of the climate, the high CO_2 accounts for most of the warmth. If there is a mismatch, the CO_2 caused a little more

Figure 10.1 *Glaciers leave behind tracks. The striations and polish on these rocks from Danmark Island, East Greenland, are only about 10,000 years old, but other, older indications tell us that glaciers have grown and melted away many times on the planet.*

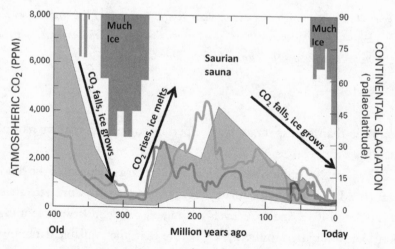

Figure 10.2 *Glaciers have grown when CO_2 fell, and melted when CO_2 rose; the last 400 million years are shown here.*

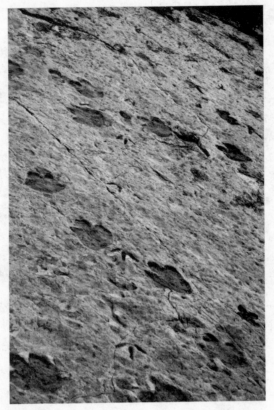

Figure 10.3 Dinosaur tracks at Dinosaur Ridge, Colorado. These tracks from the Cretaceous—the latter part of the "saurian sauna"—were formed at a time with no permanent ice on the planet, at least at low elevations.

warming than we expect, or was helped by something that we haven't identified yet.

However, 200 million years before that, distant ancestors of the dinosaurs lived on a world climatically more like today, with a great ice sheet at the South Pole[6] and CO_2 levels not hugely different from recent values.[7] And a hundred million years before that, the available data point to high CO_2 levels and warm temperatures. Correlation does not demonstrate causation, but the success of CO_2 in explaining at least most of the globally averaged temperature change, and the failure of any other hypothesis to do so, point strongly to CO_2 as the main culprit.[8]

When I started in this field, sketchy data for many times in geologic history suggested a mismatch between temperature and CO_2. Rather remarkably, additional studies with better dating, more samples, new techniques, and better instruments have reduced or eliminated those mismatches.[9] There remain a few time intervals for which disagreements among available data leave some uncertainty, but I don't know of any times showing well-established glaring mismatches between CO_2 and temperature, while agreements are found for slow changes over tens of millions of years, and for fluctuations lasting as little as a million years or less.[10] In view of the many other factors that affect climate, as we saw in the previous chapter, it is inconceivable that changes in CO_2 and climate would ever match perfectly, but the evidence shows more and more confidently that CO_2 has been the biggest knob controlling the climate, with others doing the fine-tuning.[11]

SETTING THE THERMOSTAT

So the data are now clear—atmospheric CO_2 levels and temperature have increased and decreased together over very long times in Earth's history, not perfectly coupled but closely related. Our understanding of the physics of climate shows that the changing CO_2 levels should drive temperature, and models using the physics have good success in "predicting" the climates of the past based on CO_2 levels, continental positions, and brightness of the sun, with CO_2 generally being most important. If anything, the climate changes tend to have been a bit larger than predicted by the models.

But, I am often asked, doesn't this show that warming drives up CO_2 levels and not vice versa, so CO_2 doesn't cause warming and therefore there's nothing to worry about? No! This leaves unexplained how the climate could have ignored the fundamental physics of CO_2, and what unknown cause brought the warmth instead. The equivalent would be my accountant claiming that the known interest charges on my credit-card overspending had no effect on the amount of money I owed the bank, but that some completely unknown process did—the hypothesis requires ignoring what we know, and pretending that something unknown is more important.

Furthermore, while it appears that warming raised CO_2 levels during ice-age cycles and is likely to do so in our near future (as mentioned briefly in chapter 8 and discussed more fully soon), over much longer times an increase in temperature causes a decrease in CO_2, as we will see next. This means that the high CO_2 evident during the "saurian sauna" cannot have been caused by the warmth. And the explanation of this will help show why nature cannot undo what we are doing to the atmosphere fast enough to help us.

First, though, a quick analogy. If I decide to invest my pocketful of money, will I have more or less cash to spend? There isn't a yes or no answer until you know the time interval being considered—in the next twenty days I will have less, but twenty years from now I will have more (presuming I invest wisely). In the same way, the effect of warming on the atmospheric level of CO_2 depends on the time frame we consider—warming can raise CO_2 levels over thousands of years but lower it over millions. And whether CO_2 is going up or going down at some time in response to changing temperature has no effect on the physical reality that the CO_2 in the atmosphere is a greenhouse gas contributing to planetary warmth at that time.

So, let's look at the slow controls on CO_2. We visited one important control already, back in chapter 4, with the formation of fossil fuels removing CO_2 from the air and the natural "burning" of fossil fuels returning that CO_2 to the air. CO_2 also is supplied to the air naturally by volcanoes. Our fossil-fuel burning is pumping out CO_2 50 to 100 times faster than volcanoes, but the volcanoes are persistent and have been doing their job for billions of years.[12]

Volcanoes also throw out pieces of rock and ash, while additional melted rock freezes beneath volcanoes to form granite or related rock types. Careful examination of old graveyard headstones, some of which are granite, shows that they do not last forever—rain picks up CO_2 to form a weak acid that attacks the granite and all other rocks, causing changes that we call "weathering."[13] The granite ultimately breaks down to form clay, rust, and sand that go to make our soil, while releasing dissolved chemicals that wash to the sea. Some of these dissolved chemicals are taken up to make shells, while others make the ocean salty. Many of the shells are made of calcium carbonate, which has the chemical formula $CaCO_3$; if I rewrite this in the old style, as $CaO \cdot CO_2$, you can see that these shells contain CO_2.[14] Eventually, the soil washes to the sea to pile up as sediment. And the drifting continents and sinking sea-floor plates take shelly rocks, other sediment, and salty water down to melt and feed volcanoes, closing the loop.[15]

The rate at which Earth melts rocks to feed volcanoes depends on processes deep in the planet that don't care much about the climate up here. But the rate at which the volcanic CO_2 and rock get back together depends on chemistry (plus some physics and biology) at the surface. And, as any introductory chemistry (or physics or biology) student knows, warming speeds up most chemical reactions. The result is a rock-weathering/CO_2 thermostat for Earth—when the climate is cold, the CO_2 coming out of volcanoes tends to stay in the air rather than reacting with rocks to supply shell-making chemicals to the ocean, and this extra CO_2 in the air warms the planet; when the climate is warm, the CO_2 in the air reacts more rapidly with the rocks, lowering the atmospheric load of CO_2 and cooling the planet. However, because the natural rates at which CO_2 comes out of volcanoes and goes into weathering rocks are so slow, it takes about half a million years for this thermostat to make a big difference to the planet's climate.[16]

So, unless scientists can find a way to "geoengineer" much faster reactions between CO_2 and rocks, as we will explore in chapter 23, this thermostat will not help us much over the next centuries or millennia, and it was too slow to stop the 100,000-year cycling of the ice ages that we will look at in the next chapter. This rock-weathering thermostat does show up in Earth's climate history, though, helping to explain some very big, very old climate events.

THROWING SNOWBALLS AT THE FAINT YOUNG SUN

Our physical understanding indicates that the sun has brightened by about one-third over its lifetime, as burning of hydrogen to helium has packed more mass into a smaller space in the center, giving a stronger gravitational pull on the remaining hydrogen to squeeze it more tightly and fuse it faster. With the young sun so dim, Earth should have been completely frozen early in its lifetime, but geological evidence points to liquid water and even quite warm conditions that far back. Early Earth didn't have much free oxygen in its atmosphere until photosynthesizing microbes evolved and multiplied, so it is likely that the early atmosphere had extra methane and perhaps other greenhouse gases that oxygen would have destroyed quickly. However, the available evidence still points to CO_2 as an important player, with the rock-weathering thermostat acting to keep CO_2 high enough to keep Earth warm.[17]

Any sports fan knows that a slow defense leaves a team vulnerable to a quick offensive strike, so if early Earth was relying on the slow CO_2 thermostat to defend against freezing, we might expect to find evidence of events when the thermostat failed to stop too rapid cooling. And we do![18]

A few times in Earth's history more than half a billion years ago, perhaps when rising oxygen broke down some of the non-CO_2 greenhouse gases, Earth fell into a snowball or near-snowball state, with glaciers leaving deposits extending to sea level in the tropics. The cold of this snowball would have slowed rock-weathering, allowing CO_2 from volcanoes to accumulate in the air. But the high reflectivity of such a snowball would have resisted melting until volcanic CO_2 accumulated in the air for millions of years to bring enough warmth to overcome the reflectivity of the snowball. After melting, the warmth and relatively rapid rock-weathering from that excess CO_2 would have supplied the oceans with enough shell-making materials to eventually overcome the acidity from the CO_2 and deposit rocks with the same composition as shells (calcium carbonate). And indeed, remarkable "cap carbonates" sit atop the snowball deposits.

Our knowledge of snowball Earth and the faint young sun is evolving rapidly, and almost everything we have measured about Earth's history is being refined. But our understanding of climate and CO_2 continues to work beautifully in making sense of the past. The rock-weathering/CO_2 thermostat means that the long-lived saurian sauna was warm because CO_2 levels were high, and not vice versa, because warmth tends to decrease CO_2 over long times.

RESETTING THE THERMOSTAT

However, if high temperature tends to lower CO_2 by speeding rock-weathering, how could CO_2 have stayed high? If you have shared a home thermostat with other people, you already know the answer. The thermostat's job is to stabilize the temperature at a particular "set point," but you and the other people in the house can raise or lower that set point. Your housemates may lower the set point of the air-conditioning on a particularly hot day, and you may raise the set point when you check the electric bill. Similarly, natural processes shift the set point of Earth's thermostat.

If a thick soil layer separates the CO_2 in the air from rocks beneath across much of the world, the CO_2 can't easily get to the rocks to break them down, giving a high set point and a warm world. But if widespread mountain-building causes the soil to wash or slide off, the CO_2 can attack the rocks more easily, lowering the set point and cooling the world. How much mountain-building occurs, changing the set point, depends in part on the geological "accident" of whether continents are crashing into each other or not. The rate at which volcanic eruptions bring CO_2 up from below also depends in part on slow changes in Earth's geology, and whether deep parts of Earth are "burping" (outgassing) or quiet. The high temperature of the saurian sauna probably is traceable at least in part to faster volcanism then, while the slide in temperature to the recent ice ages probably includes the effects of soil loss, allowing rapid weathering of young mountainsides.[19]

Earlier than the saurian sauna, the evolution of land plants allowed coal beds to form, taking some of the volcanic CO_2 and burying it as fossil fuels to lower the atmospheric CO_2. Land plants also allowed the "breathing" of their roots to put CO_2 against rocks down in the soil, speeding rock-weathering and lowering the thermostat. The cooling that brought ice to the South Pole about 300 million years ago followed the evolution of land plants. In fact, the ice sheet probably owes much of its existence to their influence.[20]

Drifting continents move more or less as rapidly as fingernails grow, so large changes in continental positions, mountain-building, and vol-

Figure 10.4 The layered rocks, which have caught snow to look like horizontal black-and-white lines just below the mountain tops, are from a vast outpouring of volcanic rocks—a flood basalt—in East Greenland. Much larger flood-basalt eruptions occurred longer ago, contributing to resetting the Earth's thermostat.

Figure 10.5 Fern fossil on Mt. Flora, Hope Bay, Antarctica, from the warm, high-CO$_2$ "saurian sauna." Rock hammer for scale.

canism generally require 100 million years or so. In turn, large changes in the set point of Earth's climate thermostat often take about that long, not fast enough to matter much to human planning. Big changes in evolution, such as the origin of land plants, also tend to be rather slow, or rare. So the set point of Earth's climate has been dialed up and down, taking 100 million years or so to change greatly as geology and biology have evolved, with high CO$_2$ levels giving high temperatures.

Changes in CO$_2$ much faster than 100 million years are also known from geologic history. While still not fast enough to matter much to us over the next decades or centuries, these again point to the warming effects of CO$_2$. Let's check out a warm event, and then in the next chapter we will visit some cold ones.

A BLISTERING BELCH?

Travel back to fifty-five million years ago. The meteorite had killed the dinosaurs ten million years before, and new creatures had evolved or were evolving to fill the "jobs," or ecological niches, left open by the departed dinosaurs.[21] The climate remained warm with high CO_2 levels. And then something really weird happened: the Paleocene-Eocene Thermal Maximum, or PETM.[22] The already warm climate became even warmer, from equator to pole, typically by about 9°F (5°C) or more.[23] Shells that had fallen onto the deep sea floor dissolved, leaving only a thin clay layer, which we now find sandwiched between older shelly deposits and younger ones—the ocean was more acidic during the PETM than before or after.[24] The event arrived rapidly, maybe in only a few millennia or less, although probably somewhat longer. We face frustrating uncertainty in part because the event covered its tracks by dissolving many of the sea-floor shells that otherwise would have contained the record of the event.[25] Notable changes in rainfall and soil moisture accompanied the warming.

A major extinction of certain bottom-dwelling ocean creatures occurred,[26] together with large migrations of many species on land and in the ocean, a tendency for some species to become dwarfed, and the onset of important evolutionary diversification.[27] Insect damage to land-plant leaves increased sharply.[28] After the event peaked, environmental conditions drifted back toward normal, over a couple hundred thousand years.[29]

Chemical and isotopic indicators point to a huge release of CO_2 from living or formerly living things.[30] This CO_2 may have started out in plants, or as natural-gas methane or other fossil fuels that were "burned" to CO_2, but it didn't come from volcanoes. The total amount of CO_2 released to cause the PETM may have been roughly equal to the total amount of fossil fuel or of sea-floor methane that existed on Earth at the start of the Industrial Revolution.[31]

It is difficult to figure out where that much CO_2 came from that rapidly. Sea-floor methane clathrate, as described in chapter 4, was the early favorite and probably contributed; however, the ocean was warm enough before the event to make it unlikely that such a huge amount of carbon was frozen in clathrate.[32] Other possibilities include volcanoes intruding organic-rich rocks to drive out their stored carbon, or burning of lots of land plants, or drying of a seaway to expose the organic-rich sediments to oxygen.[33]

A few smaller, although similar, events are scattered through the millions of years before and after the PETM,[34] but in size and speed the PETM looks unique. Thus, some highly unlikely convergence of several such events may have occurred—coincidences do happen if you wait long enough, and this type of event must be highly unlikely or else more of them would have occurred over hundreds of millions of years.

Regardless of the exact cause, the event provides a lot of useful information. First, the warming effect of CO_2 is again demonstrated—CO_2 explains what happened pretty well, and nothing else does. The time to remove a lot of CO_2 from the atmosphere is also confirmed—more than 100,000 years may be short in Earth's life, but it is long to people. The effects of warming on living things are also clear—species migrated, ecosystems were disrupted, and extinctions occurred.

The lack of widespread extinctions except in the deep sea might provide an optimistic note, but caution is required: most estimates of the rate at which the PETM arrived are notably slower than the ongoing human perturbation, and slower changes give more time for the living things to adjust where or how they live or to evolve in response to the changing environment. Furthermore, the climate is less sensitive to added CO_2 when the initial atmospheric CO_2 level is higher, as we saw in chapter 6, so adding a similar amount of CO_2 now is likely to have more influence than at the time of the PETM because CO_2 levels were much higher before the PETM than they were just before the Industrial Revolution. Thus, the influence of our CO_2 on living things probably will be larger than that of the PETM—yet the PETM had major impacts.[35]

The biggest difficulty with our scientific understanding is that, again, the warming during the event seems to have been a little larger than our models produce, especially near the poles. This likely indicates either that something in addition to CO_2 contributed to the event, or that the climate is more sensitive to CO_2 than the models produce on these time scales. I suspect that the greater sensitivity is more likely, in part based on cold-climate results that we will meet in the next chapter.

First, though, I have not come close to discussing all of the other cases in which CO_2-induced warmth is implicated in major climate change.[36] Times of vast volcanic outpourings have been accompanied by large warmings, somewhat slower than for the PETM but still leading to great changes for living things. The warmth has favored slow burial of dead plants to make fossil fuels while removing the extra CO_2. Sometimes, warmth and rapid rock-weathering seem to have combined to create massive "dead zones" in the ocean with minimal oxygen, producing hydrogen sulfide; the greatest extinction of all time probably was caused in this way. As we look across the vast sweep of Earth's history, we see how many things have influenced climate, but how critical CO_2 has been. And we will see this most clearly in the next chapter.

11

The Great Ice That Covers the Land

Synopsis. The ice ages of the last million years were not caused by CO_2, but by changes in the distribution of sunshine on the planet driven by features of Earth's orbit. But the growing and shrinking of ice were accompanied by changes in CO_2 and other greenhouse gases in the air. And some places cooled with increasing sunshine and warmed with decreasing sunshine, behavior that is explained by changes in CO_2 but not by competing hypotheses.

One naturally asks, What was the use of this great engine set at work ages ago to grind, furrow, and knead over; as it were, the surface of the earth?
—L. Agassiz[1]

AND SLIDE ON THE ICE

After the PETM of about fifty-five million years ago, temperatures and CO_2 began a slow but bumpy slide toward recent conditions, as the hothouse world with ice-free poles changed to our icehouse with (relatively) permanent polar ice. The large, long-lived ice sheet on the Antarctic continent probably grew slightly more than thirty million years ago.[2] The effects of drifting continents are believed to have contributed to growing this ice on Antarctica, but dropping CO_2 levels appear to have been the main driver.[3]

Once ice appeared, it grew and shrank as ice ages alternated with warmer interglacials. Note that we still have ice during an interglacial—the PETM or saurian-sauna hothouse gave way to an icehouse, and

within the long-lasting icehouse we have warmer times called interglacials and colder times called glacials or ice ages. For example, in the
modern interglacial, about 10 percent of the land is buried beneath ice
sheets, which hold enough ice to raise the sea level about 200 feet (64
m);[4] at the peak of the last ice age, about 20,000 years ago, ice covered
almost one-third of the modern land, and sea level was between 300 and
400 feet (a bit over 100 m) lower than now.[5] For the last million years,
the peaks of ice ages have been spaced about 100,000 years apart, with
smaller wiggles spaced about 41,000 years, or 23,000 years, or 19,000
years apart.

Remarkably, this spacing of the ice-age wiggles was predicted decades
before it was observed.[6] Building on important earlier work, the Serbian
mathematician Milutin Milankovitch had spent much of the early twentieth century calculating changes in the distribution of sunshine on the
planet in response to features of Earth's orbit, and had found just those
timings.

To see what Milankovitch calculated, imagine a child's spinning top.
(Better yet, go get one—they're fun! And this is a lot easier to understand if you have something to look at.) The "spin axis"—the north
pole, or the red handle sticking up out of the middle—will be tracing
a small circle in the air, even if the "south pole" isn't moving where
it touches the table. This circle-tracing is the "precession" of the top.

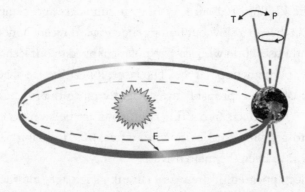

*Figure 11.1 The Milankovitch orbital variations include changes in eccentricity (E,
100,000 years), tilt (T, 41,000 years), and precession (P, 19,000 to 23,000 years).*

Earth's precession isn't quite a perfect cycle, but it repeats roughly every 19,000 or 23,000 years.[7]

Keep watching and you will see this small circle getting bigger as the top slows down and begins to fall over. The angle between the top's north pole and the vertical is called the "obliquity," or tilt. For Earth, the obliquity cycles from about 22 degrees to 24.5 degrees and back over 41,000 years, without falling over.[8]

Earth's orbit about the sun is a noncircular ellipse, with the sun off-center (at one of the foci of the ellipse), so our distance from the sun changes a little during a year. The eccentricity, or out-of-roundness, of the elliptical orbit increases and decreases over about 100,000 years, increasing and decreasing the changes in our distance from the sun during a year.[9]

If the obliquity were zero, sunshine would always just graze the poles while shining down straight on the equator. Increasing obliquity tilts the equator slightly away from the sun so it receives less sunshine, but it exposes the poles to a lot of sunshine during their summers so that they get more during a year.

Today, Earth is near its greatest distance from the sun during the northern summer, but 10,000 years ago precession had reversed this, with Earth closest to the sun during the northern summer. Thus, midsummer sunshine has dropped in the north while increasing in the south over the last 10,000 years.

Eccentricity primarily controls how much effect the precession has on summer sunshine. If the orbit were perfectly circular, then our distance from the sun would never change and precession wouldn't matter.

Therefore, characteristics of the orbit serve primarily to move sunshine around on the planet, north to south to north, or poles to equator to poles, affecting how much sunshine a latitude gets and how it is distributed through a year. The changes can be large—midsummer sunshine may increase by more than 25 percent over 10,000 years at the poles before decreasing again—even though the total sunshine received by the whole globe in a year is almost unchanged. Milankovitch calculated all of this, without a computer, in a couple of decades of serious

work before World War II, and hypothesized that the changing sunshine had controlled ice ages. And, decades later, the climate-change
spacings he predicted were discovered in histories of ice ages.

Milankovitch didn't quite get it all right though.[10] He had the idea
that the ice ages would see-saw, first at one pole and then at the other,
fleeing the corresponding see-sawing sunshine. But the ice ages haven't
done this. Instead, over the last million years or so, when low obliquity
has combined with the northern summer being far from the sun on a
highly eccentric orbit to reduce the total-year and midsummer sunshine
in the far north, ice sheets have grown on Canada and Eurasia, and
the whole world has cooled. The cooling reached to southern regions
getting extra midsummer sunshine, and to equatorial regions getting

*Figure 11.2 Franz Josef Glacier, New Zealand, is just visible from Peters Pool now,
but the glacier covered the site of the pool and extended many miles farther during the ice
age, reaching beyond the modern coast as shown in figure 11.3, with no change in total
sunshine and more midsummer sun then.*

Figure 11.3 During the ice age, the Franz Josef Glacier of New Zealand flowed out to sea along the black arrow and deposited piles of loose rocks along its side that now form the moraine labeled M. The rising sea after the ice age has eroded the moraine, leaving the boulders in the surf. These are big rocks; the 5-foot 7-inch–tall author is at the tip of the white arrow for scale.

extra sunshine for the whole year. And when the orbital characteristics combined to bring high midsummer and total-year sunshine to the far north, the whole world warmed.[11] Global temperature changes averaged about 10°F (5–6°C), despite essentially no change in the total sunshine to the planet.[12]

It is easy to see why northern sunshine has been more important—the extensive high-latitude lands of North America and Eurasia have provided lots of room for ice to grow, whereas down south the Antarctic continent has remained almost completely ice-covered, and the next land on which ice could grow is sufficiently close to the equator that immense cooling would have been needed to cover it with huge ice sheets. But why the great global temperature change?

Positive feedbacks that we met in chapter 6 helped some. Ice across northern lands reflected a lot of sunshine while grinding up rocks and causing other changes that generated extra sun-blocking dust. Grasslands or tundra replaced dark forests, reflecting more sun. But the extra reflectivity from the full-size ice-age ice sheets, the dust and vegetation then, as well as the water-vapor and other feedbacks, account for only about half of the cooling that occurred.

Furthermore, in some sense we are cheating to get half of the cooling this way. Until something else caused the other half of the cooling, the ice and dust and vegetation wouldn't have changed as much as they did and so wouldn't have caused as much cooling.

To get out of this problem, we need some help from ice-core data. As described in chapter 8, the ice cores show clearly that the temperature changes of the ice-age cycle were accompanied by CO_2 changes, as well as by changes in the less important greenhouse gases methane and nitrous oxide. If doubling of CO_2 levels contributes about 5°F (3°C) of warming or a bit more, as expected based on our best science, then the changes in greenhouse gases can explain the remaining half of the temperature change.[13]

The CO_2 must be a feedback on other Earth-system processes— we know of no way in which changing orbits could directly alter the amount of CO_2 in the air. Instead, the orbits must have affected ice and other things that then affected CO_2. But the records show that only a small amount of temperature change occurred before the CO_2 began changing, and then temperature and CO_2 changed together.

The ice ages show that changes in sunshine in a region have a huge effect on climate, but that CO_2 does too. And as for the PETM and some other events in the past, proper accounting for the ice ages, including the effects of the CO_2 on the ice sheets, suggests that if our models are in error, they are underestimating the changes that we may cause by continued fossil-fuel burning.[14]

Oddly enough, we haven't totally solved the mystery of how and why the ice ages caused CO_2 levels to change, although the picture is rapidly becoming clearer.[15] Recall from chapter 4 that in the ocean plants can use CO_2 and water to make more plants only in the lighted uppermost part of the ocean. When they die, some of these plants sink, either as dead plants or as fish poop or as the bodies of giant dead whales. Ecosystems in the deep ocean survive by "burning" this "rain" of valuable organic material and releasing the CO_2, whether the burning is done by bacteria or by bizarre "zombie worms" living on whale carcasses.[16] Deep-ocean burning of surface-grown plants transfers CO_2 downward. And water in the deep ocean takes roughly a millennium to bring its CO_2 back to the surface.

The surface ocean and the air exchange CO_2 readily—every breaking wave takes air bubbles into the water while throwing spray into the air, so CO_2 is traded back and forth. This rapid exchange at the surface means that anything that lowers the amount of CO_2 in the surface water, by causing more plants to sink or slowing the rate at which CO_2 is brought back to the surface, will also lower the amount of CO_2 in the air.

The great Southern Ocean provides an important path for CO_2 escape from the deep ocean, as water rises from the depths to the surface to be replaced by sinking of very dense water chilled by the frigid Antarctic winter. Broadly, attention is focusing on two major changes in the Southern Ocean affecting ice-age CO_2. First, ice-age cooling allowed more sea ice to grow, so the waters came up and went down beneath this floating ice without releasing all of their CO_2. Second, ice-age cooling seems to have shifted winds equatorward away from the Southern Ocean, reducing the wind-driven currents that contribute to CO_2 escape there.[17] Other mechanisms may have contributed as well, including the fertilizing effect of the extra dust blown into the ocean during the ice age, stimulating plant growth to take CO_2 out of the surface waters.[18]

ALL TOGETHER NOW

Whew! We have just completed a long tour through the long history of a fascinating planet. And we are left with a few clear messages. If the sun changed a lot rapidly, it would surely control climate changes—witness how important the regional sunshine changes were to the ice ages. But the sun seems to smile on us—its fast changes have been small, and the planet's rock-weathering/CO_2 thermostat has been able to handle the slow changes, except for the snowball Earth events far back in time. Drifting continents and evolving life have reset the thermostat, contributing to the snowballs, and to the cooling-warming-cooling trend over the last few hundred million years. And all of these bear the clear fingerprints of CO_2, which is necessary and sufficient to explain much of what happened. Shorter events, including the great PETM belch and the ice-age cycles caused by features of Earth's orbit, similarly bear the mark of CO_2.

These climate changes have been huge, have affected living things greatly, but have been slow compared to projected business-as-usual changes in our near future. Global changes of 10°F (5–6°C) in the ice ages may seem impressive, but they took 10,000 years; compared to these natural events, human-caused climate change is expected to become anomalous in size and rate.[19] (We will visit some truly weird and rapid regional events later, but globally they were not that impressive.) Nature has changed CO_2 greatly with no help from us, but we understand the system pretty well, and our understanding agrees with history, which tells us that nature cannot undo the changes we are causing fast enough to be of much use to us.

If you don't care about people or the other things living here with us, nature is highly likely to survive anything we throw at it, and a really huge, rapid climate change could be a very interesting experiment—come back in a few million years and see what happens. If you were a vast and powerful being who had many virtual Earths to play with, you might try this experiment on a few of them, just to see. (Scientists do this continually with computer models.) But if you care about the creatures that live on Earth now (including us), history shows that we are starting to cause big, fast changes that will matter a lot.

12

Kindergarten Soccer and the Last Century of Climate

Synopsis. Temperature changes over the last century have been small compared to the long-ago events discussed in the previous chapters. Because so many factors can cause such small temperature changes, analogy may be drawn to crowds of five-year-olds kicking a soccer ball in various directions during a kindergarten game. However, careful work shows that climate models are skillful in simulating these recent, small changes, and that CO_2 has had the strongest influence over the last century.

. . . playing at football. The novelty and liveliness of the scene was very amusing. —Dr. G. E. Corrie, 1838[1]

TO THE FIELD

Perhaps the finest spectator sport ever invented is soccer played by five-year-olds. The action is nonstop, even if it is sometimes difficult to figure out which team is going which way. Most players on both teams spend their time packed around the ball, although a few may be inspecting the dandelions, waving at the fans, or looking for the snacks.

In our friendly leagues, the parents and coaches usually split their time between cheering for both sides and trying to activate their offspring. ("Run! Kick! Yaaay!") The group hug around the ball lurches up and down and across the field, and at first may seem almost unpredictable. Yet, if you watch closely and long enough, you begin to see order. One of the players on a team usually figures out what is happening before the rest of the team does, and that player's kicks are resolutely in the right direction. When that player gets the ball, the lurching cluster moves toward one goal. And, a dozen years later, that player will be the All-State star on the varsity high school team.

A soccer game of five-year-olds may seem like an odd analogy for the climate of the last century. True, the five-year-olds are playing for camaraderie, teamwork, fitness, and other fine goals, while the climate is lurching toward an unfavorable future—but otherwise the comparison is apt.

In this chapter, we will meet many of the natural and human-caused climate "players." And we will learn a little more about how we know with high confidence that our fossil-fuel burning dominates the observed rise of CO_2 in the atmosphere, and that this is responsible for much of the observed rise in temperature. But we will also estimate the "kicking strength" of the other players, and how much each has contributed to the measured warming of the twentieth century.

RATING THE PLAYERS

First of all, let's identify a few players that we don't need to worry about because they didn't show up for the game of the century. The globe warmed by roughly 1°F (0.6°C) over the twentieth century, and this is expected to accelerate if we continue business as usual, so we don't want to spend a lot of time worrying about drifting continents and orbits, which "kick the climate" much more slowly—say, slower than roughly 0.1°F (0.06°C) of change per century.

Effects related to drifting continents helped to reset Earth's CO_2 thermostat to cool from the saurian sauna, and especially since the PETM. Global changes were large, and Arctic changes were especially large—perhaps roughly 50°F (say, 25–30°C)—but they took fifty million years. This gives 0.0001°F (0.00006°C) per century, not especially threatening to our near future.[2] The end of the last ice age, with CO_2 and ice-albedo feedbacks amplifying the effects of the Milankovitch orbital cycles, achieved most of its roughly 10°F (5–6°C) of warming in "only" 10,000 years, which gives roughly 0.1°F (0.06°C) per century. But the Milankovitch cooling affecting northern latitudes over the last few millennia has been notably slower, and the globally averaged change has been close to zero.[3] So even though drifting continents and tilting orbits have caused huge climate changes, we can safely ignore them in explaining what has occurred in the last century or projecting to the next one.

As I discussed in chapter 9, shorter-lived small climate wiggles bounced along the Milankovitch trend of the last few millennia, in response to big volcanic eruptions and perhaps to changes in the sun's

Figure 12.1 A glacier melting back from the Little Ice Age, East Greenland. The white glacier ice is surrounded by a light-gray "footprint" marking how big the glacier was a century or so ago.

Figure 12.2 The church at Hvalsey in South Greenland was constructed by Vikings in the 1300s and abandoned in the early 1400s, with the last documented use being a wedding in 1408. The cooling of Greenland from medieval times into the Little Ice Age contributed to the decline of the Viking settlements in Greenland.

brightness and in atmospheric and oceanic circulation. A big volcano can cause a change of a few tenths of a degree in a year, but the effect disappears in another year or so. The combined effects of volcanoes, the sun, and changing ocean and atmospheric circulation cooled the Northern Hemisphere by a few tenths of a degree (in Fahrenheit, or Celsius) over the centuries between the "Medieval Warm Period,"[4] centered around the year 1000, and the "Little Ice Age," from approximately 1500 to 1850. This cooling rate was a bit less than 0.001°F (0.0001°C) per year, but the uncertainties may be twofold or more, so we will pay attention to volcanoes, sun, and ocean-atmosphere events, such as changes in the rate at which ocean currents carry warm water into the

North Atlantic, or El Niños and La Niñas out in the Pacific. Then, who else is in the game?

The biggest player is CO_2. We saw in chapter 9 that Antarctic ice cores provide reliable histories of CO_2. The CO_2 variations during the ice ages ran between 172 and 300 parts per million (ppm; 10,000 ppm is 1 percent) over the 800,000 years prior to the Industrial Revolution, and you wouldn't be too far off to just remember that CO_2 levels oscillated between 180 ppm during the coldest times of the ice ages and 280 ppm during the warmest times of the interglacials, until the Industrial Revolution kicked in and we pushed the levels well past 380 ppm and heading upward. Records from sea-floor sediments indicate that the current levels are the highest for at least two million years.[5]

You would need to be a real believer in coincidences to imagine that the recent change is natural, and that after two million years nature decided to crank up the CO_2 just exactly at the time that humans got serious about burning fossil fuels. But we don't need to argue about coincidences, because we can demonstrate with high confidence that the rise in CO_2 has been caused primarily by our fossil-fuel burning.

The first approach is simple bookkeeping. We have a pretty good idea how much oil and coal and natural gas we are burning—how many oil tankers and coal trains—and how much CO_2 this produces, as I discussed in chapter 4. We have measured the arrival of that much CO_2 in the air and moving into the ocean. There are a few tweaks to the calculation to include the CO_2 released by burning trees, and the CO_2 taken up when trees grow back, but these amounts are not hugely uncertain.[6]

Measurements at modern volcanoes show that their CO_2 emissions are only 1 to 2 percent of human production, so a change in them cannot account for the rise in CO_2. If the CO_2 in the air were coming from the ocean, then the ocean would be losing CO_2, not gaining CO_2 as observed.[7]

Furthermore, we can check the bookkeeping by fingerprinting possible CO_2 sources, and then finding out whose fingerprints are on the CO_2 that is showing up in the air. Carbon comes in three flavors: the very rare and radioactive carbon-14 that is made by cosmic rays and decays away in about 40,000 years, and the relatively rare carbon-13 and much more common carbon-12, both of which are stable. Plants take in at least a little of all three, but "prefer" the easier-to-use carbon-12 and so have a slightly higher ratio of carbon-12 to carbon-13 than is found in shells or volcanoes.[8] Living and recently dead plants include some carbon-14, but all of it has decayed before plants are converted to fossil fuels, or before plants or shells or other things are taken deeply enough in Earth to feed an erupting volcano. As CO_2 levels in the air have increased since the Industrial Revolution began, carbon-12 has become more common in the atmospheric CO_2 and carbon-14 less common.[9] Volcanoes melt a lot of shells to make CO_2, but melting doesn't use oxygen whereas burning does, and we see the drop in oxygen that corresponds to the rise in CO_2.[10] Hence, the atmospheric shift toward CO_2 that is rich in carbon-12 and poor in carbon-14, with dropping oxygen levels, shows that the rise in CO_2 is coming from the burning of long-dead plants, beautifully confirming the bookkeeping. (We should run out of fossil fuel before we run out of oxygen, so don't add suffocation to your list of worries.)

Next, how big of a deal is our CO_2 for the climate? A useful way to find out is to use our understanding of atmospheric physics, dating back to the U.S. Air Force research and beyond, as described in chapter 6,[11] to estimate the effectiveness of CO_2 and other climate changers through what we call "radiative forcing."[12] To understand this concept, first note that Earth is heated by roughly 240 W/m^2 of sunlight that reaches the planet and isn't reflected right back to space. The heat-trapping effect of the atmospheric rise of CO_2 between the years 1750 and 2005 was recently calculated as having an equivalent warming effect to an increase in the sunlight warming Earth of 1.66 W/m^2 (about 0.7 percent),[13] so we say that our CO_2 has a radiative forcing of 1.66 W/m^2. We will turn this into temperature change soon, but first we need to identify the other players.

Methane is a greenhouse gas produced mostly by microorganisms living in places with a lot of dead plants but not much oxygen, as we saw in chapter 4. We have created many such environments, in trash dumps, below the growing plants in rice paddies, inside large numbers of cows and other farm animals, and in some reservoirs; some methane also leaks out during recovery of coal or other fossil fuels. Bovine belches[14] and other methane sources have raised the atmospheric concentration enough to have a warming effect of 0.48 W/m^2. We also have increased levels of the greenhouse gas nitrous oxide, which is produced by bacteria using the extra nitrogen in our fertilizers, and by cars and other sources,[15] providing another 0.16 W/m^2. And the refrigerants (halocarbons) we have invented and released provide a notable warming influence of 0.34 W/m^2.[16] The halocarbons offset a bit of their warming by breaking down greenhouse-gas ozone in the stratosphere, but by only 0.05 W/m^2. But some of our other pollutants have been making more ozone lower in the atmosphere, especially over cities, giving a warming effect that averages 0.35 W/m^2. Some methane makes its way to the stratosphere before breaking down, and the water released by the breakdown of this methane adds another 0.07 W/m^2 of warming; the lower atmosphere has so much water from other sources that the little bit contributed from methane breakdown is insignificant.

Gases are not the only players, though. Our activities have changed how much energy is reflected by clouds and by Earth's surface. Replacing dark forests with more-reflective croplands, and other such changes, have had a cooling effect of about 0.2 W/m^2, but this is partially offset by black carbon ("soot") from our fires, which blackens the snow and gives a warming effect of 0.1 W/m^2.

Much more important, our fires and plows have put up ash and dust, and we have emitted many chemicals that condense to form particles or liquid droplets in the air. Acid rain is a big contributor, with sulfur dioxide from coal-fired power plants and other sources reacting to form sulfuric acid, which makes small droplets in the air. The ash and dust and droplets are often discussed together as "particles" or "aerosols," and those we have put up block enough sunlight to have a cooling effect of 0.5 W/m^2. And by making clouds last longer or have smaller droplets that reflect better, particles have caused another 0.7 W/m^2 of cooling. But the very high clouds from jet contrails have a bit of a warming effect, of about 0.01 W/m^2.

Over the last few decades the sun has wobbled through its sunspot cycle with little trend or even a slight drop in output, but the sun did brighten early in the twentieth century, with an estimated increase of 0.12 W/m^2. The climatically most important volcanic eruption of the twentieth century, that of Mt. Pinatubo in 1991, had a peak cooling effect of about 3 W/m^2,[17] but the largest effects faded over a year or so, with smaller effects fading over a few years; hence, volcanoes are often discussed separately from the other, longer-lasting influences.

Pull all of these variables together in an appropriate way,[18] and you end up with a net warming of 1.6 W/m^2. The uncertainties are fairly substantial, so the net warming influence could range from 0.6 W/m^2 to 2.4 W/m^2. The central estimate is very close to the radiative forcing of CO_2 influence alone, with all the other players balancing out and the biggest kicker, CO_2, controlling the game.

The "player evaluations" provided by the IPCC include statements about the regional extent of these influences. Not surprisingly, the changes related to land use apply on land, and the particles tend to be most influential downwind of the places where we are making or stirring up particles. CO_2 and most of the other greenhouse gases, with the exception of tropospheric ozone, are globally widespread, because they survive in the atmosphere long enough to be mixed worldwide, leaving only small differences between places.

The IPCC evaluations also include an assessment of scientific knowledge on the subject. This seems to have caused a little confusion—a reader will see that the cooling effect caused by our particles changing clouds is 0.7 W/m^2, with an uncertainty range of 0.3 W/m^2 to 1.8 W/m^2. (The large uncertainty here, especially on the high side, will be important in the next chapter.) The reader will then see that the level of scientific understanding on this topic is "low," and conclude, "Oh,

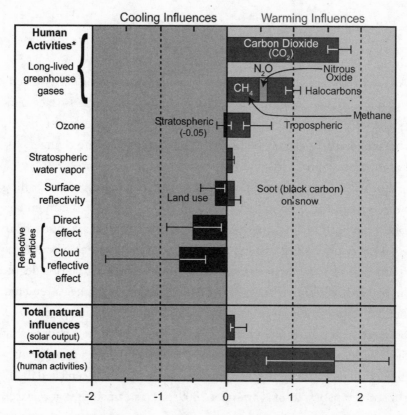

Figure 12.3 Kindergarten soccer players. These are the main influences on Earth's climate since the Industrial Revolution started (taken as roughly 1750), in W/m^2, with warming indicated by bars extending to the right and cooling indicated by bars extending to the left, together with the uncertainties (thin lines). The sun-blocking effects of volcanic eruptions are not included because they are so short-lived, and influences of orbits and drifting continents are too slow to matter. The natural influence of the changing sun (slight warming) and the net human influence (strong warming) are shown at the bottom.

then the uncertainty range must be *bigger* than stated because the understanding is low, so then we don't really know what is going on at all." Well . . . no. The uncertainty already includes our ignorance; topics indicated as having a low scientific understanding are those in which a little more research may notably improve our estimates.

KEEPING SCORE

Around 1850, enough temperature measurements from thermometers became available to allow calculation of a global average surface temperature. This is not an easy or straightforward task, because typically the temperatures weren't being collected to study climate change, posing interesting problems that have taken much effort to solve.

For example, ocean temperatures were measured on ships. Sailing vessels would pull up a bucket of water and drop in a thermometer. But some of those buckets were made of canvas, and evaporation from a canvas bucket is fast enough to cool the water measurably before the thermometer settles to the water temperature. The fraction of a degree involved didn't bother the people making the original measurements, but it is important if you're looking for a climatic warming of a degree. So modern scientists went back and tested the old buckets aboard ship and in a wind tunnel to estimate the size of the temperature bias, eventually writing a forty-eight-page technical paper that has been cited in 117 subsequent papers as of November 2009.[19] The steamships that replaced sailing vessels cooled their engines with ocean water, so thermometers were switched from buckets of water to the cooling-water inlets, raising additional issues as bigger boats pushed the inlets deeper into the sea, where temperatures may differ. These issues also can be solved, but care is required.

Weather stations on land have climate-related issues as well. If a city grows up around a weather station, the data will be biased by the extra heat that cities bring. So people assembling estimates of global temperature change eliminate the data collected by some urban stations.[20] Those to be eliminated can be identified by examining many more stations than are needed to see regional temperature trends; biased stations then show up as giving anomalous changes in comparison to their neighbor stations. To learn whether all of the stations with large biases have been eliminated, researchers compare temperature trends for still and windy nights—stronger wind mixes more of the urban heat away, so a station that reports anomalously warm temperatures only on still nights is probably giving bad numbers, but stations that report consistent data on windy and still nights are probably fine. Then, just to be careful, scientists compare temperature trends over time calculated using only rural sites and including all sites; the very close agreement that is obtained shows that the corrections have been applied properly. Similar studies address land-use changes away from cities, such as irrigation in the desert.[21]

One of the celebrated causes in climate science for a while was the mismatch between surface-temperature histories and satellite data supposedly measuring temperature trends above the surface in the middle troposphere. A data set based on satellite measurements, prepared by scientists from the University of Alabama at Huntsville and the NASA Marshall Space Flight Center in Huntsville, showed cooling or lack of significant change at a time when surface temperatures showed warming.[22] These results were widely used in public to question many aspects of global-warming science. I answered a lot of inquiries on this topic, including from high government officials, and I was far from alone. Perhaps the surface data were wrong and warming was not occurring. Or if surface and satellite data were correct, then the climate models must be wrong, because they did not simulate cooling up in the troposphere with warming at the surface.[23]

The satellites were not originally designed for climate-monitoring but for weather studies, and measurements for weather can be a bit less accurate. Many satellites have been used over the years, and calibrating one to the next is not always easy. Aging of sensors in space, changes in altitude as the satellite's orbit degrades, and other variables can influence the measured long-term trends, even though they don't matter much for day-to-day changes.

Consider the issue of satellite orbital "drift," for example. The orbits of these satellites run north-south from pole to pole, and drift slowly relative to Earth beneath. Land temperatures go from hot in the day to cold at night, whereas ocean temperatures don't change as much, going from warm to cool. If a satellite for a while mostly passes over hot daytime land (the Americas) and cool nighttime ocean (the Pacific), and then drifts to pass over warm daytime ocean and cold nighttime land, the switch from hot-cool to warm-cold will look like climatic cooling, and the switch back will look like warming, even if there is no change in climate.

As other research groups joined the field, and everyone involved worked to identify and fix these problems in the data, something highly reassuring happened: the satellite data came into agreement with the surface data and models, as well as with the balloon-borne radiosonde thermometers.[24] The warming aloft and at the surface are just about what the models yield. There is still a bit of mismatch in the tropics, but this mismatch is not statistically significant because it occurs where the data are a bit shaky.

So today, if you ask all the thermometers, or just the thermometers far from cities, or thermometers in the ocean, or thermometers sent down boreholes on land, or thermometers sent aloft on balloons, or thermometers looking down from space, you get the same answer: the world is warming. Furthermore, the great majority of the glaciers on the planet are shrinking, including those getting more snowfall.[25] Most of the changes in the places where plants and animals live and the times during the year when they do things are in the direction expected as a response to warming.[26]

PLAYING THE GAME

For the twentieth century, and into the early twenty-first century, the world warmed a bit more than 1°F (a shade under 1°C). (In case your brother-in-law told you not to believe this, just bear with me and I'll explain a little more in the next chapter.) A warming that occurred early in the twentieth century was especially concentrated at high northern latitudes, and it was followed by a near-plateau from the end of World War II into the early 1970s,[27] and then a fast rise. Lots of short-lived bumps, from El Niños (warm) and La Niñas and explosive volcanic eruptions (cold), and various other wiggles (both ways), went along with this.

To understand what happened, you need the history of CO_2 levels and particles and volcanoes. Then, do an experiment.[28] Get your climate modelers, have them turn on their models starting when we started keeping good global temperature records (about 1850), and run them to the present. Do this with just the natural forcings—sun and volcanoes—and what happens? The models driven by nature alone aren't too bad at simulating what happened in the early twentieth century; as the sun brightened and big eruptions skipped a few decades, the temperature went up. But the nature-only models are terrible at simulating what occurred during the later part of the twentieth and early part of the twenty-first century; the sun stabilized or even dimmed a bit while some big eruptions blocked the sun, but the climate ignored this slight natural push toward cooling and instead warmed.

Include the human influences with the natural ones, and the models do a little better in simulating the climate of the early twentieth century, but they already were doing fairly well, and our pushes weren't huge. But for more recent times, including the human influences really improves the match between models and reality, simulating the plateau or slight cooling into the early 1970s (natural changes plus particles from dirty smokestacks after World War II), and the warming since then. Without cheating, the models explain the history of temperature, if and only if the human effects are included.

In case you are puzzled that the industrial boom after World War II initially caused little warming and perhaps even cooling, but that industry has been causing warming more recently, recall that investing my pocket change in the stock market first leaves me less spending money but then gives me more—the short-term and long-term results are different. The same is true for industry and climate, which produces both sun-blocking particles and heat-trapping CO_2.

The particles from smokestacks fall or are washed out of the air in a couple of weeks. But some of the CO_2 we emit persists in the air for

centuries, with a little bit lasting hundreds of thousands of years, as discussed in chapter 10.[29] Thus, the particles in the atmosphere today are
controlled by the particle sources over the last couple of weeks, whereas
the CO_2 in the air today tells us about sources over centuries and longer times. When industry suddenly ramped up after World War II, the
almost immediate cooling effect of the particles was followed by a much
slower rise in temperature as enough CO_2 built up to have a notable
warming effect, so industry initially pushed the temperatures toward
cooling, followed by stronger warming.[30]

Thus, the models succeed in reproducing the history of climate in
the twentieth century if and only if both human and natural causes of
climate change are included. Furthermore, the models are successful in
reproducing other climate changes including the rise in water vapor that
has occurred with the warming, the expansion of the subtropical dry
zones, the upper-stratospheric cooling accompanying the tropospheric
warming, and more.[31]

Remember that the models are *not* tuned to match the climate changes
during the twentieth century; they are tuned to match the climate for
one time, and then tested against changes over time. If the temperature
measurements were notably in error, or the causes of climate change
were incorrectly estimated, or the models were greatly in error, then
data and model wouldn't match—because so many different tests for so
many different times are conducted, coincidentally offsetting errors are
highly unlikely to occur.

Nature does matter to climate change, and it will continue to do so.
But human influences have grown to overpower the natural ones. And,
as we will see in chapter 14, we have just started down this path. First,
though, let me provide just a little more context in chapter 13, about how
hockey is related to soccer in climate change, and how a brewer satisfied
a senator—scientifically!

13

But My Brother-in-Law Said . . .

Synopsis. Recent climate events, and recent blogospheric angst over purloined emails, have little importance to the big picture of climate change. The climate is warming, consistent with our scientific understanding.

> *"Nor all thy Tears wash out a Word of it."*
> —*The Rubáiyát of Omar Khayyám*[1]

SENSITIVE SOCCER PLAYERS

You wouldn't try to learn how hard a kindergarten-soccer player can kick a ball by watching the "group hug" of a game—you can't tell what one player accomplished when everyone else is either kicking the ball or in the way, or both. In the same way, the twentieth century is not especially central in our understanding of climate change, because lots of little players haven't kicked the ball very far.

Yet, public discussion of climate change seems to be dominated by the kindergarten-soccer game. I suspect that this is partly because we live in it, and most of us think we understand it better than physics or the Cretaceous. Also, I suspect that those people arguing that climate science is too uncertain to deserve a voice in the discussion of our energy future have focused attention on the kindergarten-soccer game. They have scant hope of convincing most people that the Air Force doesn't know how to make a heat-seeking missile, but there is much to argue about at the game—"That one is kicking a little harder than you thought . . . no, not that one, the other one, there . . . no, not right now, wait a minute. . . ."

Despite much complexity, the climate's behavior in the late twentieth and early twenty-first centuries does agree with the great body of information showing that CO_2 and other greenhouse gases have a notable warming influence, because temperature climbed despite nature pushing weakly toward cooling, our particles providing additional cooling by blocking the sun, and our other warming influences being small or zero. But recall from chapter 12 how lopsided the uncertainties are in the effect of our particles on cloud reflectivity, which allows for the possibility that a very large warming from CO_2 has been mostly offset by strong cooling from particle-enhanced clouds.

This would imply an alarmingly high "climate sensitivity"—the warming you would get if you started with a climate that isn't changing, doubled the CO_2 in the atmosphere and maintained that level, and let everything else come into balance (see chapter 7). The climate sensitivity can be estimated from the physics of the modern climate, and from use of the physics in complex models.[2] These estimates are tested against climate history, such as the ice age and the saurian sauna. Even the effects of volcanic eruptions can be used to test climate sensitivity—Earth's ice-albedo and water-vapor feedbacks are amplifiers regardless of whether the initial change is caused by sun or volcanoes or CO_2 or alien ray guns, so the cooling from the sun-blocking effects of volcanic particles tells about the feedbacks that amplify the effects of CO_2.[3]

All of these approaches, including comparisons to the last century and millennium, are consistent with a best estimate of 5 to 6°F (about 3°C) of warming from doubled CO_2 levels. Taken together, they give at least a 66 percent probability that the true sensitivity lies between 3.6°F and 8.1°F (2–4.5°C); if you want a higher probability of being right, drop the lower end a little and raise the upper end a lot because feedbacks amplify each other. However, because the range of climate sensitivities allowed by comparisons to the last century or the last millennium is larger than the range allowed by the other approaches, we don't rely on the recent changes as much.[4]

Once we know the climate sensitivity, and how rapidly the climate responds as CO_2 is raised, and how much CO_2 we are likely to put up, the future warming can be calculated with considerable confidence, as we will see in the next chapter. Notice that this path to reliable projections of future climate change does not include information on the exact temperature today, or whether this year is warmer than last year, or whether this year is warmer than the warmest medieval year. All of these are interesting, and contribute to some of the threads that form the tapestry of climate science, but they are not the biggest issues because of the uncertainties attached to all of the players in the kindergarten soccer game.

AN ANSWER BREWING

Such issues do trouble a lot of real people, though. During a meeting in 2008, a U.S. senator told me about the many claims that global warming stopped in 1998, showing that we don't need to be concerned about future warming and that climate scientists are misleading the public.

In 2009, when I typed "global warming stopped in 1998" (including quotation marks) into the Google search engine, it displayed "Results 1–10 of about 280,000." The Bing search engine displayed "1–10 of 13,700,000 results."[5] A quick reading of these "results" found many agreeing with the senator's sources, and many others debunking this claim. So what were the senator and I to do?

Perhaps check on history, or have a beer. Because in 1899, the Guinness brewery hired a statistician as a brewer who showed the way to our answer.[6]

The variability of nature often tries to fool us. If rolled dice come up double-sixes three times in a row, are we being cheated? How many times must those "boxcars" come up before we are pretty sure we've been had? Guinness wanted W. S. Gosset to find out how to separate the truly significant results from the random noise of the world.

The brewery had invested seven years of research to identify the best barley from a host of hybrids grown in many different places. Gosset

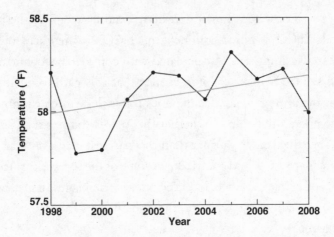

Figure 13.1 The large El Niño event of 1998 made that year especially warm. Quickly after that, some people started showing plots such as this and proclaiming "global warming stopped in 1998." The regression line from 1998 to 2008 actually shows warming, not cooling, but the confidence level doesn't rise to pound-on-the-table near certainty.

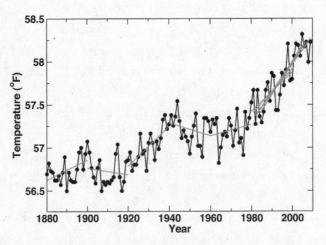

Figure 13.2 A longer view than the previous plot shows that the "weather" of El Niño does not obscure global warming, which we can conclude with high confidence has been occurring over the last few decades.

was asked to sort through measurements of yield and quality of different varieties grown in different districts and seasons, while considering
brewing results and malting assessments. Gosset found ways to show
confidently whether differences were statistically significant—whether
they really told the brewer about the quality of the barley—or were just
the usual "noise" of a complex and variable world.

Gosset's success in doing this seems to have contributed to his rise to
brewer-in-charge of the experimental brewery, while Guinness quickly
moved to put the barley Gosset helped identify into widespread cultivation to improve the bottom line. When Gosset published his new
statistical "t-test" in 1908, the author was listed only as "Student" at the
behest of the brewery—presumably, the fact that Guinness was using
statistics in brewing was considered valuable enough that they guarded
it as a trade secret.[7]

How does this relate to whether global warming or global cooling
is occurring? When the senator asked his question, the hottest year in
thermometer records was either 1998 or 2005, depending on which
compilation you chose (these years were very close in both the data set
from the NASA Goddard Institute for Space Studies—GISS—and the
data set from the United Kingdom, which I'll describe soon). A huge El
Niño had cranked up the 1998 temperature. But if 1998 was the hottest
or second-hottest year through 2008, doesn't that show global warming
stopped, just as some people argued? No! Just as a kindergarten soccer
player might be lucky in getting the bounce right for a big kick, a hot
year might be a coincidence.

To find out what is really going on, the analysis is best done with
statistical extensions of the tests developed by Gosset for Guinness.
Take any of the global temperature data sets, take the last thirty years
(a typical length used in climate studies) or any number of years long
enough that it contains a highly statistically significant trend, and you
find warming. Looking at too few years may produce trends going up
or down, but with strong indications that this is statistically insignificant
noise.

There are differences in details for different data sets, as you might expect. For example, the scientists who established the UK record didn't think that there were enough observing stations in parts of the Arctic to justify estimating temperature there, whereas GISS interpolates across the Arctic. Physical models constrained by the data indicate rapid Arctic warming, which suggests that the GISS approach is better, but you can make a scientifically valid argument for both approaches. Not surprisingly, GISS shows faster warming.[8] But whatever data set you choose, the significant trends show warming.

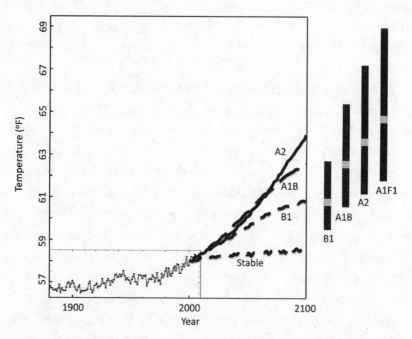

Figure 13.3 A longer view than the previous plot shows what the future may hold, as discussed in the next chapter. "Stable" is the warming that would have been expected if the atmospheric composition had been stabilized in the year 2000, with slow warming as the ocean catches up with the atmosphere. The curves B1, A1B, and A2 are commonly discussed possible futures, depending mostly on our CO_2 emissions. The bars on the far right show the best estimate (horizontal gray bar, based on a wider range of models than used in the curves) and uncertainty (black bars) for those three possible futures and for a future with still-higher CO_2 emissions.

The claim that the world switched to cooling in 1998 was so widely repeated in public that the Associated Press (AP) news wire service asked four independent statisticians to analyze the temperature data for trends, without telling the statisticians what the data were. The statisticians found no evidence for cooling.[9] More recently, Richard L. Smith, Distinguished Professor of Statistics at the University of North Carolina, told attendees at a briefing on Capitol Hill that his analyses showed no evidence of cooling from 1998. "If anything, the [warming] trend increases after this time."[10]

I am quite happy to report that the senator saw the wisdom of the Oxford-educated brewer, and acknowledged the climate statistics showing long-term warming and weather wiggles.[11] (The senator had also asked whether the climate community had flip-flopped from global cooling in the 1970s; I could confidently report that there has been no such flip-flop.)[12]

A DATE WITH THE DATA

In my experience, the folks who erroneously argued that "global warming stopped in 1998" generally relied on the slower-warming UK record just mentioned, the United Kingdom HadCRU global temperature compilation from the Climate Research Unit (CRU) of the University of East Anglia working with the UK Met Office Hadley Centre.[13] However, in the autumn of 2009 the HadCRU data set came under extreme criticism, often from the same people who had been relying on it, with many commentators suggesting that problems in this data set undermined climate science. This discussion is peripheral to the science, but was so widely disseminated that a little comment here seems wise.

Emails dating from 1996 to November 12, 2009, and involving many climate scientists were taken without their permission from a computer at the CRU. More than one thousand of these purloined emails (1073, to be exact), selected from a much larger quantity, were released publicly.

The email authors apparently believed that they were communicating in private. A few short quotes from the approximately one million words of released emails have been widely circulated in what came to be known as "Climategate." These excerpts were taken out of context, from material taken out of context, from communications between people who were not guarding their words to avoid being taken out of context.[14]

A lot of the difficulty started when some nations required their weather services to make money by selling raw data. The CRU, in trying to get global coverage, obtained data from some of these countries by agreeing not to release the raw data but to use them only in building a smoothed picture of global changes. (Other groups, such as GISS and the National Oceanic and Atmospheric Administration—NOAA—in the United States, chose not to work with proprietary data, avoiding the data-availability difficulty at the expense of having records from fewer stations in some regions; you can get everything online for free from them.) When public interest in climate data increased, many Freedom of Information requests were filed with the CRU for data that could not be released. A good bit of unpleasantness followed, leading to some emails that were publicly interpreted as indicating serious misconduct by climate scientists, who viewed themselves as being caught between a rock and a hard place.

Some of the emails raised other "hot button" issues, such as suggestions that certain scientific papers not be cited in the IPCC reports, or criticisms of some aspects of peer review at certain journals. A blog-ostorm of charge and countercharge erupted on the web, and quickly leaked into the "mainstream" press, with people questioning or defending the integrity of climate science and climate scientists.

In response, the AP set a team of reporters to read all the emails, and asked outside experts to assess aspects of the most contentious messages. The AP found that some of the climate scientists involved had stonewalled people questioning the science and had discussed hiding data from critics, but the reporters found no support for the claim that the science of global warming was faked.[15]

In light of widespread public concerns, five official investigations were conducted.[16] Two were prepared at Pennsylvania State University, my employer, to address issues involving a colleague, Michael Mann, so note my obvious conflict of interest. The various reports identified weaknesses, and offered good advice on how we climate scientists can do our job better and more usefully deal with a sometimes skeptical public. However, none of the investigations found evidence of malfeasance or of anything that would weaken the fundamental results of climate science.

If you are especially interested in the topic, I suggest starting with the report from the investigation by a committee of the UK House of Commons,[17] which explains many of the most contentious emails. For example, one email discussed using a "trick" and taking actions to "hide the decline." Now, a "trick" may be something deplorable that is deliberately used to fool others, or a "trick of the trade" that you really hope your plumber, your doctor, or, yes, your climate scientist is aware of. The House of Commons committee found that the email "trick" served to do the job rather than to deceive, and also found that " 'hide the decline' . . . was shorthand for the practice of discarding data known to be erroneous."

Regarding the email discussions of citing particular papers in the IPCC, and of paper reviewing, I have personal insight. IPCC authors were required to keep the reports short enough to be useful to policy-makers, so we had to discuss which papers should be cited. And peer review is so important to science that we would be negligent if we didn't consider ways to improve it.

The biggest issue in these emails may not be the content at all, but the tone, which taken out of context does not always cast climate scientists in the best possible light. For example, one email from a climate scientist discussed the likelihood of needing to respond to "crap criticisms from the idiots," not a warm and welcoming phrase.

Before casting your first stone, though, please consider my experience. One of the pleasures of my job is that so many interesting people contact me for information about our climate system. Tuition-paying students have first dibs, but then I get to talk to a fascinating range of folks about a subject that deeply interests them and me. I spent much time answering emails from one particular gentleman (I use the male gender purely to avoid awkward writing), a successful Penn State (PSU) alumnus, who then invited me out for lunch to discuss the matter further in a completely amicable way. Yet later, this gentleman copied me on a message he sent to our university president demanding that I be fired. The message stated in part:

> Dr. Alley's work on CO_2 levels in . . . ice cores has confirmed that CO_2 lags Earth's temperature. . . . This one scientific fact alone proves that CO_2 is not the cause of recent warming, yet . . . Dr. Alley continue(s) to mislead the scientific community and the general public about "global warming." . . . I await your prompt response confirming that an investigation into . . . Dr. Alley's activities will . . . start prior to end of this year. (His) crimes against the scientific community, PSU, the citizens of this great country and the citizens of the world are significant and must be dealt with severely to stop such shameful activities in the future.

You may recall the discussion of this particular issue in chapter 8. By the logic in the letter, interest payments cannot contribute to debt, and an egg cannot come from a chicken that came from an egg. Yet, by his demonstrated success in the real world, the letter-writer is more than sufficiently smart to recognize his flagrant error.

A few months later after re-reading some of our communications, I still believe that this gentleman is a valuable member of society who has done good things. He probably doesn't care very much about the relative timing of changes in CO_2 concentrations and water isotopic ratios in ice-core samples from roughly 20,000 years ago in Antarctica; his real concerns are with the future. He sees how valuable fossil fuels have been to us, and worries that a focus on global warming from CO_2 emissions will move us away from future fossil-fuel use that he thinks is wise. People with beliefs like his will be important in the discussion about our energy future, and should be.

But his email didn't cheer me up that morning. And his message is far from the worst I've seen. Another, for example, stated in part, "You creep . . . I hope you . . . suffer badly. . . . Eat me. . . . I am going to try to find you to expose you more."

So if you're thinking of throwing that stone at climate scientists, please ask yourself whether such morning greetings might tempt you to vent—privately—to a friend about crap criticisms from idiots. My guess is, if you can honestly say you wouldn't be tempted, you wouldn't throw the stone anyway.

WHAT IF THE HOCKEY STICK BROKE?

Many of the Climategate emails, and some of the completed investigations just noted, dealt in part with the "hockey stick" reconstruction of temperatures over the last millennium. The cooling from medieval times (about 1000 years ago) to the Little Ice Age (variously dated to one to a few centuries ago) is a tempting target for model testing, because it is so well documented in certain regions, such as Europe. Unfortunately, relatively little information is available from some other places.

The "hockey stick" papers by Michael Mann and coworkers[18] were a pioneering attempt to statistically overcome the irregularly distributed data. The authors chose to rely on especially well-dated records, and thus primarily on tree rings, although including some others. Dating to the nearest year with no errors is essential to learn the effect of a single volcanic eruption or a cluster of them. However, tree growth responds to many things in addition to changing temperature, so there is a lot of discussion in the scientific literature on just how accurate tree rings are as thermometers and how to read their record.

The tree-ring and other reconstructions yielded a relatively small overall cooling from medieval times to the Little Ice Age—a graph shows a line similar to the shaft of the hockey stick—followed by recent warming to temperatures above medieval levels—a curve similar to the blade. The notable uncertainties have been highlighted consistently in the scientific literature, including in the original papers, and prevented strong conclusions. In most discussions of the "hockey stick," the most

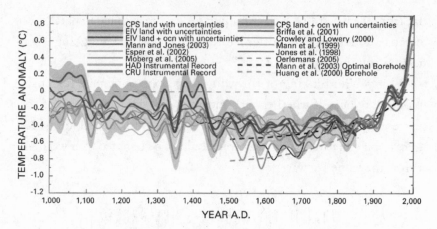

Figure 13.4 "Hockey sticks." Reconstructions of Northern Hemisphere temperatures over the last millennium, based on "proxy" evidence such as tree rings, glacier extent, borehole temperatures, and more, shown with recent thermometer records. This is presented in black and white here to emphasize the broad agreement among many different estimates, including the original "hockey stick" from Mann and coworkers.

recent and warmest temperatures are from thermometers rather than from tree rings, both because the funding and the researchers are not available to keep resampling as time passes to bring records up to date, and because the story is increasingly complicated as our pollutants, excess CO_2, and other changes affect the response of trees to climate.

Many studies since have addressed climate changes from medieval times to the Little Ice Age, sometimes including a wider range of paleo-climatic indicators and different statistical approaches. These generally agree with the original within the stated uncertainties, although perhaps with a tendency to show a bit more natural change.[19] The original work was certainly open to improvement; indeed, not only have other workers improved on it, but so have Michael Mann and colleagues.[20] This is the fate of science; bright students eventually improve on or disprove what earlier scientists have done. But an assessment by the National Academy of Sciences found that the original work basically got it right.[21]

The relatively small "hockey stick" change into the Little Ice Age in response to changing volcanoes and sun indicates a climate sensitivity slightly lower than typically found, but the uncertainties allow even slightly lower or much higher values, so the result is not overwhelmingly important in learning about future warming.[22] (Such studies are actually very important for learning about the regional patterns of climate change.)[23]

Note also that natural variability larger than what was found in the original "hockey stick" research does not mean that the current warming is natural. Satellites show that, if anything, the sun has dimmed a little over the last three decades. A couple of big volcanoes have blocked the sun not that long ago. Heat is going into the oceans, not coming out to warm the air. And in other ways, we can be confident that the ongoing warming is not natural. Surely, there might be natural cooling in the near future—another big explosive volcano, for example—but there also could be natural warming in the future—if the sun brightened, for example. Larger natural variability would not affect whether human activities are turning up the climate's thermostat now, but would mean only that the climate will stagger more on the way to a warmer future.

This scientific view of the "hockey stick" differs greatly from its popular usage, probably related to the comparison between medieval and recent warmth. We can say with moderately high confidence that temperatures have gone above medieval levels recently. But a modest change in reconstructed medieval warmth can be spun into "We are suffering from heat unprecedented in more than a millennium" versus "Our medieval ancestors enjoyed warmer temperatures." In public, the difference may appear large and important. The irony is that a warmer medieval time popularly means less concern about global warming, but to me it indicates higher climate sensitivity motivating greater concern.

My suspicion[24] is that climate sensitivity is in the mid-range, near or slightly above 5.4°F (3°C) for doubled CO_2 levels; the effect of our particles on the reflectivity of clouds has been important but not huge; temperatures now are a bit higher than those during the medieval peak; and other evidence excludes the scarily huge and optimistically small future warmings allowed by the climate record of the last millennium.

If so, then the people saying "warmer now than in a millennium" have gotten it right. But I don't think that this has risen to the pound-on-the-table near-certainty level that is associated with the physics of CO_2 causing warming, just the balance-of-evidence level.[25]

PUTTING PIONEERS OUT TO PASTURE

Our understanding of climate sensitivity and the warming effect of CO_2 rests on physics known for more than a century, clarified through well-tested military research, implemented in a great range of models from back-of-the-envelope to supercomputer-straining, and tested against the history of Earth's climate. The strong, consistent results raise concern about large future warming from business-as-usual burning of fossil fuels. Among the many parts of climate history used for testing are the cooling from medieval times to the Little Ice Age, and the warming since. The cooling is known from extensive proxy records, and the warming from proxies and from consistent thermometer data analyzed by different groups in different countries, including thermometers outside of cities, in the ocean and in the ground, and on balloons and satellites.

Against this backdrop, whether one research unit in the United Kingdom properly shared all their data, or whether one climate scientist at the University of Virginia used exactly the right statistical tests, or whether warming paused for a few years after 1998[26] are not especially important scientific issues. If you went back now and erased the CRU, or Michael Mann, or anyone else among us (certainly including me) from the history of climate science, nothing would change in the fundamental message.

When a pioneer lays down a new thread of knowledge toward the future, science does not rely on it until it has been tested and retested, by different groups using different techniques and different data in different ways. When there is such a weave of supporting results that the original could be cut out and thrown away without notably weakening the whole, the result is passed to the broader community through the assessment process, and the pioneer is put in line for invitations to give review talks and receive awards. Today, you cannot make evolution disappear by denying Darwin, nor relativity by erasing Einstein. Likewise, climate does not hang by the thread of the CRU or a hockey stick.[27]

So, before I am put out to pasture to enjoy my old age, let's see what this means for the future.

14

The Future

Synopsis. Fundamental physics shows that our CO_2 emissions will contribute to notable warming in the future. Initially, both positive and negative impacts will occur, but the negative will grow to dominate greatly. Increasing droughts and floods, sea-level rise, suppression of food production, and increasing threats of extinctions of rare species are projected.

The hypotheses which we accept ought to explain phenomena which we have observed. But they ought to do more than this: they ought to foretell phenomena which have not yet been observed. —William Whewell, 1840[1]

JUST THE FACTS

The power of misused media to polarize the modern world is huge. Climate change may not be the most polarized issue, but is certainly high on the list. Please allow me a bit of theater.

Suppose that a reasonably well-off democracy is about to hold a referendum on a proposal to reduce greenhouse-gas emissions. A hired gun with a video camera films a young but appealing victim of the latest natural disaster in a poor part of the world. The commercial might run as follows:

Sasha is an orphan. She and her family were building a better life for themselves when the drought hit. Her parents were bravely protecting the last of their withered garden from the locusts when they died heroically in the fire.

Now for the most important question—which side is this commercial on, **continue-business-as-usual** or *slow-the-emissions*? Let's do one version of the commercial for each side, with the different bits in different fonts for clarity.

"The **extremist environmentalists** tell you that this was a NATURAL disaster, but the environmentalists caused it. **They convinced government bureaucrats to pass a law that kept Sasha's family from digging a water well because their pump would have used a few cups of gasoline that emits a little CO_2 that helps crops grow better so they won't burn up or be eaten by locusts.** If you believe that people are more important than **polar bears**, vote **NO** on Tuesday."

"The *super-rich polluters* tell you that this was a NATURAL disaster, but THEY caused it. *Greenhouse gases from their smokestacks stopped the rains and destroyed the ecosystem, bringing the drought and the fire and the locusts.* If you believe that people are more important than *profits*, vote *YES* on Tuesday."

Both of these may pass a low barrier for "truth"—rising CO_2 levels may bring drought to some places and rain to others—so I doubt that a court of law would find that either version violates election laws. But neither one provides useful information to the voters. Assessed science exists to produce that useful information, to guide wise choices. Here, after a little more comment on predictions, projections, and uncertainties, I will try to present the likely future as dispassionately as possible.

FINDING THE FUTURE

We do understand a lot about climate, as described in chapter 8: we can successfully "predict" things that happened in the past without cheating, and things that are happening now were forecast in previous decades. But important uncertainties limit our vision, so certain things simply cannot be predicted, such as the weather more than a couple of weeks in the future.[2]

The climate system has "weather" too, such as El Niños and La Niñas and other wobbly modes. El Niños and La Niñas at least can be forecast with some skill six to twelve months ahead, but specifically forecasting the timing of an El Niño a decade in the future is not possible now, and may never be. If you try to predict global warming from this decade to the next, an extra El Niño or two could cause more warming than the rise in CO_2 over the next ten years, so the natural variability dominates the uncertainty.[3]

Look further than a decade or two and the effects of rising CO_2 levels grow inexorably while the "weather" tends to average out, so for the next few decades the big questions are related to the remaining uncertainties in our physical understanding, such as the strength of cloud feedbacks. Beyond that, though, the uncertainties are dominated by human choices—will we continue on a path of ever-faster fossil-fuel burning, or switch to some other path?

To address this uncertainty about human behavior, the IPCC created a range of scenarios or story lines for possible changes in economic growth, trade, income differences, and more.[4] These different futures lead to different levels of CO_2 and other influences on climate. All of these scenarios assume some sort of "business as usual"; that is, they do not include major efforts we might make specifically to reduce human-caused climate change.

So far, we have no strong reason to believe that one of these scenarios is more correct than any of the others, and the IPCC never intended that one of them would be exactly right; the intent is that reality will fall somewhere among them. The differences between the scenarios are not very large for a few more decades, so we cannot choose among them yet. Our emissions of CO_2 had been running near the highest level in the scenarios before the economic recession pulled the 2009 levels down into the middle of the range.[5] Temperature has been rising perhaps slightly faster than expected but well within the modeled range, while sea level has been rising notably faster than expected and very near the top of the modeled range.[6]

Because we don't know which scenario will actually end up being closest to correct, climate scientists cannot *predict* the future; instead, science must *project* what will happen if humans follow a particular path, and try to provide projections for enough different paths to allow wise choices. Here, I will mostly report projections from a range of scenarios for near the end of this century, as provided by the Fourth Assessment Report of the IPCC. The temperature from now until the end of the century isn't expected to be a perfectly straight line—some scenarios steepen a little toward the end of the century while others flatten, following the projections in levels of CO_2—but you won't be too far off if you think of straight lines.[7]

The climate-modeling centers of the world, including the Geophysical Fluid Dynamics Lab of NOAA in Princeton, New Jersey, the UK Hadley Centre for Climate Change, and many more, used various climate models of different types to simulate the future for the specified scenarios, and generally ran each model for each scenario many times, starting from slightly different conditions, to average over the "weather" in the models. The bigger models tax the largest supercomputers on the planet, so this is an amazingly large effort.

As described in chapter 8, results tend to be brought forward confidently if they can be understood from fundamental science,[8] appear in a range of models from different groups, are consistent with the history of climate, and are emerging in recent data. If only some of these criteria are met, the result is communicated with less certainty, or is not presented as a main result. So here is a lot of information about our possible futures meeting those criteria, with a few comments on the implications. I will delay most of the "what it means" and "what to do about it" discussion to later chapters.[9]

WARMING, MELTING, AND BLOWING

First, suppose that back in the year 2000 the CO_2 and other things in the air had quit changing, and remained at those values for the next century and beyond. The global average surface temperature in the decade 2090–2099 would be about 1.1°F (0.6°C) warmer than the average in the decades 1980–1999.[10] This additional or "committed" warming is primarily because some of the heat from the CO_2-warmed air is flowing into the huge, colder ocean, so the atmosphere hasn't yet experienced the full warming from the CO_2 already up there.

We have already increased CO_2 past the year-2000 level, with no sign of going back any time soon. For those various rising-CO_2 scenarios highlighted in table SPM-3 of the IPCC report,[11] the best estimate for global-average surface-temperature increase from 1980–1999 to 2090–2100 ranges from 3.2 °F (1.8°C) for the scenario with the smallest CO_2 rise, to 7.2°F (4.0°C) for the largest rise. Warming continues strongly beyond 2095 in all of these scenarios in response to rising CO_2 and ocean warming, and warming would continue more slowly even if CO_2 quit changing at any time in any of these scenarios, because of the delay in warming the ocean.

The estimated warming for each scenario has an associated range because of model uncertainty and internal variability. For the lowest of these rising-CO_2 scenarios, the IPCC states that it is "likely" (66 percent probability) that the warming will fall between 2.0 and 5.2°F (1.1 and 2.9°C) compared to the most-likely estimate of 3.2°F (1.8°C); for the highest of these scenarios, the corresponding range is 4.3 to 11.5°F (2.4–6.4°C), with the most-likely estimate of 7.2°F (4.0°C). (A more recent study using a slightly different methodology across many possible emissions scenarios gave a median warming to the end of the century from 1861 of 5.1°C, or 9.2°F.)[12]

Each of the IPCC most-likely estimates is shifted toward the low end of the corresponding uncertainty range. Thus, the warming is more likely to be on the high side than on the low side of the best estimate. This occurs at least in part because of interactions between feedbacks: if two of the positive feedbacks are a little larger than we thought, they amplify each other to make the warming a lot larger; if two of those feedbacks are a little smaller than we thought, they push the warming down but only toward the direct effect of the CO_2 by itself.[13] Uncertainties continue to be dominated by lack of knowledge of the future behavior of clouds. But the possibility of feedbacks in the carbon cycle may become important in pushing these projections up as we look further to the future; as the temperature and CO_2 rise, more of the CO_2 we emit is expected to stay in the air, while some of the CO_2 in the soil and the sea floor is released to the air as carbon-rich polar soils become more like carbon-poor tropical soils, sea-floor methane clathrates break down, and other changes occur.[14]

You would need to broaden the temperature ranges just given if you wanted to be "very likely" or "virtually certain" that you have correctly captured the coming warming. And that means lowering the low end of the possible warming a little, and raising the high end of the possible warming a lot—for good physical reasons, uncertainties are mostly on the big-change side.[15]

When it comes to global warming, almost all of us are above average. Because the planet is mostly ocean, and warming over the ocean is projected to be smaller than on land, the global-mean temperature changes given earlier are close to the oceanic temperature rise and smaller than the changes on land, where most of us live. Warming is expected to be largest at high northern latitudes in winter, and smallest over the Southern Ocean and a small region of the North Atlantic where ocean waters mix deeply.

Snow, sea ice, and frozen ground are expected to shrink.[16] Melting mountain glaciers will shift water from the peaks to the ocean, contributing to sea-level rise.[17] Sea-level rise is also expected from shrinkage of the great ice sheets of Greenland and Antarctica, together with expansion of the ocean water as it warms. Total sea-level rise is then projected to be roughly 0.2 to 2.0 feet (0.18–0.59 m)[18] by the last decade of the century, plus any contribution from future dynamical changes of ice-sheet flow, a subject we will return to soon.

Some of the impacts of warming are not especially surprising. More record-high temperatures and fewer record-low temperatures are projected, with more heat waves and fewer cold snaps.[19] Note that record lows don't immediately disappear with warming; many times a record has been broken by a few degrees, so changing the average temperature by one or two degrees wouldn't have prevented a record low. Thus, while it may be cute to make some smart remark such as "Global warming, huh?" on a cold morning, weather is not climate, and one record low or a big snowstorm or two in Washington, D.C., does not in any way invalidate global warming, just as one record high or a warm winter does not in any way prove global warming.

Perhaps more important, many changes in water availability are expected. Heavy precipitation events (downpours, and snowfalls where it is cold enough) are projected to become more common and heavier from the additional water vapor in a warmer atmosphere. The subtropical dry zones are expected to receive less precipitation, while the tropical rain belt and the high latitudes get more—essentially, dry areas become drier, and wet areas become wetter, but the dry subtropical zones expand toward the poles, switching some areas from being wet to dry.

The trend toward having more hot days and fewer cold ones is easy to understand, but having both more droughts and more floods may seem contradictory. However, every home gardener in Pennsylvania knows that the soil is generally damp or frozen all winter, but that in summer we're watering a week or two after a huge cloudburst. Warmer air can carry more water vapor and so can produce more rain, but can also evaporate water more quickly at a time when the plants want more water. So gardeners at least shouldn't be surprised that making the world more summerlike will give more variability in water—more floods and more droughts. These changes may be large, with a recent modeling study showing a doubling of droughts and tripling of persistent droughts.[20]

Much attention has focused on tropical cyclones—hurricanes or typhoons or whatever you want to call those big tropical storms. They are hugely important to many people, but are still rare enough and complex enough that really strong conclusions are not possible. The best estimates now suggest that the number of such storms will decrease or remain relatively constant, but that warmer waters will fuel an increase in the number and intensity of the strongest storms. Because most of the damage is done by the strongest storms, this seems likely to combine with rising sea levels to cause problems for the growing concentration of people and wealth on the coasts.[21]

The magnitude of future sea-level rise is highly uncertain, but a globally averaged rise with continuing warming appears nearly certain. In turn, this will require some combination of protecting low-lying areas through levees or other construction, or losing those areas to the sea. With about 10 percent of the world's population living on land within about 30 feet (10 m) of sea level, this seems like a major issue.[22] How-

Figure 14.1 Sea-level rise contributes to coastal erosion, and is a near-certain result of global warming, although the amount remains highly uncertain. Here, a winter storm erodes a stairway at Nauset Light Beach, Cape Cod National Seashore.

Figure 14.2 Looking north from the previous picture, at the same time, the pipes hanging out of the bluffs show that erosion has been removing human construction over the years.

ever, estimates of economic damages generally are not especially high, because economists typically assume that society will respond wisely and with foresight, although experience suggests that this is not always the case in the real world.

One way to present the economic argument is something like this. If your beach house is lost to the sea but I own the house behind yours, I get the million-dollar view, and the value of that view is not lost. And if you know early enough that your house is doomed, you can quit maintaining it so that it won't have much value when it goes. One estimate (of several given in the cited paper, using different methodologies) is that a rise in sea level by 20 inches (0.5 m) by the end of the century would cost the United States roughly $1 billion per year, with the cost to the world about ten times larger;[23] these are small numbers compared to the size of the U.S. economy.

However, very different numbers are possible if the response is not wise and careful. The disaster of Hurricane Katrina in New Orleans by itself was more expensive than this projected estimate for all of the sea-level-rise damage to all of the U.S. coasts for the upcoming century. Hence, if a poor response to sea-level rise caused even one Katrina-like flooding event of a protected city sometime in our future, the estimated costs would require a very large upward revision.[24]

I will address other, longer-term issues and uncertainties after we look at the significance of these changes for living things. Because, honestly, not many people care about the global mean surface temperature, but almost everyone cares about their dinner.

GROWING . . . OR NOT

Changing climate is virtually certain to have major impacts on plants and animals. We learn about past climate changes in part from shifts in where plants lived, as recorded by pollen and fossils. The large and rapid changes of the PETM and other past events caused extinctions, disruption of ecosystems, and extensive migration of species, as I discussed in chapter 10. Importantly, a PETM-era creature migrating northward as North America or Eurasia warmed did not need to navigate parking lots or herbicide- and pesticide-treated cornfields. Today, "weedier" species can move quickly as climate changes, but slower specialists may face grave difficulties and possible extinction.[25]

Our release of CO_2 affects living things through climate, and by changing the acidity of the ocean and the CO_2 concentration of the atmosphere. Perhaps the most optimistic feature of the whole story is that higher CO_2 levels tend to increase plant-growth rates on land. However, on careful consideration, this is not quite as encouraging as first suspected.

The plant's job is to turn CO_2 and water into more plant using the energy of sunlight, so giving the plant more CO_2 helps it do its job. And extra CO_2 also allows the plant to use water more efficiently; the stomata that let CO_2 into a leaf also let water escape, so supplying more CO_2 allows the plant to obtain the CO_2 it needs while losing less water.

The increase in plant growth obtained from adding CO_2 depends on plant type, availability of water and fertilizers, competition from other species, and more,[26] but a first-guess estimate may be that doubling of the atmospheric CO_2 level from the preindustrial value will increase plant growth by roughly 10 percent.[27] In general, raising CO_2 increases plant growth more when the plant has more of the other things it needs. In natural ecosystems, raising CO_2 initially stimulates rapid plant growth, but most of the effect disappears after a few years as the plants begin to run out of other essentials. And as we will see later, even for crops that have everything else they need, the CO_2-induced warming may reduce crop yields more than the rising CO_2 increases the yields, especially in hot places.

An additional complication is that supplying more CO_2 tends to produce plants that are enriched in carbon but depleted in nitrogen, the essential ingredient of protein.[28] Again, the outcome may be affected by the species involved, and what parts of the plants are considered, and how much nitrogen-rich fertilizer is supplied, and how much water is available, and how competitive other species are, and more. But the percentage drop in protein content from increasing CO_2 often is almost as big as the percentage increase in plant growth with rising CO_2. In a higher-CO_2 world, a young animal that needs a lot of protein to grow well may eat until it is full and still not get enough of the good stuff. Obviously, animals have lived in high-CO_2 worlds, but the evolutionary changes involved in handling such issues may be notably slower than the rise in CO_2 that we are causing, so the mere existence of animals in a former high-CO_2 world does not tell us about our immediate future.[29]

In natural ecosystems, rising CO_2 favors some species over others even if climate doesn't change. The most widely quoted effect for land plants is that CO_2 generally helps leafy vines, particularly vigorous and especially allergenic poison ivy.[30] Rising CO_2 also may favor "weedy"

Figure 14.3 High CO$_2$ favors poison ivy, and a rapidly changing climate will in general favor weeds.

behavior—some species may gain enough advantage from rising CO$_2$ levels that they invade other ecosystems.

In the ocean, rising CO$_2$ is similarly likely to cause large ecosystem changes. The increasing acidity of the oceans from rising CO$_2$ favors dissolution of carbonate shells, and while some species can grow shells in the presence of high CO$_2$ and may even do better, many seem to have a more difficult time. Higher CO$_2$ levels appear to slow coral-reef growth, which worries many biologists because coral reefs are so important as homes for diverse and valuable ecosystems.[31] Warmer surface waters tend to float more strongly on deeper, colder waters and thus to mix more slowly with them. Because this is the path that brings many nutrients for growth of algae in the ocean, warming may slow growth; this seems to be occurring already, and is projected for the future in at least some models.[32] Less growth of algae is likely to reduce the population of things that eat algae, up to and including fish that we eat, a worrisome outcome.

Back on land, the IPCC found that considering both the direct effects of higher CO_2 on plant growth and the climate-change effects, rising CO_2 levels might increase worldwide food production when warming is less than roughly 1.8 to 5.4°F (1–3°C), but reduce food production when warming is above that. Furthermore, impacts were far from globally uniform, with less warming required to reduce food production in tropical and subtropical regions, which often have difficulty raising enough food even now.[33]

Scientific work since the 2007 IPCC report further emphasizes the vulnerability of food supply to warming, especially in the tropics and

Figure 14.4 Extra CO_2 helps plants grow faster, if they are given everything else they need. However, heat stress from extra CO_2 reduces growth of our crops, such as the corn shown here. Emitting too much CO_2 is projected to do more harm than good to crops, making it harder for us to feed ourselves.

subtropics.[34] Today, anomalously high temperatures are observed to reduce crop yields, with a warming of 1.8°F (1°C) causing typical yield declines of roughly 10 percent for many important crops such as corn (maize), wheat, and rice. Climate projections for the end of the century show a more than 90 percent chance that typical summer temperatures will exceed the highest temperature ever recorded by thermometer until now across broad areas of Asia, Africa, and the Americas, primarily in the tropics but extending well beyond into the United States and parts of Asia, and with a more than 70 percent chance of this arriving by the middle of the century in many of those places. The crop-yield reduction from heat stress could easily exceed the gains from CO_2 fertilization, with the additional problems of reduced protein content in the food produced, and especially large food losses where food is now scarce.

Little is known today about how easy or difficult it will be for plant breeders to develop heat- and drought-tolerant varieties rapidly. Should this prove difficult, greatly increasing food supply to feed a few billion more people is likely to become much more difficult in a warming world. Recall that the tropics are the big belt around the middle of the planet, and the poles are the small caps on the ends, so simply shifting production poleward is not an even trade-off.

The distribution of diseases is controlled by many physical, ecologic, economic, and social factors. Knowing the mean temperature, or the minimum temperature during a few years, does not tell us whether a given disease is present in a particular area. But some diseases clearly are more prevalent in warmer places, so loss of freezing will mean loss of one line of defense. Considering temperature and many other factors, the IPCC found that

climate change currently contributes to the global burden of disease and premature deaths (very high confidence). Human beings are exposed to climate change through changing weather patterns (temperature, precipitation, sea-level rise and more frequent extreme events) and indirectly through changes in water, air and food quality and changes in ecosystems, agriculture, industry and settlements and the economy. At this early stage the effects are small but are projected to progressively increase in all countries and regions.[35]

The accompanying discussion makes clear that some health problems in some places (freezing to death, for example) may be reduced by warming, but that the net health impacts of large warming will become highly negative.

Even though warming tends to raise the global burden of disease, this does not mean that everyone will get malaria or other terrible illnesses in the future. Indeed, a comprehensive recent study showed that public-health improvements have outpaced the small effects of the warming to date, to notably reduce the incidence of malaria over the last century.[36] Human choices strongly interact with climate change to control our future, and in many cases the human choices are more important, as we will see soon.[37] For issues such as diseases, you might think of climate change as a handicap in a fight; you can win the battle with one hand tied behind your back, but the task is more difficult.

TIPPING POINTS?

All of the changes discussed here involve a climate system that "behaves itself"—no weirdness, just the predictable response as we twiddle the control knobs. But, as described in chapter 7, the climate system has switches too, or tipping points if you prefer. Is there any chance that we will tip something over? Well, maybe.

During the warming from the last ice age, regions around the North Atlantic jumped back to a millennium of cold, which ended with a remarkably large and rapid jump as a tipping point was crossed—roughly 18°F (10°C) in a decade or so in Greenland.[38]

Today, wintertime cooling causes the relatively salty surface water of the far North Atlantic to sink before freezing, and this sinking water is replaced by warm water from the south, helping Europe have relatively mild winters. As the ice age was ending, melting ice sheets supplied enough fresh water to switch the North Atlantic to wintertime freezing. The sea ice offshore made the winters very cold, changed the ocean circulation to warm the South Atlantic, shifted tropical weather patterns south, and weakened monsoonal rains in Asia and elsewhere where modern populations are measured in billions rather than millions. The global average temperature change was small—perhaps roughly 1°F (0.6°C), with southern warmth almost offsetting northern cold[39]—but the climate changed notably in some fashion pretty much everywhere. Similar events happened many times before, and a couple smaller ones happened slightly more recently.

The IPCC gives a more than 90 percent chance that global warming will not melt Greenland's ice fast enough or cause other changes fast enough to trigger such an event, but more than 90 percent is not necessarily 100 percent. And projections of future damages generally do not include a great drought clamping down on billions of people.[40]

The 2001 IPCC report highlighted the great uncertainties related to ice-sheet changes, but gave as a best estimate that the big ice sheets would not notably contribute to a rise in sea level over the next century; indeed, a slight growth of the ice sheets seemed more likely, with an increase in snowfall on Antarctica exceeding an increase in melting on Greenland and with no notable change in ice-sheet flow transferring ice from land to ocean. In 2007, the IPCC noted that ice-sheet flow had already changed in response to warming, a century early. Although the contribution of changing ice-sheet flow to sea-level rise was not large, it was unexpected. Thus, the 2007 IPCC report found that "models used to date do not include . . . the full effects of changes in ice sheet flow, because a basis in published literature is lacking . . . understanding is too limited . . . to provide a best estimate or an upper bound for sea level rise."[41]

The possibility of a major ice-sheet collapse raising sea level a lot more than what the IPCC projected in 2001 is not included in many economic assessments. We really don't expect to lose all of an ice sheet in mere decades, but we might possibly commit to such an outcome within decades, and realize the sea-level rise over centuries. And with over 23 feet (over 7 m) of sea-level rise possible from Greenland, almost as much from West Antarctica, and perhaps something similar from coastal East Antarctica over centuries, a worst-case scenario has a lot of sea-level rise.

What if warming triggers a great methane release, or collapse of a major ecosystem (the Amazonian rain forest), or huge extension of oceanic dead zones? In the next chapter we will look at economically optimal responses to global warming, which generally do not include these unlikely-but-not-impossible events.[42]

BULLDOZERS AND AC

Anyone can think of *lots* of other impacts of climate change. If you live with winter and hate shoveling snow or being stranded in an airport during an ice storm, you may cheer warming. If you love skiing or live on a low-lying island somewhere, you may not cheer the sea-level rise that comes with warming (unless you love water skiing). Climate affects most of what we do, and may change a lot.

By the last decade of this century, business as usual with the faster of the emissions scenarios is projected to cause almost as much globally averaged warming as the rise that took us from the depths of the last ice age to the preindustrial level, with greater warming continuing into the future. Most of that natural rise from the ice age was completed in 10,000 years; we may achieve a similar warming in not much more than a century. The natural change caused huge ecological shifts; the much faster human-caused change is expected to have at least similarly large impacts, with greater possibility of disruption and extinction because of the greater rate of change.[43]

For climate-change impacts on people in the near future, a useful if oversimplified summary goes something like this. If you have winter, bulldozers, and air-conditioning, too much warming will be bad but a little warming might not be too bad or might even help a little. Warming removes the damages from winter storms, while air-conditioning allows work to continue during hot summers, and bulldozers can build walls or move things in the face of rising sea level. However, if you don't have winter, bulldozers, and air-conditioning, warming is bad, bringing costs without offsetting benefits.

We are living in a time when those of us with winter, bulldozers, and air-conditioning aren't suffering much from global warming, but many of you reading this book are likely to see the day when the impacts of global warming are negative for most people. Let's explore how to assess that, and how to move forward, in the next chapter.

15

Valuing the Future

Synopsis. An economically optimal response to the rising effects of our CO_2 emissions includes beginning now to reduce those emissions. Consideration of national security, employment opportunities, possible catastrophic events, and some ethical issues favors even greater effort to reduce emissions.

> *The vast possibilities of our great future will become realities*
> *only if we make ourselves . . . responsible for that future.*
> —Gifford Pinchot, first chief of the U.S. Forest Service[1]

We have just seen that if we continue burning fossil fuels, we have high scientific confidence that climate changes over the next decades and beyond will be large. We face spreading deserts, more droughts and floods, stronger storms, rising sea levels, ecosystem disruptions and extinctions, and temperatures so high that they greatly reduce yields from our crops. So we should cut back on emitting CO_2 from fossil fuels, and save ourselves, right?

Not so fast. We also saw that fossil fuels are doing us great good. If we completely stopped burning them right now, we almost certainly would be much worse off. Instead, people all over the world are trying to get and burn more fossil fuels. So we should drill baby drill, dig baby dig, and burn baby burn, right?

Not so fast. As with other predicaments, the wise response is to weigh the good against the bad, and pick the best way forward.

I am a physical scientist, so chapters 4 through 14 are my beat, and I like history, so chapters 1 through 3 are not entirely foreign territory. But to make wise decisions while recognizing that large uncertainties remain about some things (for example, how easy will it be to breed heat-resistant crops?), we enter into the realms of economics, national security, ethics, insurance, and government budgeting. And, truth be told, I am not an expert in any of them. But I can borrow heavily from people who are. The answer that emerges is clear: there is no need for panic, but starting now to reduce CO_2 emissions and move toward sustainable energy is much more beneficial than business as usual. So, let's start with economics and move through these realms.

LOW LOW PRICES . . . BIG SALE! DISCOUNTING THE FUTURE

Probably the most comprehensively quantitative way to consider the impacts of climate change is to combine climate and economics in a single model, and calculate the optimal path for humanity. Economists have done a *lot* of work on optimal strategies—how much to invest now to maximize the net benefits. For decisions involving global warming we must consider when most of the damages from fossil-fuel burning will occur, decades or centuries in the future. Because so much of the pain and the gain of reducing global warming happen to different people at different times and in different places, we have to find comprehensive ways of making wise decisions.

The grand not-that-old man of coupling climate and economic models to provide comprehensive guidance for decision-makers is William Nordhaus of Yale University.[2] I will try to follow his general results initially. The goal is to estimate the "economically optimal" path for society to follow. It might be better to refer to this as the most "economically efficient" path, if you can think of economic efficiency as a good thing, and not as the explanation given as to why you should pay more credit-card fees and have less legroom in the coach section of an airplane. The economically optimal or efficient path in Nordhaus's work, and in many related studies, is the one that gives the most "good" from the things we buy—the "utility of consumption"—for the world's people now and in the future.[3]

Economists don't sit down over a cup of coffee and decide what is "good" for us. Instead, they do the science, building models and running experiments, testing and improving their understanding of economic activity. Economists also carefully measure what we do with our money, investing or consuming, building or binge-drinking, to constrain the numbers in their models.[4]

A central issue in many economic analyses, including those addressing global warming, is discounting, a deceptively simple-looking way of valuing the future. If I give a bank or credit union one dollar of my money, I expect to get back more than one dollar in a year. I can calculate the interest rate if I divide the extra I receive by my original investment, and I can calculate the discount rate if I divide the extra by the total I have at the end.[5]

Today, I have a personal discount rate, reflecting how much I value stuff in the future compared to now, and reflecting my current level of investments, consumption, and predilections. If I have a dollar in my pocket, but my discount rate is higher than the bank offers, then I don't value future stuff enough to wait for the little extra that the bank will give me, and I will spend my dollar now. If my discount rate is lower than the bank's, I value future consumption enough to wait for the extra money, and I will invest my dollar.

Similarly, if society's discount rate is relatively high, we collectively tend to consume more and invest less in such things as education and technologies, or in preserving the "natural capital" of the climate system by avoiding climate change. Society's discount rate depends on the three issues presented in the next paragraphs.[6] Note that inflation is not included—because inflation makes everything pricier, economists can remove those effects and report all quantities in modern dollars.

The first piece of the discount rate is the growth rate of the economy. Averaged over enough people and a long enough time, the economy has always grown, producing more goods and services—more stuff—per person, and we expect this to continue. Second, the more stuff we have, the less value we get from adding a little extra—a box of candy is worth more to a hungry person than to someone with a well-stocked, climate-controlled "chocolate cellar" in their basement. If we expect the economy to grow rapidly so we will have a lot of stuff in the future even without our investment, or we believe that our appetite for more consumption becomes satiated easily so that a little more stuff is almost as good as a lot more stuff in the future, then our discount rate is high and we tend to consume now rather than investing for the future.

The third piece of the discount rate comes from the observation that, on average, we behave as if today is more important than tomorrow, and each generation is more important than the generation that follows. Because we prefer things in the current time, this is sometimes called the "pure rate of time preference." The more we prefer things now, the less we invest for the future, and the higher our discount rate.

The information presented in the previous paragraphs upsets some people greatly. How can economists possibly dismiss the well-being of future generations by taking selfishness and renaming it "pure rate of time preference"?!? Please, for now, don't get upset, follow the economic analysis through, and then we will revisit ethics and other issues. Because, surprisingly, even this "selfish" view of the world leads to an optimal path in which we start now to head off the damages from global warming!

For the discount rate, Nordhaus adopted 4 percent, the estimated market return on capital, which is how much you might really make by investing your money.[7] The resulting optimal path, not surprisingly, splits economic activity between consumption and investment, but includes significant investment in reducing global warming. This investment could be achieved most easily by starting now (actually, in 2005) to charge a price of $30 per ton of carbon emitted as CO_2 to the atmosphere.[8] This is roughly a 10 percent tax on electricity in the United States, or nine cents per gallon of gasoline. But the optimal tax (or other instrument to put a price on emitting carbon, such as cap-and-dividend) then rises, to $90 per ton by 2050 and $200 per ton by 2100, using 2005 dollars (with the effects of inflation removed).

This ramp in carbon price reflects trade-offs among all of the economic and climatic factors involved, including the initially small but rapidly rising damages as temperature climbs, and the rising wealth of people as the economy grows from our investments. If we ignored global warming in our decision-making, we and our descendants would suffer from some damages that could have been stopped for small or negative costs—changing lightbulbs can save money now and help people in the future by stopping a little global warming. However, if we immediately took the measures needed to stop all global warming, which are much larger than just changing lightbulbs, we would suffer great economic harm. Neither panic nor business as usual is optimal.[9]

Many other scientists and economists have tried similar analyses, using a range of estimates of the damages, the discount rate,[10] and other relevant quantities from climate and economic science. A survey of over 100 peer-reviewed estimates of the *social cost of CO_2 emissions*—what the damages from CO_2 emissions cost the world, and thus what the optimal tax or fee would be on those emitting that CO_2 now—found estimates in 2005 dollars ranging from $350 per ton of carbon (emitting CO_2 really costs us) to –$10 per ton (emitting CO_2 very slightly benefits us), with a central estimate of $43 per ton (emitting CO_2 costs us some), compared to Nordhaus's $30 per ton.[11] Note the "long tail"—the central estimate is near the low end of the damages.

In response to such calculations, you may hear an argument presented as if it were real economics by people who probably should know better: global warming matters but we should ignore it anyway because more important issues affect humanity. How can anyone in good conscience recommend spending money to slow global warming when people are dying of starvation and AIDS and bad water? This argument resonates, but wrongly.

Consider that when you run your own house, you don't say, "Should I patch the roof, or should I pay to educate the kids?" Instead, you may say, "With the money I have, what is a sound strategy for educating the kids, patching the roof, and doing all the other things we want to do?" The sound strategy usually involves at least some investment in addressing many problems. Maybe you decide to nail a hunk of plywood over the hole and get new shingles later so you can pay some of the tuition while your kids get part-time jobs to help pay the rest, but you don't let the rain pour in on the piano because education is more important than roofing, and you don't force the kids out of school to patch the roof.

What you do about the roof and education is what the economic arguments by Nordhaus and others have done: choosing the best path among the possible uses of the money. "The essence of an economic analysis is to convert or translate all economic activities into a common unit of account and then to compare different approaches by their impact on the total amount."[12]

With almost any problem, there are inexpensive ways to make some progress, and the first small investment, such as switching lightbulbs, often has the greatest payoff. The optimal path allows much warming because it finds better uses for a lot of money than heading off all warming.

The discount rate is probably the most important issue in calculating the economically optimal path. However, estimates of damages remain highly uncertain, with much room for improvement from additional research. Some recent studies, including a major one by Nordhaus,[13] point to somewhat higher damages than have been used previously, and suggest that actions taken to adapt to the rising warmth cannot come close to eliminating those damages, although wise adaptation can lower the damages.[14]

Overall, hard-nosed real economic science, as applied by real economists in a classically economic approach, finds that a measured response to global warming is economically justified now (or, more accurately, is already overdue). This is not a big attack on global warming; in the Nordhaus optimization, CO_2 rises to 685 ppm in 2100, almost 2.5 times the preindustrial level, with estimated warming in the model of 5.6°F (3.1°C) for 2100 and 9.5°F for 2200 (5.3°C)[15] relative to temperature in

1900. The optimal path invests $2 trillion to stop $5 trillion in damages, but damages totaling $17 trillion are allowed to happen. Note that all of these trillion-dollar numbers are in "present value"—discounted from the future—so they represent much more money in the future.

The key to making this economic optimization work, essentially, is to convince everyone that emitting carbon will become costly in the future. Firing skilled coal miners, or throwing away cars or electric power plants, would be economically inefficient by wasting the investment that had been made in education or construction. But when the coal miners retire, or the cars or power plants get old, society is economically better off switching to jobs and infrastructure that emit less CO_2. (Clearly, firing coal miners raises many ethical and other issues that extend beyond the basic economic analysis.)

Among many other important points from the Nordhaus studies, one especially stands out: "a low-cost zero-carbon technology would have a net value of around $17 trillion in present value . . . the net benefits . . . are so high as to warrant very intensive research."[16]

Seventeen trillion dollars! Find an answer to the energy/climate problem and there is $17 trillion on the table today. Find an answer to a little piece of it, and you still may end up well off, because even a little piece of $17 trillion is real money. That's what the final third of this book will be about, after we look at a few ways to extend the economic optimization.

ONWARD, SECURELY

Economics pervades modern life, but surely is not the only way to evaluate the issues facing us. For example, the decision to defend one's country against foes is not usually based on a strict cost-benefit analysis.

Consider the following:

> . . . climate change, energy security, and economic stability are inextricably linked. Climate change will contribute to food and

water scarcity, will increase the spread of disease, and may spur or exacerbate mass migration.

These are not the words of a left-wing politician, pundit, or policymaker. They are the official position of the U.S. Pentagon.[17]

Military operations worldwide have long been tied to energy supplies. After the fateful Civil War battle of the ironsides USS *Monitor* and CSS *Virginia*, an attempt was made to move the *Virginia* up the James River far enough to help defend Richmond. Her draft was too deep for the river, and throwing most of her coal overboard did not lighten her load quite enough. Lacking passage upriver or fuel to go downriver, the decision was made to blow her up.[18]

During the lead-up to World War II, Japan had been importing about 80 percent of its oil from the United States. The de facto U.S. oil embargo in response to Japanese actions in China was a major problem for Japan, and seems to have been an important factor in the Japanese

Figure 15.1   Oceanographer of the U.S. Navy Rear Admiral David Titley provides environmental information for safer and more-effective operations, including assessments of human-caused and natural climate change.

decision to strike toward the Netherlands East Indies to obtain oil, and Pearl Harbor to prevent U.S. interference.[19] The Iraqi invasion of Kuwait on August 2, 1990, was at least in part about oil prices, revenues, and reserves.[20] U.S. President George H. W. Bush cited the danger to Saudi Arabia in his speech laying out reasons for military intervention in response to the Iraqi invasion,[21] and such an experienced policymaker was surely aware of Saudi Arabia's oil reserves and trade with the United States, among the many issues to be considered.

These national-security issues arose for the United States during the Korean War as well. President Harry S. Truman's Materials Policy Commission examined shortages of strategic materials caused by the war, and recommended avoiding dependence on imported oil, especially from the volatile Middle East, through development of alternatives— solar energy, and also synthetic fuels.[22] However, free-market policies in the following administration of President Dwight D. Eisenhower, together with objections to the synthetic fuels by the National Petroleum Council, delayed action on these security issues.[23]

Fuel supply lines are points of vulnerability, not only for nations but also for armies. Data from a report prepared for the U.S. Army from the years 2003 to 2007 show the high human cost, with 188 resupply-convoy casualties (killed or wounded) in Afghanistan and 2858 in Iraq. Because about half the load of those convoys was fuel, gains in energy efficiency or alternate-energy generation in the field would translate directly into lives saved.[24]

Added to these considerations is the emerging evidence that climate change can trigger strife, including civil wars. And strife often grows to involve militaries and national security.[25]

For the U.S. Pentagon, these considerations motivate a serious effort on energy and environment.

> Energy efficiency can serve as a force multiplier. . . . The Department [of Defense] is increasing its use of renewable energy supplies and reducing energy demand to improve operational effectiveness, reduce greenhouse gas emissions in support of U.S. climate

change initiatives, and protect the Department from energy price fluctuations.[26]

And, recognizing the need for innovation to meet these challenges:

> Solving military challenges—through such innovations as more efficient generators, better batteries, lighter materials, and tactically deployed energy sources—has the potential to yield spin-off technologies that benefit the civilian community as well. DoD will partner with academia, other U.S. agencies, and international partners to research, develop, test, and evaluate new sustainable energy technologies."[27]

A large, eager research partner and market are not likely to remove the benefits of putting a cost on emitting carbon,[28] but they will stimulate the innovations needed to reduce emissions.

I suspect that if the true experts on national security were asked to rank their worries for this week, "bad guys with bombs" would come out higher than "CO_2-induced climate change destabilizing weak countries over coming decades." But security experts really do recognize the dangers of heavy reliance of long, expensive, easily cut supply lines. And, long-term, military planners know that all too often they are called upon to fix things that others have broken, and such problems are likely to occur as climate change worsens conditions in some of the world's trouble spots. The military planners thus find that broadening the choices available for energy, including local generation and conservation, and avoiding destabilizing climate change will increase national and international security.[29]

ONWARD, CONFIDENTLY

While I was still a graduate student conducting field work in Antarctica, the commander of the military air wing was preparing for an expedition to central Greenland in support of the deep ice coring being planned.

Because of my prior experience in Greenland, he called me in and asked whether his huge ski-equipped cargo planes would fall into crevasses if he landed on central Greenland's ice. I told him that crevassing was very unlikely in the center of the ice sheet, based on several lines of evidence: a lack of crevasses in satellite images compared to the presence of crevasses in images of some more-coastal regions; our knowledge that flow conditions in central regions don't favor crevasses in the way that the coastal conditions do; a lack of crevasses where I had been, not too far from his proposed landing site; and analogy to similar sites where pilots have landed successfully.

The commander was a large and imposing military presence in uniform, and I was a small and diffident graduate student. He drew himself up a little taller, leaned in a little closer, and more loudly and forcefully

Figure 15.2 Crevasses are a danger for travel on ice sheets. Much of an ice sheet lacks crevasses, and some regions have obvious crevasses, but other regions have crevasses bridged by weak snow. These are on Daugaard-Jensen Glacier in East Greenland.

Figure 15.3 Pilots of these massive, ski-equipped heavy-lift airplanes are very careful not to fall in crevasses, for obvious reasons, and often take out "insurance" against the uncertainty of the real world by conducting a ski drag over a possible landing site.

said something very much like this (I don't recall the exact words, but the substance and tone are completely unforgettable): "If I go to Greenland and land in a crevasse, I'll come back and find you. I'll put your [reproductive organs][30] in a vise, set the table on fire, and give you a butcher knife. Now, are you sure that I won't fall in a crevasse?"

I suggested that he "buy a little insurance" by doing a "ski drag" reconnaissance first.[31] I also quivered a lot.

As posed by the commander, the setup was a bit unfair—if my science was correct (which it was) he might someday see me and thank me, but if I erred he would come after my reproductive future, not exactly an even trade.[32] Seriously, of course, he was using the threat for effect and not for fixing me. (I think.)

But, since then, I have often wondered what parts of our science are so well known that we would have the confidence to answer the commander, and what that says about buying insurance.

In his economic analysis, Nordhaus found that including normally distributed uncertainties, with "better" and "worse" equally likely, didn't affect the economically best path very much. But we have fairly high confidence that "worse" (more warming) is more likely than "better" (less warming), and this favors more action to slow warming.[33]

Back in chapter 14, we met a few "tipping points," or abrupt climate changes. If melting ice or other processes supply fresh water too rapidly to the North Atlantic in the near future, there is a chance of a really nasty drought where billions now live. Warming too rapidly may greatly accelerate the rise in sea level, or free frozen methane from the sea floor, or push ecosystems to jump to new states, or trigger "unknown unknowns." None of these is considered likely, but our scientific understanding suggests that burning fossil fuels faster is more likely to push us past a tipping point.[34]

Projections of future damages generally do not include such abrupt climate changes, but I could not assure a knife-wielding commander that such an event is impossible. And the possibility of such a high-cost event favors greater reduction in our CO_2 emissions than in the economically optimal path discussed earlier. Just like the commander making a ski drag, or me suggesting that he do so, we buy insurance in an attempt to avoid catastrophes.[35]

ONWARD, TO WORK

If those emitting CO_2 must pay for the damage they cause, employment patterns will shift, with some people losing jobs while others gain. People who will lose their jobs as the economy changes usually know who they are, whereas people who will get the new jobs don't know who they are. In the next chapter, we will visit London and Edinburgh in the 1800s, where "night-soil haulers" made a good living taking human waste to the countryside. They might have complained about a switch to modern sanitation, but the plumbers and civil engineers of their future didn't exist yet and thus could not cheer for a change.

After we visit the night-soil haulers, we will examine options for avoiding CO_2 emissions. Importantly, recent studies suggest that these options will increase total employment, with job gains notably exceeding job losses.[36]

I work in a department of geosciences, which has educated numerous people who have taken good jobs in energy companies, especially oil companies. I worked for an oil company one summer and found my colleagues and supervisors to be solid scientists and great people. During graduate school, I shared an office with a fellow who has become important in the oil industry, and our office was in a building endowed by oil money. My closest scientific collaborator enjoyed working for an oil company, and some of my undergraduate education was paid for by the foundation of an oil company. I have read with interest the scientific research done by oil-company employees. So I am deeply indebted to oil and oil companies, and I know as well as anyone that oil companies provide good jobs, as do companies recovering other fossil fuels.

But the many recent advertisements from fossil-fuel interests highlighting job creation, while true, are incomplete. For example, as the IPCC summarized in its 2007 report, a $1 million investment in wind or solar photovoltaics generated 40 percent more jobs over ten years than an equivalent investment in the coal industry,[37] while a commitment to a 20 percent reduction in European Union energy consumption by 2020 could potentially create, directly or indirectly, up to 1 million new jobs.[38] The IPCC report also noted that a higher proportion of renewable-energy jobs are relatively highly skilled than for fossil-fuel sources.[39]

A recent update[40] for the U.S. electrical power sector compared employment from coal and natural gas (which are more directly tied to U.S. electric generation than is oil) to that for possible low-CO_2 energy options. Renewable energy, energy efficiency, carbon capture and storage, and nuclear energy all were found to generate more jobs than fossil-fuel approaches. Many other studies are finding net job gains from renewable energy sources compared to fossil-fuel sources.[41]

Fossil-fuel extraction and use as practiced today are very efficient, with a few employees providing a lot of energy. One person with a joystick can run a dragline that takes apart a mountaintop to extract the coal beneath. The lowest-price producers of oil can probably get it out of the ground and onto a supertanker using very few workers for about $5 per barrel, but are being paid $70 per barrel.[42] Spend that $70 per barrel on wind turbines or solar cells, and more people are involved building and installing and maintaining the sustainable sources.

The broadest interpretation would assess the extra jobs lost elsewhere in the economy if a switch to renewable energy raises costs, plus the effects on employment of avoiding climate-change damages. One recent study for Germany included the economy-wide losses from higher-priced renewable energy, and still found that pushing more aggressively toward renewables increases employment.[43]

The study of employment implications of energy sources is far from completed, and something surprising might yet show up. But the best research so far says that while some people will lose their jobs if we address global warming, more people will gain jobs. Therefore, if increasing employment matters, a stronger push away from fossil fuels would be favored.

ONWARD, WITH A HEALTHIER ECONOMY

Economic optimizations typically do not outlaw CO_2 emissions. Instead, they place a cost on emitting CO_2 through a cap-and-trade system, or a fee-and-dividend one, or some such plan; I'll consider carbon taxes here.

In order to "provide for the common defense, promote the general welfare"[44] and do other things, governments must raise money. In the United States, various levels of government levy taxes on tobacco to raise money and to discourage smoking, so the money-raising also promotes the general welfare. Similarly, we tax alcoholic beverages to discourage too much drinking. And then we tax working and investing and owning things (income and property taxes), discouraging activities we like. Would we be better off taxing other "bad" things and reducing taxes on "good" things?[45]

To assess the costs of a proposed cap-and-trade bill to reduce U.S. carbon emissions, the U.S. Environmental Protection Agency looked at many scenarios, including one in which the money raised by placing a cost on emitting carbon was returned to the people by reducing the tax on labor. Unlike the other scenarios considered, this one "raises consumption and GDP [gross domestic product] relative to what would occur in a policy scenario without such recycling . . . over the next several decades."[46] If we tax actions we don't want (emitting climate-changing carbon dioxide) rather than actions we do want (working), U.S. household consumption of goods and services would be higher through the year 2040 than under business as usual.[47] Furthermore, this ignores the added benefits of avoided climate change.

Although much is known about these and related issues,[48] there is much still to learn. But, if we make the right decisions, the economic costs of our response to global warming don't need to be large, and might even be negative.

ONWARD, ETHICALLY

The economic approach falls short if we cannot measure the costs of important possible impacts because they have not been "monetized."[49] For example, good science shows that global warming will place many species at risk of extinction, and may make it difficult or impossible for native peoples to live their traditional lifestyles. If humanity values those species or lifestyles, but that value is not fully included in the economic analysis, then the economic analysis is incomplete. What would our great-great-grandchildren pay to have us preserve functioning cold-weather ecosystems?[50]

Some people might argue that we should give up on economics because it is incomplete. An economist is likely to argue that we should expand economics to include things that matter, and indeed, such efforts are underway. There is little doubt that some damages of global warming are not monetized in at least some studies, giving greater impetus for a response to head off global warming.[51] My impression is that the nonmonetized issues are larger for global warming than for most or all competing uses of money, because CO_2 affects almost everything. Thus, to the extent that the economic analyses remain incomplete, a greater reduction of CO_2 emissions seems justified, not less.

I have received some interesting correspondence recently from people who care about these issues based on religious grounds.[52] Warming today is being caused primarily by people with winter, air-conditioners, and bulldozers who burn most of the fossil fuels, but these people aren't being hurt very much by the climate changes; the damages primarily affect poor people in hot places who lack winter, air-conditioners, and bulldozers, and who are not primarily responsible for the changes. By almost every moral and religious code on the planet, this is wrong—if my actions hurt others, I am supposed to make it right. "Do unto others as you would have them do unto you," says that I shouldn't do my bathrooming in your front yard, and I shouldn't let my fossil-fuel CO_2 emissions change your climate in ways that harm you.[53]

In addition, recall from earlier in this chapter that the discount rate includes a pure rate of time preference, treating people who are here now as more important than people not yet born. This really bothers many people—what gives us the right to decide that our great-grandchildren are less important than we are?[54] If we don't treat ourselves as more important, the discount rate is lower, and an economically optimal response is to take more action now to head off global warming.[55]

OPTIMAL OPTIONS

With so many reasons to make larger reductions in CO_2 emissions than in the economically optimal path, are there reasons to do less? Uncertainty is *not* a reason to wait; if anything, uncertainty motivates taking more action now, because the uncertainties are so strongly on the "bad" side.

The biggest objection to action is probably from those who believe that all government actions are ethically wrong, or are certain to fail. No matter how I view it, a lot of people will need to act together to leave valuable fossil fuels in the ground, or to put the CO_2 from the fossil fuels back into the ground after they are burned. The actions might be implemented and coordinated through trade or treaties, but they are unlikely to happen spontaneously. Treaties almost surely will require verification, with data collection.[56] And all of this seems likely to require the involvement of governments. But, most of us drive on roads maintained by governments, enjoy the national security provided by government forces, vote and pay taxes, and otherwise recognize that while governments may be imperfect, they are certainly not perfectly useless. Still, future studies that more accurately include the (in)efficiency of governments may lead to adjustments in the economically optimal path.

So where does that leave us? For a hard-nosed planner who collects the best numbers and then makes the hard decisions, the best available science shows that the economically optimal future is reached by putting a price on emitting CO_2 now, a price that likely will ramp up in the future. An especially savvy planner looks at the uncertainties of climate science, sees that the most-likely costs of emitting CO_2 are on the low side of the possible range, and favors a higher price now or a faster price ramp. These planners are also likely to favor innovation through support of research and emerging markets.

What else? If you believe that the expected damages to endangered species and lifestyles have economic value that is insufficiently accounted for in existing economic analyses, then you would do even more to slow climate change. Or, if you believe that there are ethical, religious, or moral reasons to preserve endangered species and lifestyle choices going beyond the economics, and to preserve choices for poor people now and people not yet born, then you would do even more now to slow climate change. If you favor carrying insurance against large risks, including dangers to national security, then you would do more now, perhaps using taxes on carbon to offset taxes on beneficial activities.

Many large uncertainties and debatable issues remain about this topic. But solidly orthodox economics motivates taking action now, additional considerations favor taking greater action, and to the best of my knowledge these considerations are not offset by strong balancing arguments for less action.[57] In the rest of the book, we will visit the remarkably rich and broad options available to us.

PART III.

THE ROAD TO TEN BILLION SMILING PEOPLE

16

Toilets and the Smart Grid

Synopsis. Stabilizing the composition of the atmosphere within a few decades to avoid major human-caused climate change is estimated to cost about 1 percent of the world economy per year, similar to or even less than the cost of clean water and sewage treatment. We thus have solved problems this big before.

When we were finishing our house, we found we had a little cash left over, on account of the plumber not knowing it. —Mark Twain[1]

THE 1 PERCENT SOLUTION

A few professional arguers, joined by a host of amateurs, have responded to the two-headed issue of energy supply and climate change with some version of "The natural-not-human problem that isn't happening and wouldn't matter is too big to handle." We have seen how wrong most of this is, with high confidence from strong science that our continuing reliance on fossil fuels will, before they run out, change climate in ways that matter to us. But is the problem too big for us to handle? Of course not!

The optimal economic path outlined in chapter 15 would cost well under 1 percent of the world economy per year, with the benefits outweighing the costs. But that path allowed a lot of warming to occur. Enhanced national security, insurance against disasters, and fairness to others were among the additional arguments that favored more action sooner.

Several groups (see below) have attempted to estimate the costs of stabilizing the climate while still supplying abundant energy. These estimates generally ignore the benefits of avoiding climate change and present only the costs. Those costs depend a lot on how rapidly the stabilization is made. For plans that stop the warming at no more than a few degrees within a few decades, costs generally are in the neighborhood of 1 percent of the world's economy (gross domestic product, or GDP).

The Intergovernmental Panel on Climate Change (IPCC),[2] for example, assessed paths for stabilizing atmospheric composition at different levels, ranging from 1.6 and 2.5 times the preindustrial level of CO_2 (while also including the effects of other greenhouse gases),[3] and found somewhere between 0.6 percent extra growth and 3.0 percent shrinkage of global GDP in the year 2030 compared to what would happen if we were to continue business as usual. The German Advisory Council on Global Change similarly gave a range of possibilities for different rates and levels of stabilization, producing costs that are distributed around roughly 0.7 percent of the world economy.[4] Other estimates covered in a recent review[5] ranged from –1 percent of GDP—benefits, not costs— to 4 percent, with a mean of about 1 percent.[6]

In all of these projections, the economy continues to grow at a rapid rate, but one that is slightly slower than if we were to conduct business as usual—again, these projections are ignoring the benefits of heading off climate change and considering only the costs. These costs can be reported in many ways. With a world GDP of about $60 trillion per year,[7] 1 percent is a big number ($600 billion per year). If you multiply that $600 billion by, say, 100 years to give the costs over a century, you get a number that is easily shouted on talk radio ("$60 trillion dollars for global warming!?!"). You could get an even bigger number if you added economic growth and inflation over the century. But you can also say that a sustainable-energy future is ours for only "a penny on a dollar," which is equivalent to simply slowing down economic growth by a few months to perhaps a year over a century. All are basically accurate reports, except for the obvious bias that they omit the benefits.[8]

Have we ever done anything so big? World War II comes to mind as a coordinated human action that took a lot more than 1 percent of the economy. More recently, and with much less drama, a study from the International Energy Agency and other agencies found that globally, government subsidies for the production and use of fossil fuels are roughly 1 percent of the world's economy.[9] And the cost of fossil fuels used just in the United States is more than 1 percent of the whole world's economy. But even ignoring wars, because they are so different, or the costs and the subsidies for fossil fuels, we have solved a problem with costs similar to or even slightly higher than those for global warming and energy. And although the "sticker shock" was unpleasant, we wouldn't dream of going back to the old ways to save that money.

A "TERRIBLE SHOWER"

During a visit to Edinburgh, Scotland, in 1754, Edward Burt of London reported that when he "first came into the High Street," he thought that he "had not seen anything of the kind more magnificent," with the long street overlooked by well-sashed stone buildings of extreme height.[10] He was much less favorably impressed by what happened that evening though. At ten o'clock that night, "by beat of the city drum," the people dumped the contents of their chamber pots and buckets, the accumulated human waste of the day, out of the windows onto the street in a "terrible shower." Burt made it to his rooms unscathed only by having a guide go before him, shouting to the people above to "hud your haunde" (hold your hand), and he then found it necessary to hide his head between the sheets in an attempt to escape the "smell of the filth, thrown out by the neighbours on the back side of the house."

"This great annoyance," human waste dumped from as high as "eight, ten, and even twelve storeys" to accumulate in the streets, was hauled away in the morning by "scavengers"—night-soil haulers—except on Sunday, which thus was "the most uncleanly day."

When Burt considered possible solutions to the problem, however, he decided that it was "remediless." The city had been built on rock near

Figure 16.1 The magnificent, well-sashed stone buildings of the High Street in Edinburgh today are even more welcoming than they were in 1754, because there is no worry of a "terrible shower" of human waste at ten o'clock at night.

the castle for protection, in a place so narrow that construction was forced up rather than out, so people were living and excreting waste as high as the twelfth floor. Burt observed that "anything so expensive as a conveyance [for the waste] down from the uppermost floor could never be agreed on; nor could there be made, within the building, any receiver suitable to such numbers of people." He noted that there was plenty of nearby flat land with a stream, "which would be very commodious for a city," but the magistrates would not allow building there because then the people would leave the old city, "which would bring a very great loss upon some, and total ruin upon others, of the proprietors of those buildings."[11]

In other words, Burt reported that the people were stuck with dumping their human waste out the windows because the available options were too technically demanding, or too expensive for people to agree to, or involved loss of money by established businesspeople who had influence with the government.

When I first came into the High Street of Edinburgh, during a visit in 2010, I'm happy to report that I too "had not seen anything of the

Figure 16.2 Reconstruction of diners burning things to cover the smell of the "terrible shower" of human waste at 10 p.m. in Edinburgh, 1754, prepared for the filming of Earth—The Operators' Manual.

Figure 16.3 Reconstruction of the "terrible shower" of human waste at 10 p.m. in Edinburgh, 1754, prepared for the filming of Earth—The Operators' Manual.

kind more magnificent," with the same well-sashed stone buildings of extreme height that so impressed Burt more than 250 years before. But when our party came out of the pub at ten o'clock that night, we had no worry about a "terrible shower" on our walk down the long and beautifully clean street to our rooms. Edinburgh had indeed expanded into the flatter regions below, but the High Street remained the focus of the city. There, agreement had long since been reached among the people and the proprietors to build "a conveyance down from the uppermost floors" for the human waste.

Clearly, something happened between Burt's visit and mine, to turn a "remediless" problem into something that was remedied so well. To see what happened, let's visit Burt's home of London, a century later.

MORE TERRIBLE THAN THEY KNEW

Cities can be wonderful, with museums, restaurants, and universities providing great opportunities for meeting, mingling, and mating. Airlines have adopted hub-and-spoke systems for efficiency, and cities similarly provide highly efficient ways to organize our interchanges.

But cities also require a lot of cleaning up after ourselves. A million hunter-gatherers spread across much of the planet could relieve themselves behind a tree like a bear without worrying too much about the poop piling up or transmitting disease—nature can keep up with the wastes of a small, distributed, and mobile population. But pack more people than that into, say, London, and diseases take advantage of the hub-and-spoke system too.

The emergence of cholera as a killer in cities in the early 1800s was one important reason why many city dwellers suffered nasty, short, and unpleasant lives. Epidemics repeatedly swept across major cities. And one of science's true heroes, physician John Snow of London (1813–1858), was observing and learning about cholera transmission, especially in the epidemic of 1854.[12]

Against the prevailing view that cholera was spread through the air by a "miasma" of unhealthy vapors from sewers and other sources, Snow had observed in an earlier outbreak that the disease spread easily in places that the "miasmists" predicted were safe. Snow also saw that people who drank only beer in regions with bad air had less cholera than people who drank water in regions with better air. These and other inferences led him to suspect that the illness was waterborne and linked to human waste.

Snow then greatly advanced modern epidemiology, by asking how the distribution of cholera was related to possible causes. In one of his most important studies, he found that two water systems served people mixed together in one part of London, but only one of the systems used sewage-contaminated water from the Thames. Snow found that subscribers of the company with the contaminated water source were eight to nine times more likely to contract cholera.

In a second, better-known study, Snow mapped the occurrence of an especially bad cholera outbreak, and traced the source to a single water well known as the Broad Street pump. The authorities then removed the pump handle for a while to slow transmission. Subsequent studies showed that human waste was leaking into the well, especially from a cesspit that had received waste from the diapers of an infant who died of cholera. Snow even identified a woman who lived far from the pump but liked the taste of the water and had it delivered to her, and who died of cholera in the outbreak.[13]

Snow didn't do experiments by deliberately infecting susceptible lab animals to verify that some confounding factor (a bit of vapor in the top of a jug?) was not at work, and to get enough data to be statistically valid (the case of the one woman infected while drinking the pump's water far from the pump might have been a statistical fluke, given that many people became ill around the city). He didn't have tools capable of seeing or otherwise identifying the disease organisms, although he did examine the water for them. But he did succeed in forming a few of the individual threads of knowledge that have been woven into the fabric of medical science.[14] And by advancing scientific understanding of cholera and how to avoid it, Snow contributed to the huge, slow, expensive, uncertain, but ultimately successful effort to improve public health in the great cities, with much of the work focused on human waste.

In the early 1800s, night-soil haulers in places such as London, Paris, and Edinburgh removed the human output to the countryside, where it was used as fertilizer.[15] However, the use of water for flush toilets and other purposes was increasing rapidly. Flush toilets were not a new invention in Europe,[16] but their reemergence strained a waste-handling system that relied on carts and buckets. Cesspools often overflowed or leaked, with reports of basements filling with raw sewage.[17] Sewer pipes or troughs could carry the watery waste, but construction was expensive. Supposed experts promoted schemes to fund the sewers using the proceeds from selling the human waste as fertilizer, with estimates of its value running as high as 10 percent of the cost of the food eaten.[18] This source of income proved to be vastly overestimated, and with the rising realization that spreading human waste spread human disease, efforts shifted toward development of the system of sewage treatment that we have today.

This system is far from free. The Organisation for Economic Co-operation and Development estimated the cost of clean water for its members (much of the "developed world") in 2002 as roughly 0.5 to 2.4 percent of household income, with the cost in the United States being the lowest.[19] The U.S. Congressional Budget Office produced an even lower estimate for the United States, with sewer and water bills together accounting for 0.5 percent of household income, but noted that investment in the system was inadequate and that growth of expenditures to 0.6 to 0.9 percent of household income would be required to maintain the infrastructure.[20]

Connecting a new house to sewer and water systems, or installing a well and septic system, often accounts for notably more than 1 percent of the construction cost—rates vary hugely, but where I live, simply connecting the plumbing already in a new house to the sewer and water accounts for about 3 percent of the typical home price.[21] The pipes and toilets in the house cost a good bit of money, and so do the toilets at the office or the stadium, and the porta-potties at the local soccer fields. Plumbers to install and repair the system also cost money. A reasonable estimate is that the cost of our sewer-water system is similar to, or a bit higher than, the estimated cost of solving energy and global warming, representing something like 1 percent of the world economy.

Imagine for a moment that you are an elected official in London or Paris or Edinburgh in the early 1800s, but that you have the solid science on waste and water that has been woven from the threads of earlier work, including Snow's. You know what causes cholera and other diseases, with high confidence, based on epidemiology, process studies, double-blind testing, and all the other strands of modern medicine. You know that your city now needs a clean water supply isolated from waste that must be biologically and chemically neutralized at sewage treatment plants that do not now exist, by techniques that you don't yet have. Storm water will need to be separated from human waste to keep the sewage out of rivers and wells during floods. Your streets must be dug up to lay lots of pipes. Every house, every apartment, every public office will require appropriate rooms and pipes for toilets and sinks.

The night-soil haulers, who make a decent living doing an indecent job, will be put out of business, while a host of new specialists will need to be trained and hired, from educational programs that don't even exist. Every level of government will be looking for people with expertise in water and sewage. Zoning boards will be forced to decide whether population density is sufficiently high to require connections, or whether septic systems and wells are acceptable, knowing that over time increasing population density may require people who already paid for a septic system to hook their homes up to the sewage system, making them really mad about having to pay twice. Inspectors will need to be hired to ensure that regulations are followed—a few contaminated diapers apparently killed hundreds of people through water from the Broad Street pump, showing that very efficient separation of waste from drinking water is essential.[22]

If all of this were revealed to you as an elected official, at once, in 1820 or so, is there any chance that you would look at the problem and be tempted to panic? Do you suppose that the farmers might complain because they have been using low-cost human waste and now will need to import more expensive phosphate fertilizers from Chile or elsewhere? Are the owners of existing apartments going to donate to your reelection campaign when you require that they put plumbing in all of their buildings and pay to connect it to the sewer and water pipes? Might they find someone to run against you in the next election if you persist in this sewer-water plan? Will you see Joe the Night-Soil-Hauler on the next campaign ad, accusing you of bankrupting him and his family? Remember that in 1754, Edward Burt reported that Edinburgh would not clean up their "terrible shower" because the cost was too high and agreement too difficult, and Burt did not foresee the full effort required.

Cleaning up after ourselves generally is neither easy nor fun, and we usually don't like it. We want the benefit of our good ideas, but would rather not be bothered with the unintended consequences. Yet failure to deal with those unintended consequences can have huge impacts.

Sewers and cholera are in some sense local issues, but in a way they're not. In a world of travelers, all diseases have the potential to become global. And in a world economy, the tourists and businesspeople stay away from a city that does not provide modern sanitation. Essentially, standards are enforced through the economy by the trade system.

The issue of our CO_2 changing the climate has much in common with the spread of disease by our more personal waste, but much that is different. Cholera may kill you or your family this week; global warming mostly harms poor people in hot places or other species now, with the biggest impacts reserved for decades or longer in the future. "I'm gonna die" focuses attention in a way that is not achieved by "Damages will rise to a few percent of the world economy per year in a few decades." The switch to public sanitation may have cost more of the economy than the switch envisioned for the energy system, but the very large near-term payoffs of public sanitation made the economics much easier. And the political impact of the public-sanitation switch may have been diluted because the actual response was spread out over more than a century, without the foresight to add up all the costs before the action began.

When we look ahead to a sustainable system providing enough energy for ten billion people to live in comfort without changing the climate, it is very, very easy to add up all the difficulties and costs and to conclude that the task is impossible. But as we will see soon, we know that the world has more than enough sustainable energy for us, we have engineering-ready ways to generate that energy and stabilize our influence on the climate for a cost estimated as approximately 1 percent of the world economy per year, and we know that humans have done much bigger things.

WEDGES AND WAYS FORWARD

What will a solution for energy and environment look like? Many people are trying to figure this out, guide it, stimulate it, or reap the profits that come with it—recall from chapter 15 that $17 trillion is on the table for someone who can find a better way. Many well-thought-out plans are being published.[23]

These plans are diverse. Some push quickly toward a truly sustainable future, with heavy reliance on sun and wind power. Others use "bridge" technologies such as capturing CO_2 from power plants and putting it back into the ground, which surely is unsustainable as fossil fuels run out, but may be workable over at least several decades or longer. Large disagreements can be found—is nuclear fission a workhorse for the long haul, a bridge technology to help us over a rough spot to something better, or a pariah to be eliminated as quickly as possible?

The "wedge" approach provides a useful way to think about the diverse options available. Not only did ecologist Stephen Pacala and physicist Robert Socolow of Princeton University produce scholarly studies on wedges, but the popular press including *Time* magazine highlighted the advantages, and there is even a wedges game.[24] Pacala and Socolow first picked a target of keeping CO_2 from exceeding twice the preindustrial concentration through the year 2054—fifty years after their initial paper was published—figuring that we would learn enough in fifty years to make readjustments. They estimated that we could meet this goal by stabilizing CO_2 emissions at the 2004 level, by "scaling up what we already know how to do."[25]

Pacala and Socolow showed that this stabilization could be achieved if society picked seven actions, or wedges, each starting from zero in the year 2004 and growing to avoid emissions of 1 gigaton of carbon (Gt C) per year in 2054. Pacala and Socolow provided fifteen options, of which any seven would suffice; more or fewer could be used if society's goals changed. Four of the wedges involved different efforts to improve energy efficiency, one shifted much electric generation from coal to natural gas because gas provides almost twice the energy for the same amount of CO_2 released, three wedges used different forms of capturing and storing CO_2, one increased use of nuclear power, three implemented renewable energy, and two preserved carbon in forests and soils.[26] And as Pacala and Socolow stated, "Every one of these options is already implemented at an industrial scale."

I don't know which set of wedges is closest to the best answer, or whether other wedges should be included.[27] But, as we will discover in the rest of this book, a great number of practicable solutions are available—the "1 percent of the economy" generalities introduced earlier are backed up by much specific information.

I do know that two ways forward are especially highlighted in serious discussions. One we have already discussed in chapter 15: creating an economic incentive to avoid CO_2 emissions, through some combination of convincing people that emitting CO_2 to the air will become expensive in the future, and creating markets for low-CO_2 alternatives.

The second is to greatly increase research into energy alternatives, so that the ways forward are available. Energy research has simmered on the back burner for a long time. For example, the member countries of the International Energy Agency, most of the big "players" in the world's energy markets, did increase funding for energy research after the oil price shocks of the 1970s, but then they cut that research in half and held it there for a while before beginning a slow, recent increase that still has not produced sustained funding at the 1970s level.[28] And the money available from that lower level of funding was primarily directed at fission, fusion, and fossil fuels (receiving 40 percent, 12 percent, and 11 percent of the total, respectively, up to 2002); research for all of the renewable sources of energy taken together (solar heating and cooling, solar photo-electric, solar thermal-electric, wind, ocean, biomass, geothermal, and hydroelectric) received just under 8 percent of the expenditures, conservation 11 percent, power storage 4 percent, and other technologies 14 percent.

U.S. expenditures have been similar. From 1979 to 2006, inflation-adjusted research expenditures by the U.S. government in energy dropped from $8 billion to $3 billion, while industry and other non-governmental sources decreased their research expenditures on energy from $4 billion to $1 billion by 2005. For comparison, U.S. government research expenditures from 1979 to 2006 for health rose from $9 billion to $28 billion, and for military from $29 billion to $76 billion.[29] Very clearly, these numbers show that energy has not been a priority in comparison to health and defense, and that renewable energy sources have not been a priority in comparison to fission, fusion, and fossil fuels.

AGREEING TO MOVE FORWARD

When Edward Burt was considering Edinburgh's human-waste problem, he noted that some possible solutions were technically difficult, but other technically workable solutions were blocked by people or politics. The various serious plans for our energy future rely on using and improving existing technologies. But these will surely need to deal with people and politics too. In particular, the people of one country cannot solve global warming by themselves—involving most of the people responsible for most of the world's economy will be needed.

Much effort has gone into discussion of international agreements, but these have not yet met with widespread and enthusiastic endorsement. The Kyoto Protocol, for example, would not have "solved" global warming by itself. Neither would the commitments for the 2009 Copenhagen Climate Conference, or the Waxman-Markey cap-and-trade bill introduced to the U.S. Senate. But just as we had to learn how to handle sewage, and as we are learning how to handle wind and sun power, we will need to learn how to get along with each other on these issues.

Edison's first lightbulb and Ford's first car would not survive long in a modern market; they were grossly inadequate for solving the problems of illumination and transportation. I am not a policy wonk, but in my naive view of politics, demanding that the first or second international agreement solve all of the problems is just as absurd as expecting that the Wright brothers' first plane should have been ready for the Paris-to-New York route.

The good people of Edinburgh, and London, and Paris solved their problems and cleaned up the water and sewage, and people today can travel much of the world without fear of waterborne disease. The global economy helped—if you trade with other people, and welcome them as business partners and tourists, you need to meet their standards. In turn, many people considering the political future of actions on global warming are learning from the old toilet issues; if a sufficiently large fraction of the world's economy agrees on a path forward, they could use fees or tariffs on trade to encourage others to participate.[30]

A pessimist about climate-change solutions is betting against the ingenuity of a lot of smart people. We know that the solutions are out there. The problem is big, but let's go meet the bigger solutions.

Be forewarned though—there is no single solution, except the inspiration and perspiration of a lot of people. All of the paths we might follow have costs as well as benefits. If someone tells you that they have the answer, and can solve everything painlessly, you can be fairly confident that someone is being fooled. I am going to give you the pluses and minuses—protecting mountaintops from strip-mine removal might mean installing some wind turbines or solar cells instead, and running a power cable to the mountaintop to carry the power to people, and these will cost money and affect people. No one really knows the best path to a sustainable-energy future yet, in part because we have been spending so much of our effort arguing about things that don't matter very much (see chapter 13), rather than things that do matter. The rest of this book is not about giving you the right answer; it is about setting the stage for the discoveries and inventions and agreements that will give all of us good answers.

17

Sustainable Solutions on the Wind

Synopsis. Vast renewable energy resources are available, but they are widely distributed and in some cases intermittent. Wiring large areas together with a "smart grid" can offset these difficulties by reducing fluctuations, allowing different sources to supplement each other, and offering novel opportunities for storage. Such a distributed system will have much less impact on climate than the greenhouse gases from fossil fuels, but will be easier to develop if the local people affected see it favorably.

What wind will serve him that is not yet resolved upon his port? —Seneca[1]

SUSTENANCE

Some energy sources are "sustainable"—our use of the continuously replenished flow now will not reduce the amount available in the future. Other energy sources are "unsustainable"—we are extracting energy from a storehouse much more rapidly than new energy is added. We harvest sustainable sources and mine unsustainable ones.[2]

Fossil fuels are unsustainable because we are burning them a million times faster than nature stored them for us (see chapter 4). We will see that nuclear energy and much geothermal energy are also unsustainable, although the available resources are probably large enough that we don't need to worry as much as for fossil fuels.

On the other hand, the sun simply does not care whether we are using its energy to power our civilization—there is no "peak sun," no "reservoir" of solar energy that we are exhausting. Our grandchildren's grandchildren will have a steady flow of sunshine, not notably different from what we receive now.

So, let's start with the sustainable sources. Modern human energy use worldwide is almost 16 terawatts (TW),[3] or about 2400 watts (W) per person, or about twenty-four times our personal internal-energy use of 100 W.

Your favorite energy source may prove to be important even if it can generate only a small fraction of 16 TW. In 2006, energy expenditures in the United States alone amounted to almost $1.2 trillion, or almost $4000 per person.[4] Capturing just 1 percent of that market would gross over $10 billion per year—immense fortunes await even niche players. But to make enough energy for everyone in the world while saving the environment, we will need to rely on some of the big potential sources.

The maximum sustainable energy supply is given next, for various sources and in several forms: first in terawatts, then in watts per square meter (W/m^2), and how many people standing on a U.S. football field[5] would use that much internally, and how that source compares to modern human use.

Sun 173,000 TW, 340 W/m^2, 15,000 people on a football field, 11,000 times our use[6]

Wind 1220 TW, 2.4 W/m^2, 110 people on a football field, 78 times our use[7]

Plants 166 TW, 0.33 W/m^2, 15 people on a football field, 11 times our use[8]

Waves and currents 65 TW, 0.12 W/m^2, 5 people on a football field, 4.1 times our use[9]

Geothermal 44 TW, 0.09 W/m^2, 4 people on a football field, 2.8 times our use[10]

Human use 15.7 TW, 0.03 W/m^2, 1.4 people on a football field[11]

Tides 4 TW, 0.01 W/m^2, 0.35 person on a football field, 0.25 times our use[12]

Hydroelectric 1.9 TW, 0.004 W/m^2, 0.16 person on a football field, 0.12 times our use[13]

As noted, we will return to the possibility of "mining" energy from geothermal and nuclear sources. But from a first glance, it looks like the sun is the biggest of the big, with wind very intriguing, and the others powerful enough to make a dent in human use. We will also see that there is no realistic chance that we will collect close to 100 percent of any of these, but that we can get a lot more energy than humans use because grabbing even a little of the available energy can suffice.

Conservation usually doesn't show up as a sustainable energy source, but using less energy makes all the other sources look bigger, so we will consider conservation as well. And there is one way in which conservation is immediately huge. If coal is burned in a typical power plant to make electricity, just over one-third of the energy in the coal ends up in the electric line, and the rest goes to heat. This heat can be used for other purposes, such as warming houses in the winter, but the cooling towers that you typically see on power plants are just "throwing away" that heat. The amount of waste heat varies by energy source. A recent estimate suggested that switching from burning things to using wind or sun or hydroelectric sources connected in a smart grid would reduce human energy use by about one-third.[14]

AND SOME THINGS THAT AREN'T BIG

You can see from the table above that the 0.03 W/m^2 released as heat from our fossil-fuel burning and other energy sources is pathetically small compared to the energy from the sun. Although our fossil-fuel burning warms our houses in the winter, and even warms the cities a little bit, the global effect is almost entirely through release of CO_2, which strengthens the greenhouse effect. People may be out barbecuing because the weather is hot, but the heat from the barbecue fires does not measurably warm the planet.

Similarly, the 0.09 W/m^2 that globally reaches the surface from radioactivity in Earth—only 0.065 W/m^2 on land—is small compared to the sun's energy. This is why weather forecasters never include human-produced or geothermal heat in their forecasts!

Wind turbines, solar cells, and other collectors of renewable energy generate so much power that they greatly reduce release of CO_2 despite typically being constructed, installed, maintained, and discarded using energy that comes in large part from fossil fuels that produce CO_2. Wind is the most favorable, with a modern wind turbine generating as much energy in about three months as required to build, install, maintain, and discard it over its thirty-year lifetime, and thus reducing CO_2 emissions by more than 99 percent compared to a typical coal-fired power plant producing the same amount of electricity.[15] And if society moves to greater use of renewable sources, the CO_2 emissions associated with renewables will drop further because the construction, installation, maintenance, and disposal will be powered by other renewables.

Similarly, the direct effects of renewable-energy collectors on climate are very small compared to the effects of the CO_2 emissions that are avoided by using renewable energy. Solar farms covering reflective desert sands will warm Earth a tiny bit by reducing the albedo, but we will see soon that the area of low-latitude desert needed to generate all of modern human energy use with existing solar technologies is only 0.1 to 0.2 percent of Earth's area, so even a huge change in the albedo of that region would not matter much to the planet's climate.[16]

Wind may have somewhat larger impacts than solar collectors, but still less than the effects of the CO_2 emissions avoided by using the wind. Extensive use of wind turbines would make Earth's surface "rougher" (see chapter 8), slowing the winds that carry heat from equator to pole and thus causing a small low-latitude warming and high-latitude cooling, with other more local effects.[17] But, thus far, I don't know of any serious plans for renewable energy that would have nearly as much impact on Earth's climate as the CO_2 emissions that will be avoided, although careful consideration is required when we discuss certain biofuels in chapter 20.

Building enough wind turbines, or solar cells, or anything else to supply a large fraction of human energy use would be a big undertaking requiring a lot of materials. As I am writing this, there is some concern about availability of rare earth elements, which are used in many high-tech applications, including in permanent magnets in wind-turbine generators, and in catalysts for petroleum refining and catalytic converters on fossil-fueled vehicles.[18] U.S. production had been allowed to lapse, leading to dominance of world production by just one country, China. However, it appears both other sources (including reopening of U.S. production) and substitutes are available.[19]

In short, although much work remains, there are at present no insurmountable obstacles in view. Based on my reading of the extensive and very rapidly growing literature, if someone tells you that there is a technological reason why sustainable resources cannot solve our energy and climate problems, that person either 1) has discovered something new of great international importance or 2) is confused. And the second option seems more likely.

THE WAYS OF THE WIND

The sun is clearly the big resource, but I am going to start with wind here because the technologies are a little more settled (unless we opt for giant kites in the jet stream).[20] Many of us are familiar with wind turbines and have some idea of their impacts. The rapidly growing use of wind power raises many issues that affect other renewables, such as the value of a smart grid interconnecting many places with different sources, demands, and storage capabilities for energy. And wind lets us revisit Abraham Lincoln.

Before being elected president, Lincoln delivered a lecture on discovery and invention, praising the patent system and the innovations of people ("The patent system added the fuel of interest to the fire of genius"). He in particular singled out the opportunities from wind power.

> Of all the forces of nature, I should think the *wind* contains the largest amount of *motive power*—that is, power to move things. Take any given space of the earth's surface—for instance, Illinois; and all the power exerted by all the men, and beasts, and running-

Figure 17.1 The Eastham windmill on Cape Cod was built in 1680. When residents of the Cape switched to wind and sun to pump and evaporate sea water, they moved from importing salt to exporting it. Now windmills are being constructed on the Cape to provide electricity.

water, and steam, over and upon it, shall not equal the one hundredth part of what is exerted by the blowing of the wind over and upon the same space. And yet it has not, so far in the world's history, become proportionably *valuable* as a motive power. It is applied extensively, and advantageously, to sail-vessels in navigation. Add to this a few windmills, and pumps, and you have about all. That, as yet, no very successful mode of *controlling*, and *directing* the wind, has been discovered; and that, naturally, it moves by fits and starts—now so gently as to scarcely stir a leaf, and now so roughly as to level a forest—doubtless have been the insurmountable difficulties. As yet, the wind is an *untamed*, and *unharnessed* force; and quite possibly one of the greatest discoveries hereafter to be made, will be the taming, and harnessing of it. That the difficulties of controlling this power are very great is quite evident by the fact that they have already been perceived, and struggled with more than three thousand years; for that power was applied to sail-vessels, at least as early as the time of the prophet Isaiah.[21]

Lincoln had it all there—the wind is a huge resource that has benefited us greatly in the past,[22] but its intermittent nature, sometimes so weak as to be useless and sometimes so strong as to be destructive, provides real challenges for our ingenuity.

Following Lincoln's time, manufacturers did much to realize his vision, with millions of windmills erected in the United States, especially on the Great Plains, mostly for pumping water but also for generating electricity. Production in the United States may have reached close to 100,000 windmills per year from numerous companies, which also developed important export markets.[23]

In 1936, *Popular Science* reported on the windmill ideas of the first director of the British Antarctic Survey, Frank Debenham. Because he had hurt his knee playing soccer in the Antarctic snow, he was unable to accompany Robert Falcon Scott on the fatal trek to the South Pole. Debenham did, however, observe the icy rivers of wind hundreds of feet deep that howled across the frozen Antarctic wastes, and envisioned using giant windmill generators on streamlined steel towers to tap this

Figure 17.2 Windmill and water tank, 1939, Fort Apache, Arizona. Windmills were very widely used in the United States and many other countries for pumping water (as shown here) and many other uses, before largely being replaced by fossil fuels.

supply of energy, far greater than Niagara Falls.[24] In January of 2010, a little bit of Debenham's vision was working smoothly in Antarctica, as a joint U.S.–New Zealand project used three wind turbines to supply power to McMurdo Station and Scott Base, reducing fuel use by 240,000 gallons per year.[25]

But by the time Debenham was promoting Antarctic wind power between the World Wars, rural electrification, often powered by fossil fuels, offered electric pumps that worked even when the wind didn't. Wind power was losing market share, a trend that now has been reversed.[26] Let's see why.

ROPING THE WIND

Commercial wind turbines—high-tech modern windmills—now are routinely produced with a capacity of 2.5 megawatts (MW), and bigger installations are available and growing in popularity.[27] These can be placed on land or in coastal oceans, and experiments are underway to use floating installations farther offshore. The wind doesn't blow hard enough all the time to generate the full rated capacity, and usually about one-third of the capacity is achieved, but even with this, the average power production from one big turbine is equivalent to the total energy use in the U.S. economy for more than 100 people.[28]

If you speed down the highway in a car, the scenery flashes past but the dust stays on the windshield next to the dead bugs. Wind speed is zero at a surface and slow close to the surface, especially if the surface is rough. Even above trees and buildings, the wind speed still increases upward (away from the surface)—not too rapidly, but consistently. Very crudely, available wind power increases threefold for a tenfold increase in height above a relatively smooth surface.[29]

Hence, turbines in commercial wind farms are impressive machines. Towers 330 feet (100 m) high support blades spanning a diameter that large. Spinning at 10 to 20 revolutions per minute, the tips of the blades are moving at speeds close to or even above 200 miles per hour (320 kilometers per hour).

Figure 17.3   Modern wind turbines, such as these near Roscoe, Texas, are being constructed rapidly, but the potential resource is far greater than what we collect now.

Imagine spacing 2.5-MW turbines in a land-based wind farm, just far enough apart so they don't block each other's wind very much. Do this on the plains and deserts of the world, avoiding forests, cities, glaciers, and lakes, and using only those places windy enough so the turbines average at least 20 percent of their 2.5-MW capacity. Such a grid would generate more than five times the current human energy use, with an even greater excess if you take into account that this avoids the energy lost as heat while burning things.[30] Larger turbines, and inclusion of offshore sites, would generate even more.[31]

How about the economics of wind power? Different people have looked at this in different ways, and reached somewhat different answers. Regardless of how you care to "spin" it, the electricity we use from wind turbines now is not highly expensive, and probably is somewhat inexpensive,[32] although notably increasing our reliance on wind power would raise costs, as we will see.

The effort involved in wiring the world with wind power wouldn't be easy—if 3-MW turbines operating at one-third efficiency are used to generate 10 TW, we would need 10 million turbines. Installed over thirty years, that's almost 1000 turbines per day. But that's for the whole world, which is a big place with a lot of people looking for good jobs. And recall that a century ago the United States alone was producing windmills at roughly one-third that rate, albeit somewhat smaller than the ones built today. At current prices, it would cost a bit over $1 trillion per year for thirty years to install the world's windmills.[33]

I'm having fun tossing around absolutely immense numbers, but remember that the United States alone spent almost $1.2 trillion on energy in the year 2006, and the energy from wind is pretty cheap once the system is installed, so a wind-powered system up and operating could save a lot of that annual $1 trillion in the United States, and save a lot more for the rest of the world, and wind power would avoid any of the price increases that are expected as fossil fuels become scarcer. For comparison, the world gross domestic product was about $61 trillion in 2008.[34]

You would be stupid to take this analysis as the sole basis for a decision to go with wind for the world. I will raise a few more issues soon, and I won't get to all of them. Serious engineers and economists need to work these things out, and are doing so. But you can begin to smile now. We really can have sustainable energy—the task is huge, but the outcome depends on commitment.

INTERMITTENCY

Music is perpetual, and only hearing is intermittent.
—Henry David Thoreau[35]

The sustainable flow of wind and sun is a blessing but also a curse, because there is no wind well or sun mine where we can go for more if electricity demand goes up for air-conditioners on a hot day, or if supply goes down on a still, hot night. An engineer responsible for supplying enough power to meet fluctuating demand has an easier time designing a reliable system if it includes a big reservoir of stored energy that can be tapped easily and quickly.

Three approaches feature especially prominently in dealing with intermittency: interconnection (the "smart grid"), diversification, and storage.[36] The energy system can tie larger regions together to smooth out local fluctuations, link different generation techniques that fluctuate independently, and store energy in "good" times and places to be used when and where needed.

It is hard to blame an Illinois farmer for switching from a windmill to an electric pump; a few days of hot, still weather under a stalled high-pressure system can cause trouble for thirsty cattle counting on a water pump. But the wind is always blowing somewhere, and the bigger the area, the more reliable the wind.

In the United States, for example, fluctuations in wind power are very different in Texas, Montana, and Minnesota, especially in wintertime, so sharing power between these would increase reliability.[37] Estimated costs of transmitting energy from wind projects to users vary widely, but a recent summary found that additional transmission costs typically added only 10 to 20 percent to the total price for constructing a wind project.[38]

The fluctuations in wind energy are partly correlated across the Great Plains in summer, and wind energy over U.S. land in August is only 40 percent of the January value, which is problematic because electricity demand peaks in the summer to run air-conditioners.[39] However, solar energy peaks in summer, so a system blending wind and sun on a smart electrical grid will come much closer to matching demand than would a wind-only or sun-only system.[40]

Electric power grids now must incorporate "peaking" power as well as "base load" to match the changes in demand as weather fluctuates, people go to work and come home, and other things change. Natural-gas turbines are often used, but hydroelectric is already outstanding for this, and geothermal can be configured to help.[41]

Energy storage may also prove valuable. Options include pumping water uphill into reservoirs when there is extra wind power, then letting the water run back downhill to drive hydroelectric generators when the wind is slow. Or compressing air in storage chambers with extra electricity, and using the energy in the compressed air to drive wind turbines to generate electricity when the wind is weak. Or making hydrogen with extra electricity during windy times and then burning the hydrogen in power plants or cars, or using it in fuel cells to make electricity, when the need arises.[42]

We could go with hybrid cars plugged into a power grid smart enough to buy and sell power from them. Plug in your car when you park, and the electricity from the wind will charge the batteries so you are ready to drive. If the batteries run out of power before you get where you are going, the car switches to hydrogen, or biodiesel, or something else powering an internal-combustion engine or running a generator to recharge the batteries. And if the wind drops and the grid needs more electricity to run refrigerators or heart-lung machines or hair dryers, the power company can buy back some energy because you have a fuel backup. As long as there are lots of cars that need lots of batteries anyway, this appears workable.[43]

One estimate is that the fluctuations associated with generating 50 percent of U.S. electricity demand from wind could be managed if 3 percent of the vehicle fleet was under contract to buy and sell electricity this way.[44] The need for a lot of plug-in stations in parking lots, and a smart grid that can keep track of who is "buying" and who is "selling" energy, may seem daunting, but with smart people involved and trillions of dollars on the line, the problems don't seem insurmountable.

WHO OWNS THE WIND?

In the fall of 2009, the *Cape Cod Times*, a small, regional newspaper out of Hyannis, Massachusetts, was printing on average almost one article or letter to the editor per day that concerned wind power. Some of the stories involved the new turbine supplying power to clean up groundwater pollution on the Massachusetts Military Reservation. Others addressed projects being started in various towns to supply reduced-price electricity to save money for the local governments and their taxpayers. One story noted opening of a new community-based wind project on Maine's Fox Islands, across the Gulf of Maine to the north.[45] And most of the coverage concerned a long-pending project to build an offshore wind farm south of Cape Cod in Nantucket Sound.

The articles, the letters to the editor, and the discussions over coffee in the community were fascinating. Arguments favoring wind included lower energy prices, improved national security by avoiding reliance on imported oil, the economic benefits of generating new jobs, and the environmental advantages of reduced global warming. On the minus side, the noise, interference with the view, possible effects on aviation, and deaths of birds and bats were discussed.

All of these are valid issues. The noise from wind turbines is not huge, but real. The new turbines will not be so tightly packed as to intrude everywhere all the time, but they will be visible a long way off. Bird and bat deaths do occur—at speeds that may exceed 200 miles per hour, the turbine-blade tips rival a stooping falcon and outrace almost any other flying creature. But bat deaths can be reduced by perhaps three-fourths if turbines are turned off on the few nights when most migrants are streaming through, with very minor effects on power production. One recent review article even suggested that building wind farms to replace fossil fuels may reduce bird deaths by avoiding the loss of habitat caused by strip-mining and other issues caused by strip-mining.[46]

My reading of the local wind-power debate is that the community-based, town-based, and clean-up-the-dirty-water windmills all had good press and were primarily if not uniformly welcomed by their neighbors. For the project in Maine mentioned in the Cape Cod newspaper, the noise from the new turbines was "the sound of the future. . . . Now we are at the beginning of energy independence."[47]

Yet, the press coverage and the letters to the editor I saw concerning the commercial development off Cape Cod included both sides but were heavily against the wind farm. Many explanations are possible—the offshore project was the biggest of those being discussed, and the one most

likely to impact the view of tourists, who fuel much of Cape Cod's economy. But another possible interpretation is that the people were more
favorable to the community-based projects most directly helping them.
True, the cheaper electricity being touted from the big offshore project
would aid the residents of Cape Cod, and improving the environment
is good for everyone; but a large electrical utility spreads the benefits of
the cheap energy across many people not affected by the wind turbines,
and the environmental benefits of one local project, even a big project,
are harder to trace than the local costs.

I am reporting my impressions, not an assessment of scientific studies, but I am not the only one who has seen this.[48] And interesting questions follow. *Anything* we do to generate a lot of energy has important
consequences. My home state of Pennsylvania made the national news
about possibly poisoned water and exploding wells associated with
natural-gas drilling, following a finding by the Pennsylvania Department of Environmental Protection that drilling had polluted thirteen
wells in the north-central part of the state.[49] I enjoy fly-fishing, but there
are headwater streams in Pennsylvania in which the native brook trout
are absent because of acid rain, in large part traceable to the coal-fired
power plants that supply electricity to the region, including our house.[50]
I catch and release, but people who eat a lot of their catch in many
places must worry about mercury in fish, and coal-fired power plants
emit mercury that contributes to the environmental load.[51] Oil spills and
underwater gushers, photochemical smog, particulate pollution, acid
mine drainage, landscape degradation from strip-mining, and other bad
things come with our fossil-fuel burning. The costs to society of electricity generated from coal are probably more than twice as large as the
cost paid by customers for generating the electricity from coal.[52]

A big difference is that these are the problems we know about, and
have learned to live with. People have chosen where to live, and what
to do, in part based on knowledge of these problems. Drive on many
major highways across the United States and you will see places where
people have built houses commanding a view of the highway; try to put
a new highway into the view of people already there and you are likely

to have many complaints. But if the people there want the new highway, and believe they will benefit from it, and request it, the complaints may be much less common.

If wind energy is to be used widely, there will be a new turbine in or near a lot of places where people live or visit, and a cable passing more houses to connect the turbine to the rest of the electric power grid. Although some evidence indicates that the presence of wind turbines nearby does not lower property values,[53] having those people "on board" and in favor of the money and jobs coming with the wind power is likely to make conversion much easier.

WIND AND CHANGING CLIMATE

Some studies have addressed the possibility that climate change caused by CO_2 emissions from burning fossil fuel will change the winds enough to reduce (or enhance) the value of the turbines before a lot of them are built and the CO_2 emissions are stopped. However, the best results so far indicate that this will not be a big factor.[54] And, as we saw earlier, using a lot of wind power may affect the climate a little bit, but probably not a lot.[55]

Is wind *the* answer to the energy issues? No—wind surely can be part of the answer, but there are too many other bright ideas available to stop here!

18

Sun and Water

Synopsis. Solar energy is by far the biggest available source for our future, and it can be used in many ways in many places. Hydroelectric power cannot supply nearly as much energy as the sun, but it plays an important role in smoothing out fluctuations from other intermittent sources.

The sun shines and warms and lights us and we have no curiosity to know why this is so; but we ask the reason of all evil, of pain, and hunger, and musquitoes and silly people. —Ralph Waldo Emerson[1]

HERE COMES THE SUN

Fun in Philly

Frank Shuman (1862–1918) of Philadelphia surely was a worthy successor to Ben Franklin, an inventor from Pennsylvania who knew how to make money, doing well by doing good.[2] Early glass skylights sometimes broke and rained dangerous shards on people below,[3] until Shuman developed wire-reinforced glass. Dangerous breakage also afflicted windshields of early cars, but drivers were not interested in peering through chicken wire, so Shuman sandwiched clear glue between glass panes to produce the first practical automotive safety glass.

Most of the modern interest in Shuman is linked to his successes in solar-thermal applications. As a master of glass, he knew that putting glass over a box makes it hot, and he became interested in using that knowledge for much more than growing tomatoes. In 1911 he wrote in *Scientific American*, "One thing I feel sure of, and that is that the human race must finally utilize direct sun power or revert to barbarism."[4] What solar-power start-up company could resist a quote like that?!?

By 1911, Shuman had constructed in his suburban Philadelphia laboratory a special low-pressure steam engine driven by sunlight concentrated using mirrors to boil water inside glass-topped hot-boxes. This marvelous invention was described, with pictures, in the scientific journal *Nature* in 1912, which pointed out that although the Shuman design generated "by far the greatest quantity [of steam] ever produced by sun power," the Philadelphia area was clearly not the optimal place for such an installation—it was built there "simply for the convenience of being close to the inventor's house, offices, and laboratory."[5]

Shuman's next target wasn't so close to home though. He set off for Egypt, where he constructed a remarkable solar collection system. He used five parabolic troughs, each 13 feet 5 inches across the top and 205 feet long and designed to follow the sun, to focus the rays to boil water. The steam then ran a 65-horsepower engine to pump 6000 gallons of water per minute from the Nile to irrigate cotton fields in Meadi. Problems with investors and the advent of World War I quickly terminated what was until then a successful experiment, and cheap oil filled the void after the war. But solar-thermal power clearly worked. Shuman's achievement shows that powering our lifestyle with the sun is not quite as futuristic or as difficult as some people might think.

And, Shuman was not the first pioneer to make money from solar energy. Back at Cape Cod, the early European settlers wanted salt for personal use, packing fish for export, and other purposes. Early on, they "mined" wood, burning it far faster than trees grew back, to provide heat for boiling sea water to isolate the salt. But by the mid eighteenth century, no large timber remained from the woodlands that had covered 97 percent of the Cape when the Pilgrims landed, and the local people were importing salt.[6] After the American Revolution, entrepreneurial inventors developed an integrated wind-water system, with small windmills pumping sea water into solar-powered evaporators. This system was so efficient that the Cape switched to being a net exporter of salt.

Using the Sun

We saw in the previous chapter that the sun is by far the biggest potential source of energy. Some is reflected from the planet, and some warms dust or clouds in the air, but capturing just 10 percent of the sunshine reaching the surface in sunny places such as the Sahara and Yuma, Arizona, would provide at least 20 W/m^2 of electricity.[7] At this value, the world demand of 15.7 TW could be supplied from a square area of just over 500 miles on a side (just under 900 km). A recent estimate is that the desert Southwest in the United States alone has enough land suitable for solar-power generation to supply more than 80 percent of this world use,[8] with places such as the Sahara and the Australian Outback offering vast additional resources.

Shuman's technique is called "concentrated solar-thermal energy"—focus the sun's energy to make something hot, and use the heat, most simply by making steam to run turbines and generate electricity. A hybrid plant might use natural gas or something else to power the turbine when the sun isn't shining. Or some of the heat can be stored in giant vats of molten salt or in other ways, and then extracted at night. Conversion of

Figure 18.1 Concentrated solar-thermal power near Seville, Spain, part of a larger project updating Frank Shuman's work to an output of more than 300 MW, with built-in storage capacity to extend operations after sunset.

more than 15 percent of the solar energy to electricity can be achieved commercially, and since electricity avoids the lost heat of burning, technologies already exist to double human energy use from a square 500 miles on a side.[9] Other options include using the heat to drive chemical reactions, perhaps turning otherwise-unused biomass wastes into useful fuels, or even converting CO_2 or water to fuels using catalysts.[10]

The sun also can be used to make electricity directly, avoiding the high temperatures altogether. Adding a little bit of the right chemical to silicon creates mobile electrons that can be knocked free by sunshine. A second layer with a different chemical additive gives a destination for the electrons, and you can run those electrons from this photovoltaic cell or solar cell through your electricity grid on the way to power your air-conditioner or heart-lung machine.[11]

Once you have gotten this far, you may be asking whether you have to use silicon, and whether there are better substances for "doping" the silicon than the chemicals that people are using now, and how much to use in doping, and how thick the layers should be, and whether it matters that the silicon is really pure or a little impure, and much more.

To a scientist or an inventor, this is a playground, with a seemingly endless list of exciting things to study that matter to the real world. For example, solar cells might simply be painted on a surface.[12] Or living organisms such as tobacco viruses might be tricked through genetic engineering into growing solar cells.[13] And scientists and engineers are looking to nature, including polar-bear fur, to gain insights for building better solar collectors.[14]

There are no guarantees that any of these, or any of many other ideas, will be the answer we are looking for, but the remarkable cleverness, the sheer diversity, and the passion of the people involved surely are encouraging to those of us watching and cheering, and additional progress is almost guaranteed. Furthermore, we already know that we can make solar-thermal systems, and photovoltaics that make electricity from light, even without new inventions.

A Costly Decision?

The big issue with solar power remains price. And estimating the price may be almost as difficult as building the system. Are you calculating for a really sunny place, or a gray one? Just as for wind, adding a little sun-produced electricity to the grid is easy, but adding a lot requires additional investment in backup or batteries or other options discussed in the previous chapter. Solar energy is like nuclear energy in that it has relatively low operational costs but high construction costs, so even a small change in interest rates charged on loans can have a huge influence on economic viability. Government or societal subsidies can be estimated in various ways—recall from the previous chapter that the cost of getting electricity from coal may be less than half of the true cost to society. Similarly, it may be very difficult to accurately compare the value of tax breaks for solar power to the value of the government carrying the "catastrophic insurance" for nuclear power (see chapter 21). Overall government subsidies to well-established fossil fuels are much larger than those provided to sustainable energy such as solar power, but they are less per watt being generated.[15]

To those hoping for solar power, it is encouraging that at least some ways of looking at the costs indicate they are closer to the competition than you might suspect. For example, a recent study found the then-modern (2007) cost of electricity produced by solar (both photovoltaic and solar-thermal) power in the U.S. desert Southwest to be comparable to the more expensive electric rates paid by U.S. household consumers, although solar power cost twice as much as the cheaper rates.[16] Similarly, the *Christian Science Monitor* reported that solar-thermal electricity from the desert Southwest is barely more expensive than electricity from combined-cycle natural-gas plants, although it is slightly more than double the cheapest prices of electricity generated from coal.[17] However, a less favorable analysis from 2005 placed the cost of solar energy at or above the estimates just given, listed roughly a doubled cost for solar power if the price of storage is included, and quoted a lower cost for coal, making solar power quite far from being competitive at present; the differences arise primarily from the way the numbers were viewed rather than a change in price over the intervening few years.[18]

Thus, if we ignore the damages of global warming and the other damages from fossil fuels that are not included in the costs, solar energy may be almost competitive with a barely perceptible extra cost, or roughly double the cost of the competition, or off by a factor of 10. Within the assumptions made, all of these appear accurate; there are places and times when solar power makes sense already, but much improvement in relative cost will be needed to allow much more market penetration.

With solar, though, the prices have been dropping steadily for a long while, with no sign yet of leveling off. This does not prove that the drop will continue forever, but a lot of the analyses of solar power foresee major cost reductions in the future. Furthermore, these are not wild-eyed wishful-thinking analyses—they are based on engineering and science.[19] In light of the limited investment in solar research—just under one-tenth of the investment in nuclear-fission research from 1987 to 2002, averaged across the International Energy Agency member nations—the progress in solar energy does not represent some prejudice in its favor by the funding agencies.[20]

Warming to the Sun

Solar energy also is used directly, without first making electricity. A greenhouse warms plants, and our cats bask on our sun porch before racing around the house. Our "window quilts" shut out the sun to help keep us cool in the summer, and can be opened easily to let the light stream in to warm us during the winter. Constructing buildings to maximize such uses can save a lot of energy, as can solar water heaters—heating water uses almost 10 percent of household electricity in the United States.[21] The highly effective "Passivhaus" (Passive House) building standard in Germany combines extra insulation, features to allow the sun to provide winter heat, a solar hot-water collector system, and more.[22] Wintertime heating is reduced by 90 percent, and the houses save money: depending in detail on interest rates and construction and energy costs, the extra mortgage payments to cover the added construction costs are more than offset by the lower energy costs.

I have a twenty-five-year-old solar-powered calculator, one of the better investments I ever made. I took it with me to Greenland, and it worked fine in the pale blue light of a plywood-roofed snow pit at −20°F (−29°C) while I was processing cores in the summer of 1985. Today, when I pulled it out of the dark case on my desk in the fluorescent light of my office, the numbers appeared and the calculations worked. And that may be a lot of the appeal of solar power—it is *not* a single, centralized, one-size-fits-all answer. The tiny solar cells on my calculator could not possibly power my car, but they have saved me several sets of disposable batteries by now, and they should have plenty of life to go— similar cells in the workaday world of intense sunshine and changing temperatures are expected to last thirty years or more.[23]

Solar energy underlies Project Surya, in which the great climate scientist V. Ramanathan and colleagues are supplying sun-powered ovens to people in India so that they no longer must endure the labor and the bad air of dung-fueled cooking fires.[24] My colleagues Sridhar Anandakrishnan, Don Voigt, and Peter Burkett have run scientific instruments through the dark of Antarctic winter and the blazing sun of summer by using a combination of wind and solar power, collecting important data that otherwise would not have been available.[25]

In the modern world, a backyard inventor has an uphill climb to outdo well-organized university, government, and industrial laboratories. But the chances are almost certainly better with solar than with, say, nuclear—someone trying to build an improved nuclear reactor in their garage risks unpleasant interactions with anti-terrorism squads and concerned neighbors, but a better solar hot-water heater is unlikely to cause the same distress. The odds of bringing in an oil well in the backyard are not high, but there are plenty of backyards where the sun shines.

Of course, the sun shines only during the day and is not especially evident when it is raining, but to see how that rain helps solve the problem, let's head down the river.

HYDROELECTRIC POWER

The summer of 1830 I . . . blasted the tunnel through the rock to take water
from the dam above the falls for the mill. . . . In 1831 we lowered the tunnel
four feet, and built a new dam across the creek. —Ezra Cornell[26]

Ezra Cornell was a successful U.S. entrepreneur in the middle 1800s, founder of the important telegraph company Western Union, co-founder of Cornell University, elected member of the New York Senate and Assembly, and in his early days a hydropower engineer. And this combination is not the least bit surprising.

Hydroelectric power generation has dominated renewable energy for decades, although the recent rise of wind power is changing this. The

Figure 18.2 Glen Canyon Dam, with transmission lines carrying away the hydroelectric power, viewed from the Colorado River below the dam. With a full reservoir, hydroelectric power can be turned on rapidly to follow changing demand or changing production from other sources.

Figure 18.3 The water stored above the Glen Canyon Dam on the Colorado River can power the electric grid at a moment's notice. However, evaporation from the surface loses valuable water that could irrigate crops downstream or have other uses.

total potential for hydroelectric power is sufficient to supply only about 15 to 20 percent of human use. But the power from a dam can be turned on really fast. Open the valve, the water zips through the turbine, and there is electricity on the line. If the wind isn't blowing hard enough, or the sun isn't shining brightly enough, or the coal plants or oil burners are off and cold, they may have trouble meeting a sudden rise in demand for power, but a full lake above a hydroelectric dam is ready to go at a moment's notice.[27]

So, even though there isn't enough hydroelectric power for all the world's people, it can play a critical role, calming the mood swings. Not all dams now in place have hydroelectric generators, so there is room for additional generation. The possibility also exists to build off-stream storage, or to build new dams close to existing ones upstream or downstream, using excess power from wind or sun at "good" times to pump water uphill, ready to flow downhill and generate electricity when needed.

Greatly expanding hydroelectric power could cause some problems, however, and already is contributing to cross-border tensions and "saber rattling" in certain parts of the world. So let's take a quick look at how we got here, and where we might be going.

Eager-Beaver Power Engineers

Before European settlement in North America, the beaver population may have been approximately 100 million animals, with dams spaced as closely as 300 feet (100 m) in favorable habitats.[28] Beaver dams, like their human-built counterparts, have large and varied impacts on the environment. Beavers use the resulting ponds as protection from enemies and as a place that doesn't freeze solid to stash winter food. The pond behind a beaver dam traps the rocks that streams use to help erode,[29] slows flood waters, and forces the stream to spend most of its power tumbling over a dam so that floods don't destroy beaver lodges.

True, the sediment trapped in ponds causes them to fill up, so the beavers may be able to raise the dams for a while, but eventually they must be abandoned. The stream then cuts through the dam, the trees killed by the pond or eaten by the beavers grow back, and the beavers return to start over. A single dam thus is not sustainable, but a system of dams moving up and down the stream can be.

Humans didn't take too long to become eager beavers for dam-building. The oldest known dam is probably at Jawa, in Jordan, estimated to be 5000 years old, and our building of dams became widespread thereafter.[30]

Early settlers in North America often killed off the beavers and then replaced their work with mill dams. By 1840, there were more than 50,000 mill dams in the eastern United States.[31] Nearly every stream was dammed that was big enough to drive a mill wheel but small enough to be spanned by the dam-building technologies of the time. Big rivers tend to slope gradually toward the sea, while smaller streams are steeper, so these mill dams captured much of the elevation drop and thus much of the available power of water.[32] This water power was used for many purposes, including blasting air into furnaces and forges, and especially for milling of grain. The technologies had been refined in Europe for centuries beforehand, where similarly extensive damming of streams occurred in many places.

The people of the time knew the bad and the good of mill dams. The dams interfered with valuable fish migrations; one dam in the eastern United States had been removed to allow fisheries as early as the 1700s, with others at least generating petitions for removal. Laws were passed to reduce crowding of dams and provide compensation for lands flooded by mill ponds, while still encouraging dam-building.[33]

Most of the hydropower used today is from big dams that could not have been built in the 1800s, with little streams that once hosted mill ponds now drowned beneath vast lakes, and the elevation drop shifted to the single big dam. We are probably using only roughly one-sixth of the total available resource, which means that a lot of dam-building could still be done.[34] But because the big dams have been centers of controversy, there is also a lot of arguing left to do. As of July 2010, for example, there were 190 indexed scientific papers with "Three Gorges Dam" of China as their subject, and millions of "hits" on major Internet search engines for that subject in quotes.[35]

On one hand, the dams help control floods, allow recreational boating and fishing, and supply power, all of which look great to many people, as well as providing water for irrigation, which we will discuss later. But the interruption of the pre-dam river services makes a lot of people unhappy.

Dams and their reservoirs drown prime lands and cause loss of some forms of recreation. The dams may block the natural passage of fish, shellfish, and other species. For example, there are long-lived mollusks in the upper Mississippi River basin that have persisted for decades without being able to reproduce because their offspring must spend part of their lives attached to the gills of particular migratory fish that cannot pass dams to reach the upper part of the river.[36] Sedimentation in reservoirs tends to "back up," burying fields upstream of the reservoirs. The clean water released from a dam generally can't wash big rocks to the sea anymore, but will wash away sandbars, with large impacts on ecosystems. This has happened in the Grand Canyon below Glen Canyon dam, for example, where the switch from muddy to clean water has endangered some native fish species.[37] Once those sandbars are gone, the presence of dams reduces the sediment supply to the sea. Normally, beaches lose sand that is carried into deep water by waves, but gain sand carried along the coast from rivers; dams that interrupt the supply of sand from rivers can cause beach erosion far away.[38]

Some of these issues are quite large and important. For example, in some regions dams can increase the spread of schistosomiasis[39] and may harm human health in other ways.[40]

Because many rivers cross borders, dams also can create international tensions. Evaporation from reservoirs may be important, depending on local conditions, leaving less water to flow across the border, and use of dammed water for irrigation clearly increases water loss by evaporation. Recent discussions, including warm words and "saber rattling over water security," have focused on plans by India to generate power from a river flowing into Pakistan,[41] while other discussions centered on Chinese plans to dam flow into India.[42]

The United States and some other countries are discussing the removal of dams, and have lost a huge number of old mill ponds to dam failure or burial by sediment.[43] But elsewhere in the world dam-building is proceeding rapidly—a recent report noted 33 hydropower projects under construction in India and almost 300 more planned, and China is planning perhaps 750 projects in Tibet alone.[44]

The biggest issue with dams may be related to food. Where I live in central Pennsylvania, the rainfall per year of just over 3 feet (about 1 m) is very close to the global average. With this much rain, very little of our agriculture is irrigated. But almost everyone with a backyard garden has a hose nearby, because we really do get more from our tomato patches if we water them. Irrigation would increase yields during many years for almost any farmer, and we would irrigate more if the demand for food were high enough to make the effort worthwhile. As the world adds billions of people and tries to feed the billions already here, including the billion or so who now go hungry,[45] the demand for food is likely to grow a lot, maybe nearly doubling. For much of the world, river water that reaches the sea is food that was not grown. And, historically, dams have been an important part of a lot of irrigation schemes. A pioneering study turning this common sense into numbers estimated that vigorous enhanced construction of small reservoirs to supply supplemental irrigation water to regions with below-average farm yields could increase global production of cereal crops by about 35 percent, although this would notably increase evaporation and would completely dry up some rivers seasonally.[46]

Some parts of the world have relied on snow and ice as reservoirs, with winter precipitation held in snow cover or glaciers until summer warmth releases it to maintain river flow.[47] If warming continues, causing a reduction of snowpack and glaciers, this storage will be lost, impacting water availability during some seasons.[48] Building new dams can help hold this water until it is needed for irrigation, or fish, or power, or whatever else we want to do with it.

Dams, in short, are intrusive, bringing major benefits but major costs, and getting a lot of people excited. We do know how to live with them, and have done so for a long time. Despite the world having almost one million dams more than 50 feet (15 m) high, and capable of holding back about 15 percent of global annual river runoff, many more dams could be built, and some of them are being built.[49] A lot of hydropower is available—not enough to solve all of human energy needs, but enough to smooth the swings of other renewable energy sources. However, as the world population grows, the time may come when we will be looking for other ways to smooth those swings, because food has first call on the water. Fortunately, there are other swing-smoothers, as we will see next.

19

Down by the Sea,
Where the Water Power Grows

Synopsis. Tides and currents move immense amounts of water in the ocean, while waves race across the surface. The energy in these can be extracted, and at least a little is being extracted already. Large engineering challenges exist, but some places will be able to get much energy from the ocean in the future.

Time and Tide Wait for No Man. —Chaucer[1]

EXPANDING THE ENERGY MENU

We have already seen that we have enough energy sources to power the planet's people, replacing all of the fossil fuels and still providing extra to allow for growing demand from those who aren't using much energy now but would like to. Sun and wind, wired together with a smart grid and including some hydropower, or some plug-in hybrids, or any of the other storage options or peak-power providers discussed in the previous chapters, really can make a sustainable solution for humanity. The challenges are immense, and building the required collectors won't be easy. We would be happier if the price of sustainable energy kept coming down, but we know that we can do this for no more than the cost of toilets and a lot less than the cost of our national defense or major wars.

Given the scale of the problem, we are unlikely to just jump right into the fully integrated, fully renewable energy system envisioned in so many serious plans for our future. We may rely on a lot of "bridge" technologies to get us over rough spots on the way. Furthermore, a "smart grid" that allows local or distant power generation and energy storage in many different forms and places would also enable extensive entrepreneurship. If you live in a cloudy place, or have light winds, you don't have to purchase all of your power from others because there are many more options that can contribute to the overall solution, and you almost surely live near some of them.

In this chapter, we will look at the tides, waves, and currents out to sea and along the coast. Then, in the next chapter, we will visit plants and hot rocks for the landlocked.

Figure 19.1 Waves breaking on the coast of New Zealand. The ocean offers a wide range of energy alternatives, if we can learn to harness them.

TAMING THE TIDE

When our daughters were just old enough to go kayaking by themselves on Cape Cod's Nauset Marsh, we played the parental part and warned them, loudly and often, about staying home from Portugal. The mile-wide marsh is blocked off from the North Atlantic Ocean by a barrier beach, with a narrow inlet through which enough water rushes to raise or lower the water level in the marsh by 5 feet or so (1–2 m) in a few hours. Even a strong kayaker may be unable to make headway against a rapidly ebbing tide, which would never reach Europe but might carry a young paddler into uncomfortably high waves beyond the barrier.

Of course, Janet and Karen were fully aware of these issues and did not need the lecture. But parents and professors sometimes cannot help themselves, and when the parent is a professor . . .

Tides are produced by the gravitational pull of the moon and the sun. We will focus on the moon for simplicity, because it has more influence than the sun (and vastly more than the other planets). The closer you are to something, the stronger the pull of that thing's gravity on you. The ocean water closest to the moon thus is pulled toward the moon more strongly than is the center of Earth, and the center of Earth is pulled more strongly toward the moon than the water on the other side of Earth. This causes two bulges in the water—one almost beneath the moon, and one on the other side of Earth as Earth is pulled away from that far-side water. If Earth didn't rotate, these bulges would just sit there, a couple of feet high in the ocean (much smaller on land). But Earth rotates under the bulges, so we, sitting on Earth, see the bulges moving around the planet, keeping the moon nearly overhead.

When a tidal bulge runs into the edge of the continental shelf, it mixes the water, bringing nutrients up toward the surface to grow algae, feed fish and those who fish for them, and otherwise make interest-ing ecosystems.[2] Mixing driven by tides and winds also helps drive the circulation of Earth's oceans, taking less-dense waters down into the depths and so causing the homogenized water mass to rise and be replaced by dense, oxygen-rich water sinking in the North Atlantic or around Antarctica.[3]

As the tidal bulge moves shoreward, the water may focus into a "funnel" such as the Bay of Fundy, giving spectacular tides. And if water is focused into some places, it must be focused away from others, so the tide varies a good bit along the coast. Coasts aligned in a north-south direction tend to get the largest tides by damming the motion of the bulge.

The tidal bulge is not directly under the moon, but is dragged just a bit off center by Earth's rotation. The gravity from the water in the bulge gives the moon a tiny tug along its orbit, while the moon's gravity pulling back on the tidal bulge tends to slow Earth's rotation. The net result is that Earth's day gets longer while the moon moves away from Earth.[4]

The tides make currents that can sweep a kayak out to sea, and raise and lower water and floating logs and boats, eventually losing energy to make heat. The heat energy is very tiny compared to the sun's input, so we don't worry about tidal heating when we are calculating weather or climate.[5] But with an estimated 4 TW globally, tides could supply a non-negligible fraction of human energy if we exploited them vigorously.

The energy for tides really is left over from formation of the solar system—Earth started out rotating, picked up some more rotation from the Mars-sized body that collided with us to form the moon, and the energy of this rotation is gradually being converted to heat. Just as slowing the rotation of your car's wheels with the brakes makes them hot, in the same way slowing the rotation of Earth with tides creates heat. Fortunately for us, the tides do this much more slowly than the brakes!

If we develop tidal power extensively, we will shift some of the heat generation from where it now occurs to where our generators sit. A tide-power generator in the inlet of Nauset Marsh, for example, would slow the rate of inflow and outflow of the water, making the tidal change a little smaller and slower with less turbulence releasing heat across the marsh, and with the difference turned to electric power sent away on the grid. Putting out lots of tide-power generators along the coasts also might "roughen" them, increasing the total conversion of Earth's rotational energy to heat and slowing the planet more rapidly than natural. So, in some bizarre technical sense, tidal power is not absolutely perfectly sustainable because eventually Earth will run out of rotational energy. But if you are not too worried about changes that take longer than many millions of years, this won't matter enough to merit much discussion.

People were quick to realize that they could use tides to power mills for grinding grain or sawing lumber. Tide mills were operated at towns named Yarmouth on both sides of the Atlantic in the 1600s, for example.[6] Because the time of high tide shifts an hour or so per day following the moon, a miller who used the tide had to adopt sometimes-strange working hours—an early example of "shift" work. But the power was highly predictable and didn't disappear even in a dry year, as water power could.

Tidal power can drive turbines placed in natural flows such as in New York's East River.[7] Or dams can be used to restrict the tidal flow and direct it across turbines, as at the Rance River plant in France, which has been operating since 1966, reliably supplying enough energy to power a city of a quarter of a million households.[8]

Of the roughly 4 TW of tidal energy available worldwide, perhaps only about 1 TW is in places where development is likely to be practical, and developing even that much would face notable difficulties; thus far, the power we generate from tides is miniscule compared to what we use.[9] No one I know has seriously proposed putting a power generator in the inlet to the Nauset Marsh, presumably because the value of the natural setting to tourists and endangered species is now much larger than the value of the power that might be generated. Underwater turbines such as in the East River may affect shipping, fishing, and diving, but they don't affect the "view" for most tourists, whereas tidal dams would. However, if structures are required to protect cities from rising seas or sinking lands by holding out the highest waves and tides while allowing the sea to flow in and out, adding power generation might prove practical and could even improve the protection.

OTHER OCEAN ENERGY—WAVES, CURRENTS, AND COLD FEET

Wave at the Camera

The tide pools of Acadia National Park in Maine, or of Olympic National Park in Washington State, or of many other parts of the world, are wonderful places to explore. Anemones and starfish, urchins and barnacles are washed by the swirling waves at high tide, and are tough enough to survive the still water, with its low oxygen and other unfavorable conditions, at low tide. What they cannot survive is being buried by mud or sand. So while tides get most of the credit for tide pools, one of the secrets is to have waves that wash away mud or sand without bringing more in. The rocky coast of Acadia is great for this—the hard granite doesn't break down easily to make sand and mud, so bare rock is left exposed as waves wash sediment into deep water.

Visitors to Acadia usually stop at Thunder Hole. As described by the National Park Service, "Thunder Hole is a familiar example of the awesome power of the sea. When the wind is strong, the rising tide surges into the narrow chasm, compressing the captured air, and resounds with a boom that is felt as well as heard."[10] In fact, it is waves riding in on the high tide that really do the job, and that same wave energy that washes the silt away from the sea urchins offers renewable energy to us.

Recent estimates of total wave energy include 2 TW in near-coastal sites,[11] 2.7 TW in high-energy sites, and perhaps 0.5 TW available with known technologies.[12] While not huge in the global picture, these are still large enough numbers to interest entrepreneurs.

Surfers long ago figured out how to get a little of the energy from waves to make a great ride, but turning that energy into electricity is not quite so easy. One option being tested is to build human versions of Acadia's Thunder Hole—each a large structure with a hole just below water level. When a wave hits such a structure, the high pressure beneath the wave crest forces water into the hole, compressing air inside to blow out the top and drive a turbine, generating electricity rather than "oohs" and "aahs" from the crowd.

Figure 19.2   Blowhole, near Wollongong, Australia. Waves crashing into a crack compress the air there and force it out a hole at the top. Human-built versions use turbines to generate electricity from the escaping air.

People have used breakwaters for a very long time to protect harbors and cities from waves. Like beavers with their dams, we build structures to keep nature's power away from our more valuable constructions, but we do not use that power for anything else. Adding wave generators to existing breakwaters, or to new breakwaters that might be needed to deal with rising sea levels, would get some good out of energy that otherwise is just thrown away. Floating versions might be constructed, to calm waves a bit while capturing their energy offshore.[13]

If you and I were sitting on surfboards, with you a little farther out to sea than me, a wave would be raising you while I was sinking into a trough. If we each were holding onto a rope, my hands would get hot as the increasing distance pulled the rope through them. Replace us by floats, attach the rope to a generator, and you can make electricity. You need to be clever to allow the rope to be shortened when we move closer together before the next wave moves us apart again, but engineers can handle this. Or a float at the surface rocking with the waves can be attached to another one below the wave base, again using the motion between the two floats to generate power. Or devices can be built to "corral" waves, funneling them up a ramp so the water can flow down through a turbine.[14] Research is continuing on other possibilities, including work at the U.S. Air Force Academy using turbines driven by lift rather than drag, drawing on the knowledge of wings that has proved important to the Air Force in the past.[15]

Go with the Flow

British Customs officials in the 1760s were puzzled why British packet ships heading for New York took two weeks longer to cross the Atlantic than Yankee ships bound for Providence.[16] Loading, ship type, and distance didn't explain what was happening, so the officials turned to Benjamin Franklin, Deputy Postmaster General for the American colonies. Franklin was a practical man, and he contacted his cousin Timothy Folger, a Nantucket whaling captain.

Folger described the "Gulph" or Gulf Stream, a great "river" flowing in the ocean. Nantucket whalers were familiar with it, tracking the whales that skirted the edge of the flow but that were not found within it. The Rhode Island merchant captains knew of the current from the whalers and rode its flow heading east but avoided it on the westbound journey. The British packet captains should have known about it too; in crossing from one side to the other in search of whales, the whalers often met the packet captains and told them about the current. But the British packet captains "were too wise to be counseled by simple American fishermen."[17]

The map prepared by Franklin and Folger to help the captains proved to be an important contribution to the early scientific understanding of the ocean's currents. Franklin successfully identified the role of the trade winds in moving water and helping drive the Gulf Stream. He even provided guidance on identifying the Gulf Stream, taking thermometer readings while crossing the Atlantic and noting that the Stream's temperature was notably higher than that in surrounding waters, showing that a thermometer is a useful instrument to a navigator.[18] But Franklin's chart and advice didn't help the British packet captains, who resolutely ignored good science based on practical experience and continued to sail into the teeth of the current, which at up to three miles per hour did add a couple of weeks to their transit.

Many other currents similar to the Gulf Stream exist, including the Kuroshio, Brazil, East Australian, and Agulhas currents. This flowing water has kinetic energy, which nature eventually converts to heat, but

Figure 19.3 Diagram of the Gulf Stream by Benjamin Franklin. Although it flows slowly, the immense size of the Gulf Stream means that it is powerful and could drive turbines generating electricity.

Figure 19.4 In the 1700s, whaling ships such as the Amelia, *and other U.S. vessels, used the Gulf Stream to their advantage, but British packets ignored existing knowledge and wasted valuable time sailing against the powerful flow.*

we can convert some of it to electricity first. Estimates of how much energy is available in these currents remain highly uncertain, but a few terawatts seems possible.[19] Put a turbine in the moving fluid, and you can extract power—we have done this with hydroelectric and wind, and we can surely do it with the Gulf Stream.

Many questions remain about how best to capture the energy—horizontal turbines or vertical ones, or machines that oscillate something like a fish, or some other approach—and how to get the energy to shore, and whether there will be effects on the currents and thus on the climate or on fish.[20] But the energy is there, people tend to be rather clever, and extracting the energy doesn't seem to be that difficult or that intrusive.

Cold Toes

Also on the drawing board, and with some success in field tests, although not widely commercially available, is Ocean Thermal Energy Conversion, or OTEC. Estimates of available power include 1.8 TW for the Atlantic thermohaline circulation, and 3 to 10 TW for the world ocean. The idea is not a new one, but traces back to Jacques-Arsène d'Arsonval in 1881.[21]

Once scientists and engineers figured out the thermodynamics of steam engines, it became evident that energy could be extracted from any large difference in temperature. The deep ocean is filled with cold water, barely above freezing, that sank in the polar regions, while a relatively thin layer (a few hundred yards or meters) of warm water floats on top of this in most areas. The temperature difference can be used to run a very-low-pressure steam engine to make power. The temperature difference is so small compared to that in a coal-fired or nuclear-fired power plant that the efficiency of OTEC is very low, but there is a *lot* of water out there, so useful power can be extracted.

Climate change might affect the temperature difference between deep and shallow waters, which would affect OTEC. Generating a huge amount of power from OTEC might affect climate too, by changing ocean temperatures and currents; some careful consideration will be needed if OTEC becomes economically important, although initially this is not likely to be a big deal. Some versions of OTEC can produce fresh water, and it might be possible to combine OTEC with extraction of valuable things from the sea water, perhaps including uranium. Nutrient-rich waters can be brought up and used to fertilize aquaculture, and the cold can even be used in air-conditioning of nearby cities. A lot of engineering will need to be done, but the potential to help power society while growing more food is intriguing.

MOVING AHEAD ON MOVING FLUIDS

Electricity from moving wind or water offers great advantages. More than enough energy is available to power the world, especially because so many of the technologies avoid much of the energy wasted (as heat) when we get power by burning things. Energy from wind and hydroelectric power is well proven, economically competitive, and sustainable.

Producing energy from tides and currents faces additional development hurdles, but many people are working on these topics, and demonstrations or fully working plants exist. Some of these face problems of intermittency, and others have not-in-my-backyard issues. But the wind blows, the water flows, and we know how to tap in and go.

20

Power from the Land

Synopsis. Earth continues to produce plants that we can burn, and we can get much more power if non-food plants or parts of plants are exploited, but much care is needed to avoid unintended consequences. Geothermal energy is locally important, with some chance for much more widespread use.

After all our discoveries and inventions . . . a pile of wood . . . is as precious to us as it was to our Saxon and Norman ancestors.
—Henry David Thoreau[1]

People living on the coasts get the beautiful view, the good seafood, the romance and adventure of the sea, as well as the potential to tap into the ocean's waves, tides, currents, and thermal energy, as described in the previous chapter. What does that leave for those of us who are landlocked? Two big possible sources are heat in the ground and plants growing in the ground: geothermal power and biofuels.

BIOMASSIVE

Our very humanity may have required biomass burning, as we saw in chapter 2. We have relied on burning plants for energy, outside our bodies as well as in, for a very long time. The amount of energy available from plants worldwide—maybe 166 TW, or eleven times the energy humans use today—says that plants can make a big difference. Depending on just how you do the calculation, our plant-burning now may be supplying about 10 percent of our energy mix. Biomass is like an old friend we already count on and can call for more help if things get bad.[2]

But asking for too much from an old friend can spoil a beautiful relationship. Friendships should be tended and nurtured. So rather than simply assuming that biomass will supply what we want when we want it, we might want to proceed carefully.

The biomass we burn includes garbage, dung and wood, plants we have converted to liquid fuels, and more, ranging from the oldest of our technologies to the latest genetic engineering. Discussion in the United States, Brazil, and other places has especially focused on conversion of biomass to alcohol to reduce gasoline use.

Probably the biggest concern with some forms of biomass energy is the ethical issue of burning food for fuel on a planet with a billion hungry people and a host of alternative fuels that nobody can eat. We face the task of greatly increasing food production in the future—perhaps almost doubling it—to keep everyone well fed, with the reality that we have our best lands in production already, and in many places the soil is eroding, the groundwater level is dropping, and the fertilizer is being depleted. So we are already planning to ask our old friend for a lot of help, when our friend is getting a little tired of our continual requests.

The potential problem is obvious if we burn food directly. The issues are just as real but harder to see if we grow fuel using land or fertilizer or water that otherwise would have been used to grow food.

If we raise more crops destined to become fuel by first cutting forests and burning them to make fields, the wood releases CO_2 to the air to warm the climate. When we humans aren't disturbing forests, they tend to reach a balance between the rate at which they take CO_2 out of the air to grow more wood, and the rate at which they return CO_2 to the air by being burned in forest fires or inside termites or bacteria.[3] If you start a new forest, for a while much of the CO_2 taken up will be stored in growing tree trunks rather than being burned, until some trees get old and the balance between burning and growth is reestablished. The plants of the world now store almost as much carbon as is in the air, and about as much carbon as is used for a decade of plant growth (presuming that none of the plants die—springtime without fall), with much of this storage in forests.[4]

If we cut down and burn a mature forest to make room for biofuel crops, it will take many decades before the crops have produced enough biofuel to replace enough fossil fuel to avoid enough CO_2 emissions to equal the CO_2 released by burning the forest. Almost surprisingly, even if you build houses or other things with the wood, so much of the carbon in the forest still is released to the air (from leaves and twigs, sawdust and more) that a few decades of biofuel production on the cleared land will be required to reach the break-even point.[5] The biofuel farming may increase carbon loss from soils to the air to raise atmospheric CO_2, and globally soils contain more carbon (maybe three times more) than the air or the plants, so releasing much of this could be a real problem. However, these difficulties can be greatly reduced if abandoned and nonforested former agricultural regions are put into fuel production.

An additional issue is that nitrogen fertilizer applied to grow biofuels and other crops increases the rate at which the potent greenhouse gas nitrous oxide is produced in the soil. Preliminary estimates indicate that the extra warming from the nitrous oxide caused by biofuel growth competes with the cooling from reduced greenhouse-gas emissions, and in some cases the nitrous-oxide warming exceeds the avoided-CO_2 cooling.[6]

Biofuel crops are often planted with a fossil fuel–burning tractor, and plowed and fertilized and cultivated and harvested and shipped and processed with power from fossil fuels. The amount of fossil fuel eventually replaced by the biofuel ranges greatly. The energy output in biofuel from sugar cane may be eight times the energy input from fossil fuel, but for corn (the kernels, or food part, not the stalks and cobs) the ratio is only about one and one-fourth—as it is done now, producing enough alcohol from grain corn to replace 1¼ gallons of fossil-fuel gasoline may require roughly 1 gallon of fossil-fuel gasoline.[7]

Discussion of these issues can become very heated and overlap into the political arena.[8] The science is evolving rapidly, the amounts of money involved are potentially large, and the numbers are notably uncertain and depend on where, when, and how the fuels are generated, whether the leftovers from the fuel production are then used instead of other grain to feed animals and improve the payoff, and many other issues.

In general, sugar cane grown in tropical or subtropical regions seems to produce much more biofuel from a given area of cropland and given input of fertilizer than much of the competition, especially if that competition is produced at higher latitudes. Growth of biofuels on marginal croplands that have been left fallow can have notable net benefits, but there isn't enough land of that sort to make a huge difference to global CO_2 emissions.[9] It helps to switch to using the nonfood parts of crops or to crops that grow where food doesn't, as well as to use the leftovers to feed animals that otherwise would be eating the main crops. A lot of very exciting work is ongoing to produce cellulosic ethanol (from corn stalks and many other plant parts), and the infrastructure built for handling grain corn might be converted with relative ease to handle corn stalks.[10]

Interest in biofuels, including ethanol but extending to many others, is huge and growing rapidly. For example, a recent search in one of the major indexing services focusing heavily on refereed scientific papers found none on the subject "biodiesel" before 1992, but a total of 3700 by late November 2009, a remarkable increase addressing aspects ranging from economics to genetic engineering of novel biodiesel-producing microbes.[11] A search on "sustainable forestry" found 1498 scientific papers, the oldest in 1988. "Corn and ethanol" yielded 1722 scientific papers, and "food versus fuel" found 134 papers, including one from way back in 1981 titled "Food versus fuel: The moral issue in using corn for ethanol."[12]

Moving the search "Food vs. Fuel" to the broader worldwide web yields an amazing array of vigorous opinions, including a lengthy entry in Wikipedia.[13] If anyone considers this book to be important, people may be gearing up to complain about the bias I've shown here either for or against biofuels. The Organisation for Economic Co-operation and Development (OECD) in 2008 found that, regarding biofuels, "the

Figure 20.1 Biodiesel can be produced from soybeans, one of the many possible applications of biomass to future fuels. However, this is also one of the applications that raises questions of fuel competing with food.

high level of policy support contributes little to reduced greenhouse-gas emissions and other policy objectives, while it adds to a range of factors that raise international prices for food commodities."[14] And one scientific paper even addressed the benefit that New Zealand's pasture-based agriculture receives from the run-up in prices caused by the U.S. use of corn for alcohol.[15]

Remember "peak whales" and "peak trees" from chapter 3. Nature does not grow things nearly fast enough to support us the way we used them in the past—biomass was insufficient then, and there are many more of us now.

Furthermore, the damage of burning biomass inefficiently can be huge. Recall from chapter 18 that Professor V. Ramanathan is running a project to supply solar cookers in India. As detailed in *Science*,

Ramanathan recalls his grandmother "cough[ing] endlessly over her smoky indoor cooking fire of sticks and dung."[16] Fires just like that still feed many people in Asia, harming them locally and sending a brown cloud of pollution far beyond their borders, while other people breathe the output of similar cooking fires in sub-Saharan Africa or of pig-iron production with charcoal from Brazil. Ironically, many of these people must exert heroic effort just to find the fuel to cook their dinners.

Summary? We can get a lot of good out of biofuels. We know how to use them in many ways already, and are learning rapidly. Enough energy is available to make a big difference. But there are also many ways to go wrong with biofuels. Using them inappropriately may increase hunger, climate change, and other problems. Like any old friend, biofuels deserve respect and consideration and never should be taken for granted.

GEOTHERMAL

There is also a number of places where the pure suphor is sent forth in abundance one of our men Visited one of those wilst taking his recreation there at an instan the earth began a tremendious trembling and he with dificulty made his escape when an explosion took place resembling that of thunder. During our stay in that quarter I heard it every day.
—Daniel T. Potts, 1827[17]

The Icelandic and English languages share many words through their common Germanic ancestors, but not many words have jumped directly from Icelandic to English. Not surprisingly, given the spectacular geology and long history of geologic study in Iceland, geologic terms have led that jump. The flood that bursts out when a volcano erupts beneath a glacier, or any other flood from a glacier, is a *jokulhlaup*, or glacier burst, and the flood races down the sandy *sandur* river plain in front of the glacier. The best-known word in English from Icelandic, though, is the name of a gushing or erupting hot spring in the Haukadalur Valley of Iceland: *Geysir*. We English-speakers know it as a geyser.

Tourists flock to the geysers in Iceland, and other geysers can be seen

Figure 20.2 Natural geothermal power is easy to see in a few places, such as Yellowstone National Park in the United States (with Cliff Geyser shown here) or Rotorua in New Zealand.

in places scattered around the world, including the Kamchatka Peninsula of Russia and the high Andes of Chile. Outside of Iceland, probably the two leading tourist destinations for geysers are across the planet from each other: Yellowstone National Park in the United States and Rotorua in New Zealand. The quote from Daniel Potts that opened this section is an early description of a geyser in the Yellowstone country.

Geyser formation requires rather special conditions. First, hot rock needs to be close to Earth's surface, often so hot that it has melted into a pool of magma waiting to feed a volcanic eruption. Then, plenty of water from rain or melting snow must percolate downward through the rocks to be heated. And there must be a narrow, sturdy natural pipe, perhaps formed in the space where two cracks intersect, to bring the heated water back to the surface. If the pipe is too small, not enough water can be pushed out to make a good show; too large, and convection cells will form in the pipe and transfer heat to the surface smoothly without an eruption. Just right, and the fun begins. It helps if there is a lot of silica in the rocks to be carried by the water and deposited in cracks, to seal them and keep the pipe from breaking under the high pressures of an eruption.

Magma or sufficiently hot rocks will quickly warm the deep water, but the weight of the colder water above initially will prevent boiling. When a little of the water at the bottom finally succeeds in boiling, it expands and drives out a little of the water at the top of the pipe, reducing the pressure and allowing more of the water at the bottom to boil, pushing out more water on top, and whoooooosh, the geyser erupts.[18]

Yellowstone has about half of the world's active geysers. New Zealand used to have more than are currently active, but development of geothermal plants to supply electricity has drained the pressurized hot water from some of them.[19] Once this problem became evident, the country acted to preserve the remaining geysers for future generations of tourists while maintaining power generation.[20] This problem of draining hot water from geysers is not unique to New Zealand, and geothermal development elsewhere, including in Nevada (near the former Beowawe geyser field), has been linked to geyser loss.[21] One estimate has more than 250 geysers lost in the United States, Iceland, and New Zealand, primarily from use of geothermal power.[22] But the technology to provide geothermal power while minimizing impacts on geysers also opens a much greater potential resource, as we will see soon.

Any energy engineer watching a geyser erupt would immediately know the possibility of power generation. Naturally, the water jets upward, falls down, and runs away to a river, dissipating its energy to heat the air above. Put a turbine with an electrical generator in the way of the water, and the eruption will spin the blades and power the grid. Geysers are rare in part because the right holes are so rare—not too big or small, not too tightly or loosely sealed. But humans can make holes the right size rather easily. Drill the hole in relatively solid rock. Then make the right number of fractures. Oil companies and water-well drill-

Figure 20.3   Humans can put geothermal energy to work, although a little care is required if we want both electrical power and geysers.

ers have known about hydrofracturing or "fracking" for a long time, to let oil or water flow into their wells, and this has recently been gaining widespread use for extraction of natural gas from shale beds.[23] Drill a hole, put in a special device (an inflatable packer) that seals the hole with a hose going through, and use the hose to pump up the pressure below the packer until the rock cracks.

This geothermal-energy technique works fine if there is a lot of hot water just sitting there waiting to erupt from any holes you drill, but then nature must get enough water down rapidly enough to make enough steam available to come back up and keep the power company happy. If there isn't enough steam available, you can drill two holes, pump water down one into the hot rocks, make cracks between the two, and let the steam come up the other. But if you let the water from the steam run away on the surface after the steam drives the turbine, then you need a new water source, so condensing the steam coming up and putting the water down the other hole keeps the power coming without drying out nearby geysers. Unfortunately, the steam comes up with sulfur compounds that can stink up the neighborhood and peel paint on adjacent houses, and with CO_2 that contributes a little more to global warming.[24]

Technological advances now allow everything coming up to be kept in pipes and sent back down, with the heat transferred to a low-boiling-point fluid in other pipes to drive the turbines. By mastering this "binary" technology and using a fluid that boils at a lower temperature than water to drive the turbine, we don't need the rocks to be as hot to make the system work. Just a little water is needed to get the system started, and then the water is reused, and heat is turned into electricity, with nothing escaping.[25]

Power engineers like geothermal energy—the power doesn't stop when the sun goes down, or when a weather system with very little wind parks over your state. Instead, like hydroelectric and nuclear energy, the engineers can be confident that geothermal plants will be online whenever they are needed, and can design the geothermal plants to fill in the "gaps" left by the intermittent nature of wind and sun. Geothermal power can be quite inexpensive too—Iceland has become a center for power-intensive aluminum smelting in part because of the cheap electricity available there from geothermal and hydropower installations.[26]

The ultimate heat source for geothermal energy is primarily radioactive decay in Earth, mostly from radioactive uranium, thorium, potassium, and other elements that were incorporated in Earth when it formed and that are slowly being depleted as they decay and release their energy (very slowly—you will need to wait a billion years to notice a difference). A bit of the geothermal heat comes from cooling of the planet, and is left over from Earth's formation, with a little more released as the liquid outer core freezes on the solid inner core and as denser bits of the planet sink downward and then stop. A little even comes from the tides bending the solid Earth.

Geothermal energy is not a huge part of human power production now (less than 9 gigawatts [GW], well under 0.1 percent of total human energy use),[27] although geothermal is important in some of the places mentioned earlier, including Iceland, New Zealand, and parts of the United States. Recall that the total geothermal flux to the surface is only 44 TW, or a shade less than three times the total human energy use, with most of that geothermal energy supplied beneath the sea where we would have great difficulty in collecting it, and with almost all of it fairly difficult to collect. Geothermal-energy development so far has been focused on those few places where magma or very hot rocks sit close to the surface and collecting the energy is relatively easy.

Even in these special places, though, geothermal plants remove heat faster than nature supplies it, so over time the rocks begin to cool off and lose their oomph for the power plants.[28] The flux of melted rock to the surface is not large enough to supply humanity with all the energy we want, and the conduction of heat through nonmelted rocks to bring up energy from deeper is simply too slow.

All is not lost, though. There is an immense amount of very hot rock deep below our feet, and the environmental impacts of making that rock cooler are likely to be very much smaller than the environmental impacts of burning a lot of fossil fuel. A report from the Massachusetts Institute of Technology estimated that the energy stored in hot rock beneath the United States and above a depth of 6 miles (10 km) is equivalent to 130,000 years of U.S. human energy use at modern rates, with enough of that energy technically available that the unsustainable aspects of the exercise would not become important until much, much further in the future than for fossil fuels.[29] As described earlier, drill two holes in hot dry rock, fracture the rocks between, pump cold water down one, take the warm water that comes up the other, transfer the heat to a different fluid in pipes to drive a turbine, and put the now-cooled water back down to warm up again. When the rock is too cold to supply energy economically, drill deeper, or move to a new place. The energy used in drilling is small compared to the energy that can be generated. The colder rocks left behind are, well, colder rocks.

A potential shadow over generating geothermal energy from hot dry

rock is the possibility that the process of making the cracks will trigger earthquakes large enough to matter to people living nearby, or will
upset their water wells or otherwise change things in ways they don't
like, with earthquakes especially important. We have long known that
injecting fluids into Earth, for whatever reason, can trigger earthquakes.
One famous series of quakes in the early to mid 1960s near Denver,
Colorado, with many having magnitudes of between 3 and 4, was triggered when people tried to dispose of waste fluids by injecting them
under pressure into deep rocks.[30] A rapidly growing scientific literature
addresses such issues for purposes of carbon sequestration (see chapter 21), enhanced oil recovery, natural gas extraction, and geothermal
power, among other purposes.

Quite simply, we like to think of the ground beneath our feet as rock
solid, just sitting there and behaving itself. No way. We are all raft-
ing across the surface of the planet, with the Atlantic spreading as the
Pacific narrows, India still burrowing into Asia to push up the Hima-
layas, the west coast of New Zealand's South Island and the western
slice of California sliding north past their neighbors, and much more.
Slabs of dense rock are sinking into the mantle, tugging on the continents or sea floor behind them, while mountain ranges and the high
undersea ridges where sea floor is made are spreading under their own
weight and pushing neighboring rocks ahead. Pretty much everywhere
all the time, the pushing and pulling on the rocks have stressed them
enough that they either are breaking occasionally or are not too far from
breaking.

Furthermore, almost all rocks contain mostly-closed cracks from
when they broke in the past. Any of this cracking can make earthquakes,
but most are tiny—the big ones that people worry about occur when
the rocks on one side of a long crack (called a fault) slide past those on
the other side.[31]

If the old cracks are oriented such that today's stresses are trying
to reopen them, then the "fracking" from gas extraction or waste disposal or geothermal-power generation will just help reopen the old
cracks, with not too much chance of a big earthquake. But if the rocks

have rotated so that Earth's stresses now are trying to slide the rocks across their neighbors along the bumpy crack, "fracking" may trigger an earthquake.

A commercial geothermal project that was being built in Basel, Switzerland, in 2006 triggered a number of earthquakes, including one of magnitude 3.4, just enough to cause cracks in plaster. Many homeowners filed insurance claims, the insurance company had paid out US$7 million as of the summer of 2009, and "one of the directors of the project (was) facing prosecution before the criminal court of Basel."[32]

The MIT expert report suggested that with more research and care, the danger of a damaging quake is low.[33] I have to admit to a not-so-secret hope that a lot of geothermal energy proves safe and economical, not only because of the good we will get from the energy, but also because we will learn so much fascinating information about how Earth works.

Meanwhile, back in Rotorua, and Yellowstone, and Iceland, the geysers erupt, the hot springs bubble and blurp, a little natural CO_2 seeps out to mingle with the vastly greater supply from humanity, and the heat suggests yet another way that we can power our society without changing the environment very much.

21

Put It Where the Sun Doesn't Shine: Nuclear Energy and Carbon Sequestration

Synopsis. Nuclear fission power contributes to our current energy mix and could be increased. Capturing and storing CO_2 are technically feasible, which would allow fossil-fuel use without associated global warming. But escape of the waste from either of these power sources could cause serious problems.

> *Through caverns measureless to man*
> *Down to a sunless sea . . .*
> *And from this chasm, with ceaseless turmoil seething . . .*
> *A mighty fountain momently was forced . . .*
> —Samuel Taylor Coleridge[1]

WE DIDN'T MEAN IT

Our good ideas almost always have unintended consequences. We deal with these in many different ways, including switching to an even better idea, cleaning up the unintended consequences, or just covering up the problem somehow.

When certain refrigerants were depleting our ozone shield and risking a rise in skin cancer and other bad things, we replaced those chemicals with safer ones—a better idea. But earlier, when we found that human sewage spreads disease, we didn't quit going to the bathroom—we cleaned up the unintended consequences by killing the disease organisms in water- and sewage-treatment plants, or isolating drinking water from the waste. And when faced with an unpleasant odor, we sometimes use an "air freshener" to make a more pleasant smell.

For example, back in chapter 16 we met Edward Burt during his visit to pre-sewage-treatment Edinburgh. At 10:00 p.m. when the "terrible shower" of human waste was dumped out of the windows, Burt's new friends in the pub "began to light pieces of paper, and throw them upon the table to smoke the room." They said that they did this "to mix one bad smell with another," but it is safe to assume that burning paper was a crude air freshener that smelled better than the human waste raining down outside.[2] Similar reasoning applies to our energy issues and global warming. We can substitute new energy sources for the fossil fuels to eliminate the CO_2 problem, as we discussed in the previous chapters. Or we can capture the waste CO_2 and isolate it from the environment, or replace the fossil fuels with nuclear energy and isolate its waste, as we will see in this chapter. Or we can try to cover up the CO_2 effects with geoengineering, which we will explore in chapter 23.

These very different paths have different costs and benefits. New energy sources are the sustainable solution for the long term, but some of the others may prove useful on the way. All of the new energy sources have their own unintended consequences that bother some people, but in most cases these are small enough that the public discussion is focused more on costs. For nuclear, carbon sequestration, and geoengineering, though, the unintended consequences are important for public discussion.[3] Here, I will use "nuclear" to refer to nuclear *fission*, the splitting of big atoms to release energy, as is done now in the many nuclear power plants; we will visit nuclear fusion in chapter 23, but I will call it "fusion."

GONE FISSION

In my experience, the "nuclear question" is the most polarizing issue in energy. I have met many people, in venues ranging from Capitol Hill to community gatherings, who will end the discussion if you give the wrong answer to the nuclear question. Unfortunately, any sizable gathering of interested people may include some who disagree passionately on what is the right answer. If you offer the opinion that nuclear has a role in our future, some people hear you endorsing terrorism, and sickness and death for unborn generations. Say instead that nuclear has no role in the future, and other people hear you advocating black helicopters to take over the world and establish a UN dictatorship, because no one who honestly fears global warming could possibly oppose nuclear. And these are not nutcases, but intelligent people who have thought about the issues.

First, please be assured that I strongly oppose terrorism, sickness for future generations, and world domination enforced by black helicopters, and you can quote me on that. Second, let me be clear that I am not going to solve the nuclear dilemma for you or anyone else. Nuclear can be used, it is not a silver bullet, and simplistic answers ("Nukes forever!" or "No nukes anywhere ever!") will not get us very far.

Now that I have addressed those basic issues, here's a little more perspective on nuclear, starting with cats in our nuclear family.

Sometimes our cats are strongly attracted to each other, forming a warm puddle on the bed. Sometimes, they are non-interacting, passing on whatever mission has caught their cattish fancies without changing course. Sometimes, they are strongly repelled, purposefully scattering in different directions.

Physics similarly displays attractive, neutral, and repulsive interactions, although things rarely change their minds the way the cats do!

Atoms have nuclei with protons and neutrons. The positively charged protons repel each other electrically, but they are attracted to each other and to the neutrons by the nuclear force.[4] The neutrons attract protons by the nuclear force without repelling them electrically, helping stabilize nuclei. But nuclei with too many or too few neutrons tend to be unstable and change spontaneously to lower-energy forms by emitting particles and electromagnetic radiation. We call these changes "radioactivity."[5]

The nuclear force that holds nuclei together is strong only over very short distances, but the electrical repulsion between protons works

Figure 21.1 Attractive interactions between cats (Coral, at left, and Prancer). Cats and elementary particles can exhibit attractive, neutral, and repulsive interactions, but particles don't change their minds, while cats do.

over longer distances. The natural wiggling of a nucleus may randomly reach an arrangement in which the repulsion of the protons exceeds the total attraction of the protons and neutrons, causing the nucleus to split apart. This is "fission," and it often spits out a couple of big pieces, two or three neutrons, and perhaps others.

Fission also can be induced if a neutron or something else hits a nucleus that is prone to splitting. And if one neutron splits a nucleus that spits out two or three neutrons, and many of the nearby atoms also are prone to splitting and spitting out two or three neutrons, the result can be a big bomb with a mushroom cloud. However, if the situation is tuned properly so that some of the excess neutrons are soaked up before they split another atom, a sustained chain reaction can be produced that will run for a long time and release an immense amount of energy from not much stuff, taming the atom bomb and turning it into nuclear power.

Uranium-235 (which occurs intermixed with the more common

and not-useful-in-this-way uranium-238) is the usual target for fission. Plutonium-239 would also work, and others are possible.[6]

Natural and Not-So-Natural Nuclear Reactors

A modern nuclear reactor seems about as far as possible from being natural. Yet, as with so many energy options, nature "discovered" nuclear fission reactors before we did. In the country of Gabon at Oklo and Bangombé, rocky remnants of natural nuclear reactors from more than two billion years ago may even help guide us on how to deal with the waste from our reactors, as we will see later.[7]

Uranium is generally quite water-soluble, but only when oxygen is present. Many of the uranium ores that we mine today were deposited where oxygen-rich waters carrying uranium met oxygen-poor waters, which often were migrating with petroleum.[8] On early Earth, the appearance of photosynthesizing organisms created oxygen-rich oases expanding into the primordial oxygen-poor environment. Uranium-bearing minerals would have dissolved in the oxygen-rich zone and then been deposited around it, locally reaching a high enough concentration to ignite the nuclear chain reaction, as in Gabon. These were not bombs, just sources of heat and radiation. The chain reaction is revealed in the rocks by their shortage of uranium-235, which decays in fission reactions, but enrichment in isotopes produced by those reactions.[9]

In the natural setting, heat released from the fission probably pushed water slowly through the rocks. In a modern nuclear power plant, the heat drives steam turbines that make electricity. As of 2006, a total of 443 commercial nuclear reactors were generating electricity around the world, supplying about 6 percent of global primary energy and 16 percent of electricity.[10]

The proven reserves of uranium available for mining at modern prices are quite limited, enough for only about thirty years at current production rates.[11] This is a shorter time than for coal, oil, or natural gas,[12] causing some worry about expanding nuclear greatly.[13] However, stockpiles of uranium from the cold-war years are being used, which has reduced mining exploration. Uranium is sufficiently common on Earth that there is optimism that more would be available if sought. We might even learn to recover enough of the uranium dissolved in the ocean to be useful, perhaps employing ocean thermal energy as mentioned in chapter 19.

Safety Today

We are continuously bombarded by radiation, both electromagnetic and particles. The longer-wavelength electromagnetic radiation, ranging from visible light through infrared and down to radio waves, generally doesn't have enough energy to harm us, as we saw in chapter 5. But the shorter wavelengths, from ultraviolet light through x-rays to gamma rays, can trigger cancer. Similarly, being hit by many high-energy particles increases the risk of cancer.

Natural and human-induced radioactive decay release cancer-causing radiation. Natural sources dominate the radiation exposure for most of us, and medical sources dominate the human-caused dose. Less than 0.01 percent of the average U.S. resident's exposure to radiation is estimated to be from nuclear power.[14] Coal-fired power plants may spread more uranium across the landscape than nuclear plants,[15] because the low-oxygen environment in which coal forms tends to precipitate uranium from water. However, existing technologies can prevent the release of this uranium in coal.[16]

Many people are more worried about nuclear accidents than about day-to-day operations, including unplanned leaks,[17] or a partial meltdown such as the one that occurred inside the Three Mile Island nuclear plant in my state of Pennsylvania in 1979, or a far greater disaster such as the 1986 accident at the nuclear power plant in Chernobyl.

A huge, authoritative study of the Chernobyl disaster found that in addition to the twenty-eight recorded deaths from acute radiation poisoning, Chernobyl's excess radiation ultimately will cause several thousand cancer deaths among those most exposed.[18] This total is based on models, using estimates of radiation exposure and of the rate at which low-dose radiation causes cancer. In the affected group, there are expected to be roughly 100,000 cancer deaths from other causes, a number that could vary greatly based on smoking and other factors, so it would be very difficult to actually detect the few thousand extra cancer deaths from Chernobyl.[19]

The Chernobyl nuclear power plant was a type that no one would build today, with widespread agreement that modern plants could not fail the way Chernobyl did. Comparing death rates from nuclear power to those from other power sources such as coal, with its mining accidents, shows that nuclear is on the safe side.[20] Still, the aftermath of Chernobyl can be "spun" easily to influence people—"no detectable effect on death rates" and "thousands killed" are both consistent with the science.

Safety Tomorrow

Nuclear power brings up many other safety issues, however. Nuclear reactors break down uranium-235 but build up other radioactive elements including much plutonium. In some countries, some used fuel is or was reprocessed to extract the most useful isotopes for additional power generation. Concern exists because reprocessing can also extract plutonium for atomic bombs that release a terrifying amount of energy, or for "dirty bombs" that could spread poisons far and wide. The most dangerous isotopes might be transmuted to safer things, perhaps using neutron beams, but no commercially mature technology exists yet.[21]

Dealing with the radioactive waste is probably the most contentious issue for nuclear power. We might someday mine the waste to generate more power. But for now, nations are trying to isolate the waste from terrorists, "rogue states," and the environment for a long enough time to be safe and politically acceptable, which might be 1000 years[22] to more than 100,000 years.[23] Ways to do this are generally not in place around the globe, and are not even especially close in many countries, although many technically trained people will tell you that this has much more to do with politics than with technical barriers.

The existence of deposits in Gabon shows that nature has left some radioactive waste in one place for two billion years. Those wastes are bound to certain organic chemicals, clays, and phosphate minerals, providing one line of research for us.[24] However, some of the radioactive material in Gabon did migrate, so that natural setting was not a perfect waste-isolation scheme. And it remains possible that geological history has been kind to these deposits but moved other deposits that we will never know about because nature spread them around so well.[25]

If we think about long times, to beyond 100,000 years, then we may need to worry about volcanoes erupting through the waste, or glaciers cutting into it, or many other issues that will be very hard to predict with confidence.[26]

Looking Forward

The contradictions of nuclear power continue to impress me. It works, but we have no accepted way to deal with the waste. I have met people who argued that nuclear energy is the most important source for our future, but that "we" should consider bombing "their" nuclear plants to stop nuclear proliferation. I have met others who greatly distrusted centralized government, but who advocated having the government guarantee the safety of the nuclear plants and their waste against terrorism and leakage for a very long time.

Nuclear power is highly likely to continue contributing to our energy mix, with some new plants being built in the coming years. But I suspect that the proponents of nuclear energy have a large task in front of them if nuclear is to grow greatly and replace a notable fraction of our fossil-fuel use. Safer reactors (so-called fourth-generation technologies), more-efficient fuel cycles, reduced opportunities for accidental leakage or terrorism in the mining-refining-generating-disposing pathway, and more advances are technically possible, but in general are not market-ready and often appear further from being ready than many competitors.[27]

Yet nuclear has been receiving a much larger fraction of the limited funds for research than its competitors—five times more than renewables from 1987 to 2002, and still substantially more.[28] That so many questions remain about nuclear, and that gains in renewable energy sources have been so rapid, raise interesting questions.

There is much more to be said, and much to be argued, about nuclear. Nuclear is almost certainly not *the* answer, because there almost certainly is no single answer—some parts of the world today rather clearly lack the infrastructure or the political stability that accompany successful nuclear installations. And the not-in-my-back-yard (NIMBY) problem remains a large obstacle to expanding nuclear power.

Note, however, that I recently joined two distinguished local elected officials in observing the capstone presentations of advanced high school students addressing the energy future of the United States. The different student groups promoted a great range of options, but just two were chosen by all the groups: federal mandates for higher gas mileage of cars, and more nuclear power plants. These are politically strange bedfellows in the United States. Thus, the modern stances of political parties on issues may not be as natural and obvious as one might at first suspect. And this may give some comfort to nuclear supporters.

CARBON CAPTURE AND SEQUESTRATION

At the American Geophysical Union meeting held in San Francisco in December 2009, Klaus Lackner of Columbia University and his colleague Allen Wright used manipulators to carefully drop an Alka-Seltzer pill into the water bowl in their sealed aquarium-like chamber. CO_2 came bubbling off, driving the CO_2 meter upward. But the concentration soon plateaued even as the bubbling continued, and then dropped as the bubbling slowed. The secret was the "plastic" tree-like structure in the chamber, which was collecting the CO_2 on its surface. Then, with a little more manipulation, Klaus and Allen convinced the plastic tree to release its CO_2 where and when they wanted.

Our understanding of the Earth system shows that the CO_2 we release will be removed from the air over a few hundred thousand years by reacting with rocks or by being buried to start replacing the fossil fuels we used, as we saw in chapter 10. And carbon has very interesting behavior that a lot of chemists understand pretty well, contributing to all the organic molecules of life, plus shells and gases and beautiful minerals.[29] If we could use our knowledge of chemistry to really accelerate the natural removal of CO_2, we might not need to worry about global warming, ocean acidification, and other problems from the CO_2. This is the broad field of *c*arbon *c*apture and *s*torage, or *c*arbon *c*apture and *s*equestration, or just CCS.[30]

Clearly, CCS cannot be the whole answer because it doesn't create energy—indeed, it uses energy—but it might allow us to continue using fossil fuels until we solve the energy problem. That may buy a few decades or longer for research addressing permanent solutions.

The first step is to remove CO_2 from the air. Plants do this well, and some people are enthusiastic about growing more plants and keeping them as forests, or turning them to charcoal and burying it, or in some other way using plants to sequester CO_2. But fossil fuels hold many times more carbon than modern plants or soils, so we would need to make a huge change to the amount of carbon in plants or soils to really matter. And because plants are so tasty for people and caterpillars and termites and bacteria, it is hard to imagine getting enough more plants to really make a difference for a long time in a hungry world. Keeping our forests and even growing some back have many advantages including holding down CO_2, and are part of most plans that I have seen for dealing with CO_2, but they would not capture nearly enough to allow us to burn all the fossil fuels and still stabilize the climate.

Reactions of rocks with CO_2 give us the half-million-year thermostat that has helped maintain Earth's equable climate over billions of years, as discussed in chapter 10. If the reaction could be accelerated by 10,000 times or more at a reasonable price and energy investment, then our CO_2 problem could largely or completely disappear, because the carbonate minerals produced are rather friendly and easy to live with. Research is ongoing,[31] but we do not seem to be close to a market-ready solution.

A related idea is to neutralize the CO_2 by reacting it with shell rock (limestone) mined on land, rather than letting the CO_2 dissolve in the

Figure 21.2 Water and CO_2 react with limestone to dissolve caves, transferring the CO_2 to the ocean in a form that does not increase acidity there. If we could geoengineer this process to go faster, it would neutralize some of our CO_2.

ocean and react with shells there.[32] In principle, a lot of CO_2 could be taken up this way, potentially at least 10 to 20 percent of U.S. point-source CO_2 emissions at relatively low cost. The dissolved materials discharged to the ocean are nearly harmless, and might even prove beneficial. However, a number of issues remain to be resolved, and scaling up to handle most or all CO_2 emissions would be notably more costly.

More widely discussed is the possibility of directly collecting the CO_2 from power-plant exhaust, or from the air far from power plants, then liquefying the CO_2 and putting it somewhere that the sun doesn't shine. Almost surprisingly, taking the CO_2 out of the air is not too much more difficult than taking CO_2 out of power-plant exhaust.[33] The energy cost of collecting the CO_2 and liquefying it is vaguely 30 percent of the energy released by burning the fossil fuel—large enough to be worrisome, but not so large as to be unworkable. And the search is on to find even better ways to do the job.[34] The CO_2 might even be regenerated into liquid fuel, again using "extra" energy from wind or solar farms at times of peak production, closing the loop.[35]

For now, though, the CO_2 has to go somewhere. Putting CO_2 into the ground is occurring already, commercially. Some oil and gas are naturally associated with high concentrations of CO_2 in Earth that are commercially unacceptable for gas sales. Oil companies strip off this excess CO_2, liquefy it, and pump it back into oil fields to force more oil to their wells. And beneath the North Sea off Norway, CO_2 is being put back into the ground not to recover more oil, but because Norway's

Figure 21.3 Carbon capture installation for coal-fired power plant, Shanghai, China. Photographer Andy Quinn for scale. Designed to capture up to 100,000 tons of CO_2 annually, the plant will initially supply the CO_2 for carbonating drinks and for chemical processes, but the CO_2 could also be sequestered.

taxes on CO_2 emissions to help slow global warming make CCS economically favorable.

Unfortunately, while a little CO_2 is easy to put down, sequestering a lot may prove harder. The expected CO_2 production from the United States over the next fifty years would cover the entire country in a liquid-CO_2 layer 2 inches (5 cm) thick.[36] Doing the equivalent of raising the country 2 inches, sticking CO_2 underneath, and not having leaks, is a very challenging problem. Very much leakage would defeat the effort.[37]

And leakage can be dangerous. Although CO_2 is normally mixed rapidly through the atmosphere and doesn't build up much near the main sources, a really big CO_2 leak or very unfavorable still-air conditions can produce dangerous concentrations. In 1986, a catastrophic natural release of volcanic CO_2 and perhaps other gases stored in the deep waters of Lake Nyos, Cameroon, killed over 1700 people.[38] Natural CO_2 springs in Italy have killed many people, including three during the 1990s at one site.[39] If an expensive effort to reduce CO_2 in the atmosphere ended up killing people with leaked CO_2, I speculate that public reaction would not be favorable. Indeed, protestors have vigorously opposed plans to sequester CO_2 underground near population centers.[40]

We also might put the CO_2 into the ocean. One way to do this would be to fertilize the water to turn more CO_2 into plants that eventually die and sink, hiding CO_2 in the deep ocean for a while (see chapters 10 and 11). But the CO_2 won't stay down forever. Also, this approach might increase toxic blooms of algae and oceanic "dead" zones without oxygen, which would produce greenhouse gases more potent than CO_2.[41] Enthusiasm for this idea is not completely gone, but seems to have dropped over the last few years.

Instead, we might extract the CO_2 from the air or from power-plant exhaust and pump the CO_2 into the deep ocean. However, this also would acidify the ocean, and the eventual escape of that excess CO_2 back to the shallow ocean and the atmosphere would raise the CO_2 level in the air and keep it raised for a long time. A better option may be to put the CO_2 into subsea sediments—there is a lot of ocean, a lot of sediment, and very few human neighbors to complain about proximity to their backyards or the possibility of walking into a dangerous puddle of leaked CO_2.[42]

Many geoscientists, including some of my colleagues, are very excited about CCS. If it works, geoscience students could continue getting good jobs with energy companies, and more jobs putting the waste CO_2 back down after it is produced, while our studies of global biogeochemical cycling and rock-weathering kinetics would be of great economic importance, students would flock to our departments, and the world could enjoy its fossil fuels while looking for alternatives. However, there is still a lot of work before we can declare this a done deal, with very large difficulties to be addressed, and the possibility that when the costs are all calculated CCS will prove not to be the way to go. Regardless, CCS is not the long-term answer; it is simply a way to buy time, albeit perhaps enough time to be worth doing.

22

Conservation—Why Saving Energy Doesn't Mean Sitting Around in the Dark

Synopsis. Saving energy may be the cheapest and fastest way to make initial progress in reducing CO_2 emissions. Deep cuts would notably impact quality of life, but modest cuts can be made with little, zero, or even negative reduction in standard of living.

. . . you will reproach us, not for what we have used, but for what we have wasted. —Theodore Roosevelt[1]

GLOW, LITTLE GLOWWORM

"Glowworm" sounds a lot more appealing than "fungus gnat" or *Arachnocampa luminosa*. Whatever you want to call them, though, they're amazing creatures.[2]

The wormlike larvae of fungus gnats, or fungus flies, and in particular the famous glowworms of New Zealand, are among the many living things that generate light. Widespread across the planet, fungus flies make their livings in many different ways. The larvae of some, especially those of the north, are vegetarians, often munching on the spores of fungi. Others are carnivores, and even turn to cannibalism on occasion. Some trap their prey in sticky webs, much the way spiders do, and those in Sweden build their webs on the undersides of fungi. And many glow. The Swedish web-builders glow all over with a rather dim blue-white light, while those in the Appalachian Mountains of the United States have blue "lanterns" at either end.

The well-known glowworms of New Zealand and Australia live in damp spots, in caves or mines or back in the woods along streams. The worms grow to more than an inch in length. A New Zealand glowworm first spins a tubular nest of silk and mucus, and then lets down "strings" of silk, extending as little as half an inch or as much as a foot and a half (roughly 1–50 cm), and beaded with droplets of sticky mucus. (The resemblance of these silky bug-catchers to spider webs led to the spidery "arachno" in the name *Arachnocampa*.) The glowworm lurks in its overhead lair until the twitching of a string shows that a bug has been trapped in the goo. Then the glowworm reels in the thread and consumes the hapless prey.

That would be the end of the story for the larvae of some fungus flies. But, like those in Sweden and the Appalachians, the New Zealand and Australian types have added a lure for their prey, which also lures the tourists. Watch an electric light during a summer night and you will see the many types of bugs buzzing around it.[3] Add an electrical device to your back-porch light, and you have a noisy bug-zapper. The glowworms are a lot quieter, but they turn on their personal light, and dinner flies into their diner. In certain New Zealand caves, masses of glowworms on the ceiling look like stars in the night sky, with beaded strings hanging down toward the water or ground beneath in a truly amazing display.

Many other species have figured out how to make light too, probably for many different reasons. Perhaps the most fascinating are the beetles commonly known as fireflies.[4] The flashing taillights of some species are advertisements for mates. In the common American *Photinus*, for example, the male will fly along flashing out, in his own version of Morse code, "I'm a *Photinus* male on the make." If Ms. *Photinus* is down below and interested, and she likes the brightness and length of his display, she will flash out an appropriate reply. He will land so they can discuss the matter further, using flashes that carry meaning in their length and pattern. If all looks good, they will mate.

However, like someone who posts a personal ad in a public space, Mr. *Photinus* can never be sure who will answer. Occasionally, "Ms. *Photinus* interested" proves to be "Ms. *Photuris* hungry," another type of firefly that is a *Photinus*-eating flash-linguist capable of "speaking" the language of other species to lure them in. In addition to a meal, Ms. *Photuris* gets certain chemicals from her prey, called "lucibufagins," that make her unpalatable to the *Phidippus* jumping spiders that otherwise would eat her.[5] Ms. *Photuris* is so set on eating flashing beetles that she will attack decoys made with light-emitting diodes pretending to be *Photinus* males. Somewhat similarly, the lights of the New Zealand glowworms may attract long-legged harvestmen, who dine on glowworms.

Not all living lights are as large and charismatic as fireflies, but many are at least as intriguing. For example, many planktonic dinoflagellate algae can be bioluminescent—stir the water, and they light up. One hypothesis is that they are motion-detector lights to warn the world about robbers. Small fish can pig out on the plankton under the friendly cover of darkness, securely unseen by the bigger predatory fish that would love to dine on their smaller cousins. But what if the plankton respond to the small fish enjoying planktonic hors d'oeuvres by flashing an "Eat Here" sign for the big predators, lighting up the skulking small-fish diners so that they suddenly are bait? Like humans with sensor-triggered lights on their houses, the dinoflagellates call in the cops with a motion detector, and the cops clean up the neighborhood before the dinoflagellates are cleaned out.[6]

Our family participated in such a display a few years ago, when bioluminescent dinoflagellate algae washed into Nauset Marsh on Cape Cod. We were transfixed, as are so many observers, by the sheer beauty of the night lights. Kayaking through the dark marsh, the bow-wave of the boat glowed, as did the drips from the paddle and the whirlpools spun off the blades. Marshes are wonderfully productive, and schools of bait fish went skittering away from the boat, often skipping along the surface to make dotted lines of light. But following those dotted lines were the great swirls of striped bass, slashing through the crowds of smaller fish. We stay at a house without television, but you can be confident that no one missed the tube that night!

Control of light thus is highly valuable. You can use it to attract dinner, or to see dinner, or to scare away something that wants to turn you into dinner, or in many other ways. My early camping experiences required our use of rapidly generated light to keep bears away, as described in chapter 3, and our distant ancestors surely did the same. Whalers put their lives on the line because people wanted to read at night, and whale oil gave them that ability without the stink of animal fat or the danger of explosive camphene.

I used to make light on my bicycle to warn away motorized "predators," with a little generator turned by the front tire that powered a flashlight-sized incandescent bulb on the handlebars. Even that little light took a lot of "oomph," notably increasing the difficulty of climbing hills—recall that the total average power output of a human is just enough to run one large incandescent bulb. Dinoflagellates can't be using energy nearly so rapidly. For the *Photinus* fireflies, flashing is quite inexpensive, increasing their resting metabolism by only about one-third[7] compared to the more than one-half increase in resting metabolism for walking. Comparison of bioluminescent and non-bioluminescent species from the same family suggests that the cost of maintaining the capacity to make light is undetectably small. And measured against the effort going into courtship flights by males, or growing eggs by females, the energy used in flashing is minor.

So why is light production apparently so easy for glowworms and fireflies and dinoflagellates, but so hard for me on my bicycle? Mostly because I was doing it wrong—if a firefly got as hot as the incandescent bulb on my bicycle, the bug would be burned to a crisp. But before we learn what glowworms can tell us about saving energy and reducing global warming, let's take a detour to learn about light production.

Where does light come from? Most is produced by charged particles accelerating or jumping. For millennia, humans have relied on light from hot sources—the sun, or fires, or electrically heated wires in lightbulbs. The rapid motions of hot atoms and molecules, with frequent collisions with neighbors that cause decelerations and accelerations, produce electromagnetic energy from their charged particles (visible light, and infrared, and other wavelengths, which really are the same thing). Because of the random motion of hot particles, with some moving faster and others slower, the light emitted covers a spectrum rather than being at a single wavelength.

Thus, the hot surface of the sun sends many different wavelengths to Earth, which sends many wavelengths back to space. In principle, both Earth and the sun radiate all wavelengths, but Earth is so cool that its production of high-energy shortwave radiation is pathetically tiny compared to the sun. For practical purposes, all of the radiation from Earth, or a human, or a doorknob, is at infrared wavelengths too long for humans to see, whereas the hotter sun emits some infrared but radiates most strongly in the visible band that we can see.

We saw earlier that CO_2 blocks some of Earth's outgoing infrared radiation, but only at certain wavelengths, and this is a clue to what the glowworms are doing that we can mimic. The CO_2 molecule can wiggle in certain ways, with certain energies, but not in other ways with other energies. The molecule absorbs infrared light carrying the energy needed to make the CO_2 molecule wiggle in one of the possible ways, while light of other energies passes without being absorbed. Absorbed energy may be given to another molecule in a collision or radiated away; if radiated away, the wavelength given off will be the same as (or very close to) the wavelength originally absorbed.

The next part of this discussion will be easier if you recall that light exhibits characteristics of both particles and waves. The particle of light is a photon, photons come in different energies, and a high-energy photon corresponds to a short wavelength of light. Atoms have negatively charged electrons, which hang around in certain orbits with particular energy levels. An incoming photon may be absorbed by an atom if the photon has just the right energy to knock an electron to a higher level that has space for a new arrival. A photon of that energy will be released if the electron jumps back down to where it started, fully analogous to the story with absorption and emission for CO_2.

Next, suppose that a high-energy photon knocks an electron to a higher level, then the electron moves down to an intermediate level by transferring some of its energy to neighbors through collisions or in other ways, and finally the electron jumps down the rest of the way. The photon emitted in this final jump will have less energy than the original incoming photon, because of the intermediate energy-loss step. This is called "fluorescence." If you have seen a black-light mineral display at a museum, with high-energy ultraviolet "black" light stimulating certain minerals to glow with spectacular colors, you know what is going on, and this is the basis for fluorescent lightbulbs. In some materials, the final jump is delayed long enough after the initial photon is absorbed that we notice the mineral glowing after the black light is turned off; in these cases, the name is changed from "fluorescence" to "phosphorescence."

In other cases, the initial jump to a high-energy state is achieved not by absorbing a photon, but by a chemical reaction that puts interesting molecules into high-energy states. The "decay" to a stable end state often involves emission of a photon, so the reacting chemicals glow. This is called "chemiluminescence" if it happens without living things, or "bioluminescence" if done by living things, and is what is going on with the glowworms and fireflies and dinoflagellates.

The essential difference between the hot light sources—sun and fire and incandescent bulbs—and the cooler ones—glowworms and fluorescent bulbs and others—is the brute force of heat versus the finesse of transitions. Make something hot enough and you get light, but you also get infrared and other wavelengths, while you may be losing a lot of energy to conduction and convection—you are making more heat than light and only a tiny bit of your energy actually lights up your life. Or you can find a way to make just the wavelength or wavelengths you want by carefully tuned jumps between states. Then not much heat is involved to fry you or your device, or to lose energy to convection or conduction or other wavelengths—you are making more light than heat.

In the United States, lighting consumes more than 20 percent of all electricity generated, with similar usage in the European Union.[8] For an incandescent light, the standard from Edison until recently, a modern bulb puts 2 to 3 percent of the energy used into light. This is almost ten times better than Edison's first bulbs, a bit more than ten times better

Figure 22.1 Andy Quinn (left) and Richard Alley, Nile River Cave, Paparoa National Park, New Zealand. The incandescent bulb in the flashlight turns 2 to 3 percent of the electricity from the battery into heat. That is more than 100 times better than the candle, but several times worse than the LEDs in the headlamps and far, far worse than the glowworms in the next picture.

than the best mantle-style gas lanterns, almost one hundred times better than earlier lamps, and well more than one hundred times better than candles. Burning and heating just aren't very efficient at making light. But compact fluorescent bulbs are running four to five times better than incandescent bulbs, and LEDs (light emitting diodes) promise to double this.[9] A lot of people are not overjoyed with the color of compact fluorescents, especially for early bulbs, but LEDs may fix this as well.

So, by trying to follow the lead of the glowworms, we may get high-quality light while saving almost 20 percent of our electricity in the United States. And advances in organic light-producing devices are coming rapidly,[10] so some day we may catch up with the glowworms.

LESS, OR SMARTER?

I believe that when you discuss "conservation" of energy in a room
of normal people, they are thinking different things. Some think
efficiency—getting more goods and services from the energy we use.
Some think reduction of demand—using fewer goods and services.
Which one people prefer may say a lot about their politics, but we will
leave that for late-night TV and the blogosphere.

In reality, these are opposite sides of the same coin. An in-ground heat
exchanger for your heat pump may save 25 to 50 percent of the electricity while doing the exact same job as an in-air heat exchanger.[11] This is
surely efficiency—you get the same results, but you pay less for them

*Figure 22.2 The roof of the Nile River Cave looks like a starry night above one of the
cave entrances, which is reflected in the subterranean river below. The cool light of the
glowworms on the roof lures dinner into their diner.*

(depending on installation costs and interest rates). Moving in with your parents so that you don't need to pay for heat may be a loss of goods and services—you are still alive, and have a climate-controlled place to live, but you may have a lot less room in which to live, and be giving up privacy and other things you value. Turning the thermostat down and wearing a sweater in the winter falls somewhere in-between, a bit of efficiency, a bit of lost goods and services. You are still doing what you want, but you have to put on the sweater, and that will increase your laundry some day, and you may just feel a bit clunky in the sweater. (Unless you feel toasty and snug, so you're gaining rather than losing . . .)

Local fire departments across the United States have stories about houses so crammed with stuff that they were fire hazards, with only tiny pathways winding through the flammable materials. In my town, some people make their living running businesses to help customers throw away their excess "stuff," things that they don't need but can't make themselves discard. In a world with a billion hungry people, some of us still had a bit too much to eat for breakfast, and are trying to slim down a little to avoid health problems associated with obesity. One way to do that is to reduce meat-eating—our ancestors surely were cooking carnivores, but not necessarily for all meals every day.

Thus, some of us in the developed world use more goods and services than are good for us. Reducing this excess use would be a win-win outcome, helping us directly while reducing our energy consumption and the rate of global warming.

I suspect that some of us will reduce our demands in the future. Many of us are trying to bring our meat consumption into line with our doctors' recommendations, and to clear out the extra stuff from our houses. But I doubt that this will be a lot of any answer. When we must hire experts to help us throw away things, it is clear that we are fighting very deeply rooted human needs. And you are welcome to take away some of my knickknacks, but you may not take away the ambulance that is waiting to haul me to the hospital in case I have a heart attack from the extra meat I ate. Furthermore, anyone planning to solve the global-warming problem by denying things to people who want them may be flirting with "political suicide" rather quickly—solutions that people choose for themselves tend to be much more popular.

Clearly, if we cannot produce or pay for the food and stuff we want, we will reduce our consumption, and this will contribute to reducing global warming. But, except for the overconsumption just noted, loss of food and stuff would mean an economically unhealthy reduction in our standard of living. Like the glowworms, the conservation goal for most people is to be smarter with what we use, following Teddy Roosevelt as quoted at the start of this chapter, by eliminating waste, not by eliminating honest use.

In the autumn of 2009, the U.S. Environmental Protection Agency (EPA) was asked to project the costs of bills being considered in the U.S. Senate and House of Representatives for reducing U.S. CO_2 emissions.[12] The EPA analysis found that in response to the costs of a cap-and-trade program with a rising cost for CO_2 emissions, the economy would respond in many ways, but that conservation would be a major player because it is so inexpensive initially. By the year 2050, with a cost of $70 per ton for emitting CO_2, the laws would have decreased the average household economic consumption by less than 1 percent, while increasing the share of low- or no-carbon energy sources to 38 percent from 14 percent, which is the modern share and the expected share in 2050 under business as usual. Energy expenditures for the average household would increase 21 percent, much less than the 35 percent increase in electricity costs, because reduced demand for energy, and increased efficiency of use, would reduce usage by 12 percent in 2050.

Thus, good economics says that a cost of $70 per ton for CO_2 is projected to conserve 12 percent of U.S. energy consumption. For comparison, a cost of $70 per ton for CO_2 is approximately $0.60 per gallon of gas; during 2009, in the absence of any such law, the price of gasoline in the United States rose by $1.00 per gallon.[13] To a consumer, $0.60 per gallon may seem expensive, but a rise of $0.60 per gallon over forty years is economically rather small when year-to-year variations can be much larger. Yet, such an economically small but sustained influence is expected to reduce consumption by more than 10 percent. A reasonable conclusion is that it is easy to reduce U.S. energy usage significantly, but it is not easy to reduce U.S. energy usage grossly. As a general rule, the easy and inexpensive things are done first, so greater energy savings would have greater costs.[14]

A somewhat similar insight comes from the observation in chapter 2 that, across the major developed-world economies, the economic activity produced per watt of primary energy used varies by twofold or so, but generally not by tenfold. At least some of the differences are undoubtedly explained by the local circumstances (e.g., energy used for transportation may be greater in a more sparsely settled country that must move things farther, and some countries are making things for other countries), although some of the differences reflect policies aimed at greenhouse-gas reductions and other energy-related issues. If conservation were really easy, the extra effort that some countries have devoted to reducing emissions would have given larger differences.

But, in some aspects of life, very large savings may be possible or even desirable. We saw in chapter 18 that the German "Passivhaus" can reduce heating bills by 90 percent. Recently, I have been avoiding flying across the country and the world, instead taking very-low-energy "trips" via web links. I visited a high school in Pennsylvania, and a university-sponsored lecture in New Zealand, and a whole bunch of universities spread across eight time zones, without leaving the comfort of my building, while saving much more than 90 percent of the energy I otherwise would have used. True, some of the people I visited were probably quietly doing their email or surfing the web rather than paying attention to me, but the discussions afterward in each case indicated that a lot of people were paying attention and even enjoyed it (and I know that some of the people at in-person talks are surfing the web too).

Viewed broadly, the opportunities for conservation are manifold— everything we do that uses energy can be done more efficiently. And because we use a lot of energy, the savings will mount. Some conservation is going to be very inexpensive, or free, or paying us, such as switching from incandescent lightbulbs. Conservation does not have to mean a reduced standard of living. And as for solar energy and the smart grid, conservation is likely to open great opportunities for small businesses, entrepreneurs, and start-up companies. Because energy use is everywhere, conservation opportunities are everywhere. No one thing can be the big solution, but a whole lot of things are parts of the solution.

The glowworms of New Zealand don't worry about start-up companies or standards of living. They just go about their business, enticing dinner company with their pretty and highly efficient lights. Conservation for them isn't a political statement; it's survival. And they make it work. They get light without burning themselves up. Efficiency cannot possibly solve all of humanity's problems, but it is almost guaranteed to be an important part of a solution. And, done with vision, we can have fun and make some beautiful things on the way.

23

Game-Changers?
Geoengineering, Fusion, and . . .

Synopsis. Among the options available for dealing with
our CO_2 emissions is "geoengineering," to treat the symptoms rather than the cause of the problem. In addition,
fusion or other radically new technologies might emerge.
These are "unknown unknowns"—nothing is close to being
engineering-ready, we are not sure that they will work or
work well enough to be part of the mix, but they are interesting. If some game-changing technology should arise, an
optimal path will allow changing the game.

Give me a lever long enough, and a prop strong enough, and with my own
weight I will lift the world. —Archimedes[1]

SUNSCREEN

We have discussed ways to make energy without making CO_2, and
ways to capture the CO_2 after burning fossil fuels for their energy.
But what about simply dumping the CO_2 into the air and then
counteracting its effects—treating the symptoms and not the disease.
We load up on painkillers and anti-itch creams and throat-soothing loz-
enges at the pharmacy because sometimes our bodies will cure the disease and we just want the comfort of avoiding the symptoms until we
are healthy again. Nature eventually will draw down the CO_2, so why
not just cover up the effects until then?

Figure 23.1 The 1980 eruption of Mt. St. Helens wasn't big enough to have much effect on global climate, but it was spectacular.

This idea is usually called "geoengineering." Different people use the term in different ways—to some folks, carbon capture and sequestration is a form of geoengineering, and I will mention a few others soon. But the most widely discussed idea is to block some sunlight, cooling Earth just enough to offset the warming from the CO_2.

We have known for a long time that this is possible. The particles from our smokestacks, and especially the reflective sulfuric-acid droplets that form from sulfur dioxide and eventually feed acid rain, have offset much of the warming effect of our CO_2 over the last century, although with large costs to health and environment. Much of our CO_2 will remain in the air for centuries, causing warming, while the cooling particles fall down in a couple of weeks, so eventually the influence of the CO_2 is expected to greatly exceed the effects of the particles.

However, we have been really inefficient with our particles. A big, explosive volcano puts sulfur into the stratosphere, high above the rain that washes the lower atmosphere, so the sun-blocking effect lasts a year or two, rather than a week or two. Much less sulfur is then needed to cause a given amount of cooling, and less acid rain results. A fleet of specially designed planes, or giant artillery shells, or rockets, or a pipe might deliver enough sulfur dioxide to the stratosphere to do the volcano job and cool the planet.

A lot of other, similar ideas are percolating around, such as putting sun-blocking things in space, or seeding reflective clouds by using fleets of ships to stir up sea salt.[2] Creative thinkers have little difficulty in coming up with clever ways to influence the climate.

Some of the world's best climate scientists have been involved in generating or exploring these ideas including the president of the U.S. National Academy of Sciences, and a Nobel laureate in chemistry.[3]

Interest in this topic is growing very rapidly. In the scientific literature, the oldest paper listed by a major indexing service[4] under the subject "geoengineering" that used the word in the modern sense is from 1993, and it was already looking at geoengineering as a way to allow continuing CO_2 emissions if necessary, after careful study.[5] Without using the term "geoengineering," the same author and coworkers in

1984 had already conducted an analysis for offsetting warming using particles distributed by commercial aircraft.[6] A 1992 report of the U.S. National Academy of Sciences discussed geoengineering among many other issues related to greenhouse warming, and cited older, pioneering work by M. I. Budyko addressing effects of sulfur gases.[7] The number of scientific papers published on the subject of "geoengineering" remained at or below just four per year through 2005, but reached forty-five in 2009.[8]

This scientific interest piqued the curiosity of the popular press. Reporters were further stimulated by the collaboration between Lowell Wood, promoter of "Star Wars" technology and protégé of Edward Teller of H-bomb fame, and Ken Caldeira, occasional antinuclear protestor. *Rolling Stone* asked, "Can Dr. Evil save the world?"[9] *Wired* wondered, "Can a million tons of sulfur dioxide combat climate change?"[10]

These and other popular treatments led to some public statements that global warming is no big deal and can be solved easily at low cost

Figure 23.2 The cataclysmic eruption of Mt. Mazama that produced Crater Lake, Oregon, about 6,600 years ago blasted ash to Greenland.

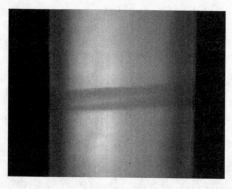

Figure 23.3 Fallout from a volcanic eruption, from 1706 m depth in the Divide core, West Antarctica. The core section is a few inches high, oriented the way it would have been in the ice sheet, with the volcanic layer running across the picture.

by the application of well-known engineering.[11] In turn, something of a backlash developed among climate scientists who understood the issues much more deeply than did many of the public commentators, and who knew that many issues must be addressed before any "solution" is engineering-ready.[12] At that point, I received an inquiry from an old friend and distinguished hydrogeologist, asking why climate scientists were being so rude to the geoengineers. We bounced some ideas back and forth, and I hope that our discussion will be instructive in evaluating the issues.

GEO-HYDRO-ENGINEERING

We never know the worth of water till the well is dry. —Thomas Fuller[13]

Hydrogeology is one of the many fields in which Earth scientists are valuable to the real world. A first-rate hydrogeologist is rare and valuable. If your water well is dry, or contaminated, or your yard won't drain, having someone who knows about water and Earth, and about engineering and the law and some other things, is essential. Hydrogeologists often don't win major scientific awards, because they're too busy being useful, but they have the satisfaction of watching their students go off to good jobs.

I learned early in my "career" mowing lawns to raise money for college that gasoline tastes lousy, and if you get a little on your hands before lunch, applying soap and water in abundance is wise. Some chemicals in gasoline may be carcinogens or otherwise unhealthy. And, gasoline has a tendency to burn or explode under certain conditions. Based on this, if gasoline were to show up in your water well, you almost certainly would be unhappy. And, you might involve a hydrogeologist rather quickly, perhaps with a lawyer in pursuit. Yet, we have gasoline storage tanks sitting beneath filling stations across the world. Older tanks can rust and leak; newer ones might rupture if a refueling truck drives over improperly applied pavement, or fittings might come loose. So gasoline does get into the ground sometimes. Once there, the gasoline moves, and eventually may spread out rather widely into many water wells. Furthermore, gasoline is far from the only problem of this sort. Diesel, and pesticides, and many other liquids or chemicals that dissolve in water can escape and do harm as they leak into wells or streams or coastal waters.

Hydrogeologists have a few tools to use in such cases. Contaminated water can be pumped out and treated, or made to flow through treatment zones where chemical or biological reactions attack the contaminants especially rapidly, or special waste-eating microbes or chemicals can be added to the site of contamination.

One neat trick is "air sparging"—if the contaminant evaporates easily, clean air can be pumped down to the contaminated zone and then back out, picking up the contaminant on the way.[14] Air sparging is not necessarily the most important tool for a hydrogeologist, it is inappropriate in some applications, but it works, and it is probably the hydrogeological tool with the best name.

Anyway, air sparging, and all of the other tools available to hydrogeologists, are not perfect. Contaminants spread out a lot, so while the hydrogeologist is trying to get the contaminant back out of the ground, the contaminant is running away. Then costs go up, and more people get unhappy.

In response to these difficulties, a lot of hydrogeologists have lent their weight to efforts to keep pollutants out of the ground, because keeping them out is often a lot easier than getting them back out. But suppose an economist showed up, or a cold warrior, and said, "Don't worry about passing laws to require sturdy underground gasoline storage tanks and frequent inspections to keep gasoline out of groundwater. If there are leaks, we can build a super-air-sparger and remove all the gas. We don't technically have one now, and no one has ever built one, but we know from existing air spargers that air sparging works."

The hydrogeologists are likely to point out that it isn't that simple. For example, air sparging should not be used where "... nearby basements, sewers, or other subsurface confined spaces are present at the site. Potentially dangerous constituent concentrations could accumulate in basements unless a vapor extraction system is used to control vapor migration." (Boom!)[15]

GEOENGINEERING BITES BACK

So, can we trust a "super sun blocker" that doesn't exist and therefore has never been tested under real conditions, but that seems technically possible and uses principles that have been tested in nature? Or are there issues to be investigated before we spray sulfur dioxide into the stratosphere, or otherwise geoengineer the climate?

Maybe the first issue is that the cooling induced by sun-blocking particles does nothing to counteract the acidification of the ocean, or the ecosystem changes on land. Thus, at best geoengineering treats one symptom, not the whole set. Sick people sometimes take a lot of pills to treat a lot of symptoms, sometimes taking one pill to treat the effects of another one. Likewise, a geoengineered "super sun blocker" might cause other symptoms.

Particles and CO_2 are not exact opposites. We saw that in general the feedbacks in the climate system amplify temperature change, regardless of the initial cause of the change. But, in detail, blocking the sun coming in, and blocking Earth's radiation going out, are different things. Particles may cause wintertime warming in some high-latitude places, for example, even though the overall effect is cooling, because there is outgoing energy from Earth but no incoming sunshine to be blocked in the dark of a polar night.[16]

More important, particles are projected to reduce precipitation. By blocking incoming sunlight up in the air, particles reduce the heating of the surface by the sun, and surface heating supplies energy for evaporation.[17] Research on this topic is in early stages, with estimates ranging from a small effect[18] to much larger impacts, including disruption of monsoonal rains.[19] The natural experiment of the major eruption of Mount Pinatubo caused not only cooling but also drying, so concerns about drought from geoengineering in this way seem warranted.[20]

Adding sulfur dioxide to the stratosphere is also expected to contribute to ozone loss, increasing depletion in the Arctic and delaying recovery of the Antarctic ozone hole.[21]

On a slightly optimistic note, one model experiment found that only part of projected warming would need to be blocked to protect the ice sheet on Greenland from melting away greatly, because of the way the atmospheric circulation and other things change.[22]

The prospect of geoengineering raises a lot of other issues. If the effort is sufficiently easy, any nation could do it. What is the legal framework in which this can be done? At the moment, such actions might be illegal internationally—ENMOD (Convention on the Prohibition of Military or Any Other Hostile Use of Environmental Modification Techniques) is a UN treaty that entered into force in 1978 and prohibits military use or other hostile application of techniques for environmental modification.[23] Many countries might view an externally imposed drought as a hostile act.

Much of the CO_2 we put up will remain there for centuries to millennia. In turn, any geoengineering scheme must either continue to function for millennia or be replaced by a different scheme, such as taking the CO_2 back out of the air, or else at some time the avoided warming will happen. If geoengineering failed suddenly (perhaps by terrorism or war), the resulting climate change could be especially abrupt. And because faster changes generally are harder to deal with, this could be very bad.[24]

Many climate scientists, including me, have publicly affirmed the possible value of research on geoengineering. Recall the immense damages from warming that are allowed to happen in the Nordhaus optimization,[25] and it is easy to be attracted by the possibility that we could inexpensively head off much of the trouble. But the dangers of ozone depletion, drought, continuing ocean acidification, changes in weather patterns, international tension, huge abrupt climate changes if the geoengineering fails in the future, and more, are real and as yet poorly quantified.

I have grave doubts that any international agreement would be possible without a greatly improved ability to model not just the global but the regional and even local effects of an intervention—people are likely to demand that we know how any plan affects them where they live, to help guarantee fairness. Climate modeling is quite skillful globally and somewhat skillful regionally, but I doubt that it is close to being up to the task that would be asked of it by the policymakers of the world if we adopted geoengineering.

A serious effort on geoengineering looks like a wonderfully secure employment program for climate scientists, yet many climate scientists are very worried about geoengineering. Some pundits and other people express doubts about the quality of climate science and our understanding of global warming, while expressing confidence that we can easily and safely geoengineer our way out of global warming if it becomes a big problem. However, these two ideas are highly inconsistent—confidence in geoengineering requires confident climate science.

So, is geoengineering potentially valuable? Absolutely. Is geoengineering easy, safe, and secure engineering? Not now, and we don't know whether it ever will be. Would hard-nosed engineers or businesspeople place great faith in a technology with this many poorly addressed questions and this much uncertainty? You can decide, but my friends in hydrogeology surely would not stake their professional reputations and the drinking water of their customers on a technology that doesn't actually exist and hasn't been properly tested.

FUSION

In 1958, the year after I was born, the British nuclear pioneer John Cockcroft wrote about our desire to emulate the sun through controlled fusion.[26] He noted the many difficulties to be overcome, but in those days there was a lot of optimism that fusion was coming soon.

That optimism is still around for at least some people, but fusion isn't. This gives rise to the oft-repeated old joke that fusion is the energy of the future, and always will be. Physicist David Goodstein observed that for the last fifty years the solution for fusion has been twenty-five years away, and it still is.[27]

This twenty-five-year estimate may be optimistic. A recent review[28] included an update on a 2004 study from the United Kingdom Atomic Energy Authority,[29] which described a path to commercial fusion reactors in forty years or more, beginning in 2004. However, the review noted that progress had fallen behind this "accelerated" path to fusion.

Back in 1976, the U.S. Department of Energy's Fusion Energy Advisory Committee released a report with a pathway to commercial fusion power, and an estimate of the research funding needed to get there. Subsequent analyses have argued that failure to reach that level of funding is responsible for the slow rate of progress.[30] Yet, until recently, fusion received a larger share of the small energy-research budget than did renewable energy sources.[31]

Many people (including me) remain fascinated by fusion and hopeful about its future. If we could master fusion, economically and effectively, the prospects for supplying everyone as much power as they want would be bright indeed.

The goal is to mimic the sun, causing hydrogen to fuse to helium and release energy. Hydrogen nuclei must be squeezed very closely together for this to happen; if they are too far apart, the repulsion of the proton charges exceeds the attraction of their nuclear force. The great gravity of the sun, because of its great mass, does this job, but we can't match that. Our first step is to use the rare forms of hydrogen that fuse more readily, either tritium:deuterium (the favored approach now) or deuterium:deuterium (harder to do). There is so much deuterium in the ocean, and it is sufficiently easy to separate from "ordinary" hydrogen, that deuterium:deuterium fusion could supply our electricity for millions of years. Tritium is rare on Earth, although it can be produced by neutron bombardment of lithium. There is probably enough lithium to supply us for centuries, although there is a little more worry here than with deuterium.[32]

Even using deuterium or tritium, fusion remains difficult. A very high temperature makes the nuclei move so fast that some of them can speed through the electrostatic repulsion to get close enough for the nuclear force to dominate, if some extra compression is supplied. The key is to squeeze deuterium and tritium nuclei together while heating them to 180 million °F (100 million °C; or to 540 million °F, or 300 million °C, for deuterium:deuterium), and to hold them long enough to fuse and for us to get the heat out, without allowing much leakage of heat or nuclei to places we don't want them to go.

The favored approach is to squeeze the hydrogen in a magnetic bottle,[33] although lasers also can be used to achieve the squeezing.[34] Whether these or other approaches will work and be economic, we just don't know. At present, the facilities required are very large, expensive, and complex. Radioactive materials are involved, although their half-lives are much shorter than those for fission materials, yielding smaller but still potentially important "not-in-my-back-yard" issues.

I believe that most people contemplating fusion energy subscribe to old wisdom. If our ship comes in, great. But rather than hanging around the dock, we're better off doing something useful while we're waiting.

This wisdom can be applied to other things as well. We might develop really cheap space-launch capabilities and build giant orbiting solar farms. We might even come up with something that no one has dreamed of yet. Many new technologies have surprised people in the past. But energy is so important to us, and the possible damages of climate change so large, that sitting here wishing is not going to keep the hard-nosed folks happy.

MID-COURSE CORRECTIONS?

Way back in chapter 1, I suggested that we might "prepare to come about," steering a new course to a more prosperous future. But what if we embark on a path to produce more energy sustainably while reducing CO_2 emissions, and then something totally unexpected comes up? Suppose cheap commercial fusion does arrive. Or the sun suddenly dims, and in some way tells us that it will stay dim for a long time, so we need the heat from CO_2 to avoid an ice age. Or, somehow, the science of global warming is shown to be wrong, and CO_2 doesn't warm things up. We know how unlikely these are, because the available data show that the sun hasn't changed much rapidly, we can't forecast it with confidence, and we know that strong multiple lines of evidence support climate science including global warming. But science is never 100 percent absolutely certain, so just suppose. What then?

The answer is that we can always make mid-course corrections. The actions to slow global warming will have slowed growth of the economy a little bit, perhaps unnecessarily.[35] The slowdown in the economy, compared to business as usual, is often noted as representing a few months of growth over a few decades—the time to reach a given level of consumption is delayed a few months—and levels of consumption for these cases are virtually indistinguishable on a graph.[36] Real costs are involved, but not end-of-the-world broke.

Furthermore, if sometime in the future we realize that burning fossil fuels is wise, the fossil fuels are still there. Certain conquerors may have salted the fields of the enemy to prevent future use,[37] but we are not "salting" the fossil fuels to make them unusable in the future; I haven't the vaguest notion how this could even be done. With the potential for carbon capture and storage, we are likely to keep using some fossil fuels for a while, keeping the technologies active. Overlap between some techniques for fossil-fuel extraction and geothermal applications can maintain the knowledge base even longer.

If everyone approaches the reduction of CO_2 emissions assuming that we are not serious about it, we won't succeed. But if we get serious, and then very strong reasons for backtracking emerge, we can backtrack, having lost a few months of economic growth.

In short, just as we are not locked in by decisions of our ancestors, our descendants will have the ability to modify our decisions. In the United States, we revere the Constitution, and we are very careful about amending it, but we do on occasion make corrections. Virtually every U.S. citizen, as well as most of the rest of the world, recognizes that a brilliant and forward-looking document contained some very bad ideas about slavery. Fixing the bad ideas did not require throwing everything away; it required a few amendments.[38]

24

Ten Billion and Smiling

Synopsis. Solid science shows that our fossil-fuel burning will warm the climate and affect us in many ways. Fortunately, we have a wealth of options to power ten billion sustainably smiling people, and we know that we can reach that goal.

None of us is as smart as all of us. —Japanese proverb[1]

LEARNING WHAT WORKS

am a scientist, supported in part by public funds, and I hope you have figured out by now that I love my job and am thankful for the opportunity to do it. Science is done by real, imperfect humans, yet it produces something bigger than all of us who do it or pay for it.

Being a scientist means learning the most successful ideas of humanity, respecting those ideas but working to improve them by testing new and old ideas against nature and keeping the best ones, following procedures that make it as hard as possible to fool ourselves and as easy as possible for others to see if we are fooling ourselves anyway. By supporting science, tightly coupled to its application in medical practice, engineering, and agriculture, society really does get cures for diseases, more food, computers that calculate, and planes that fly.

You may have a smart phone in your pocket now that can access much of the knowledge of humanity, tell you where you are, identify the bird singing outside your window, show you whether a storm will keep you away from your doctor's appointment, and even make a phone call.

All of that comes from a mere handful of carefully configured chemicals, a little bit of power from batteries . . . and a world of scientific and engineering discovery and support.

Those who study Earth and its environment have contributed to this success, helping find the key chemicals for the phone, the energy to power its battery, and the knowledge of Earth's shape that is used in the global positioning system (GPS). Environmental scientists produce the weather forecast, and guidance on promoting the health of the bird's ecosystem. You and the phone rely on Earth, and you have been paying scientists to assemble an "Operators' Manual" for the planet, to help keep it and you running well.

Despite frequent jokes, weather forecasting is a useful, successful part of science. The outcome of wars has hinged on the skill of forecasters,[2] so military authorities support weather research and include forecasters in their forces. So do many industries, paying good money for accurate predictions.

If you pay attention to weather long enough, you are studying climate, an increasingly useful effort. For example, La Niñas and El Niños shift patterns of drought and floods, hurricanes and snowfall. Forecasts of El Niños, while imperfect, have enough skill to guide water management, crop planting, and planning for the next fire season, saving money and lives.[3]

The governments of the world have invested literally billions of dollars to study weather and climate—with satellites often costing one-third of a billion dollars apiece, the billions add up quickly. The money spent on other aspects of climate research is more modest, but modeling centers, student traineeships, and ocean-sediment coring still cost real money. The knowledge gained from the investment of those billions fills in important parts of the Operators' Manual. And one of the most important parts is the high scientific confidence that our energy system endangers our future.

OUR SEARCH FOR ENERGY

We may be human because we figured out how to use more energy than the 100 W that burns within us from the food we eat. Certainly, we hadn't been human for very long before we learned to benefit from extra outside energy, with some of the wealthier of us now using more than 10,000 W apiece.

The switch from a hunter-gatherer lifestyle to agriculture gave us more food, but we have remained hunter-gatherers of external energy, collecting and burning wood, whales, or whatever we could get. We have always burned more of these than nature could supply. Even with Ben Franklin's amazing stove, trees didn't grow fast enough to fuel his fellow citizens of Pennsylvania. Burning whales for light was wiping them out when there were far fewer people than now, and when many of those people were illuminating with other, cheaper sources.

Earth spent a few hundred million years storing fossil fuels that hold much more energy than the available trees or whales. As we mastered hunting and gathering of coal, oil, and natural gas, we used more energy than ever while allowing trees and whales to grow back.

Good things almost always produce unintended consequences, and fossil-fuel use has been no exception. Sherlock Holmes sought criminals through the "opalescent London reek"[4] or the "greasy, heavy brown swirl . . . condensing in oily drops upon the window-panes,"[5] a result especially of coal-burning in the city. The witches in Shakespeare's *Macbeth* chanting, "Fair is foul, and foul is fair: Hover through the fog and filthy air" may have been referencing the coal-fired pollution of London as well as the coming tragedy.[6] The burning of soft-bituminous "sea coal" grew rapidly in London as wood disappeared near the city, beginning probably in the 1100s. The smoke from this sea coal was so bad that King Edward I banned its burning in 1306. This didn't work—in the chill of a London winter people wanted warmth, and with wood and cleaner-burning anthracite coal too expensive, the people found something to burn.[7] Later, the huge increase in burning that came with the

Industrial Revolution generated worse coal-smoke fogs that sometimes remained for months.[8]

Nearly a millennium was required to solve the smog/fog problem in London, which finally cleared in the second half of the twentieth century. The history of our burning is even written in the layers of Greenland's ice sheet, which show levels of acid-rain sulfate and black carbon rising greatly during the Industrial Revolution, and first black carbon and then acid rain largely being cleaned up in the North Atlantic basin in the twentieth century.[9] Clearly, we can pollute, and we can clean up after ourselves.

Cleaning up the worst of the immediate problems from fossil fuels has allowed us to burn more with less damage. We still suffer the damages from acid rain, soot, mercury, uranium, and other by-products of the burning, and from coal-mine cave-ins, oil-well blowouts, acid mine drainage, and mountaintop removal that accompany our hunting and gathering, but we as a society have decided that the benefits outweigh the costs.

The material released by fossil-fuel burning is grossly dominated by CO_2. At first, CO_2 may not seem like an especially alarming by-product—it is essential for life, after all, and is around us all the time. Yet, as the early physician Paracelsus stated in the sixteenth century, "All things are poison and nothing is without poison, only the dose permits something not to be poisonous."[10] This is often simplified to "the dose makes the poison." If you were to chat with someone who barely escaped drowning and then with someone who has faced the desert without water, you would surely learn that too much as well as too little of something can be bad.

WARMING TO THE UNINTENDED CONSEQUENCES

More than a century ago, the great chemist Svante Arrhenius calculated that human burning of fossil fuels could warm the world through the enhanced greenhouse effect of the CO_2 released.[11] Scientific progress on Arrhenius's hypothesis was slow[12] until U.S. Air Force research just after World War II greatly clarified the physics of radiation in the atmosphere. As this improved understanding made its way into atmospheric science, the realization of the warming influence of CO_2 became almost unavoidable. The 1979 report of the U.S. National Academy of Sciences really nailed this.[13] The Academy committee found that warming would increase water vapor in the atmosphere, amplifying the warming effect of the CO_2. Despite its best efforts, the committee could not find anything that appeared capable of undoing this. The committee estimated that doubling the atmospheric concentration of CO_2 would cause total warming of $5.4 \pm 2.7°F$ ($3 \pm 1.5°C$). This number is still often quoted, and is not far from our best estimate, although we now understand that in expressing the uncertainties, the warming is more likely to be on the high side of the central estimate than on the low side.

The U.S. National Academy of Sciences has returned to this topic dozens of times since, assessing the likely climate changes, what they might mean for economies and ecosystems, and what might be done. Through these studies, the understanding has simply strengthened without changing—the fundamental physics show that adding CO_2 to the atmosphere causes warming, this is enhanced by water vapor and other feedbacks, and the resulting amplified climate change will affect us.

The United Nations developed its own, extended assessments through the IPCC, with the first report released in 1990. The story is the same, and scientific confidence has increased as study after study strengthened rather than overturned our basic understanding. My email in-box has seen a steady stream of announcements over the years on the latest study, or talk, or revelation that supposedly will overturn "global-warming theory," but the warming effect of CO_2 has been untouched. Most commonly, the correspondent is promoting a paper that addresses some interesting aspect of the climate without hitting the fundamentals. In a few cases, the paper really did present a hypothesis that would change our understanding, but the predictions of that hypothesis failed in several subsequent studies. Some good news has come in—our fears have decreased that global warming will trigger a North Atlantic "shutdown" that might chill Europe's winters and dry out Asian monsoons, for example. But the fundamentals have remained solid: CO_2 warms, we are raising CO_2, and this will influence us.

And while alternative hypotheses have failed, several potential problems with this understanding have disappeared. For example, we used to worry about the "fact" that satellite-derived temperature trends disagreed with data from surface thermometers and results from climate models. This apparent problem with our understanding of climate was very widely discussed in the highest halls of government, in scientific meetings, and in public. Eventually, as described in chapter 12, errors in the satellite data were detected by several research groups, including the scientists who originally thought they had discovered the anomaly. Surface data, satellite data, and model results now agree within their uncertainties. Some apparent mismatches between the history of temperature and the history of CO_2 in the paleoclimatic record also have disappeared with further research.[14] Furthermore, climate science has been running long enough that we now know that early climate-change projections have proved to be accurate, as described in chapter 8—this is not just an explanatory science, but a successfully predictive one.

Based on our best understanding, we expect future climate changes from our business-as-usual CO_2 emissions to become much larger than the changes to date. The leading projections include: wet areas will tend to get wetter, dry areas drier, and the subtropical dry areas will expand poleward; ice will melt in many places, and sea level will rise; extreme cold will diminish, and extreme warmth will expand; both drought and floods will increase; the strongest storms may get stronger; and one of the important limits on the expansion of many diseases will be removed in some areas.

Learning what this means for living things has been more difficult, but our knowledge is advancing rapidly. The questions are harder. To estimate change in global mean surface temperature, a worldwide community of scientists addresses the big-picture behavior of the climate. To know what that means for agricultural productivity, scientists must learn the global picture first, and then the implications for averages and extremes of water and temperature in individual regions. We would really like to have some idea of how well humans will figure out solutions, to add to estimates of how much the extra CO_2 will fertilize the crops, and more. With lots of questions about diseases and agriculture, transportation and biodiversity in many places, the community of scholars is spread rather thinly. A huge amount of progress has been made, but our understanding of climate-change impacts is evolving rapidly, with much left to do.

Nonetheless, the big results are increasingly firm and reliable. Rising CO_2 fertilizes land plants, but in natural ecosystems something else tends to become scarce and limit the extra growth. Giving crops plenty of other fertilizers lets them use the extra CO_2, but heat and water stresses under business as usual are projected to become large enough to outweigh the fertilization and reduce our ability to feed ourselves. CO_2 also increases the acidity of the ocean, probably with large impacts on life there. Rising temperatures and other changes will force shifts in ecosystems worldwide, favoring weeds that can migrate rapidly across human-dominated landscapes to track favorable climatic zones, while endangering specialists, those that have difficulty migrating rapidly, and those happiest in cold conditions.

Taken together, initial warming will be bad for poor people in hot places, with mixed and not especially expensive impacts at higher latitudes, and overall impacts that are probably economically negative but might even be positive in colder climes. With increasing warming, the costs are projected to rise and spread until the effect is negative for most people in most places. Damages of a few percent of the world economy per year are expected before the end of this century, amounting to trillions of dollars per year. "Inertia" in the climate system, mostly from the slow warming of the oceans but also from the slow melting of ice sheets, means that we will have committed to many damages before we experience them—waiting until the damages are evident to everyone will mean that even larger damages will occur.

Military planners in the United States and elsewhere are already gearing up for such challenges as environmental refugees, regime destabilization, and opening of the Arctic to greater ship traffic. Economic analyses indicate that for an optimal economy, some investment should be made now to head off warming. However, these economic analyses usually indicate that more investment should be devoted to other uses, leaving our wealthy descendants to deal with most of the damages from warming. These economic analyses often don't include some damages, including the value of species that become extinct, the lop-sided uncertainties with more chance of greater warming, and the slight chance of really catastrophic events, all of which point toward doing more now to limit warming.

FUELING THE FUTURE

Fortunately, we are awash in solutions. Putting a wind farm on the windiest 20 percent of the plains and deserts of the world would supply more than the current worldwide human energy use, and the turbines themselves could be constructed over thirty years for a worldwide cost less than current U.S. expenditures on energy. The wind captures just 1 percent of the sun's energy, so mastery of solar power would vastly increase the possible supply. Nuclear fission is known to work on a large scale, capturing CO_2 from fossil fuels and putting it back in the ground appears workable, and biomass, tides, waves, currents, hydropower, and geothermal offer additional options. Geoengineering to block some of the sun's light and cool us a little remains possible, if untested. Conservation is better than free for the first savings, and can save a lot more at low cost. Wild cards such as nuclear fusion might possibly come up in our favor.

Many organizations and people have looked at the options, and while there is widespread agreement that we can solve the problems, there is much disagreement on the preferred path. Conservation may be a "hard sell" to people who hate being told what to do. Nuclear energy makes some people very nervous because of radioactive waste, terrorism, and nuclear proliferation. Leakage of stored CO_2, and competition between food and fuel from biomass, are also worrisome. The intermittency and distributed nature of sun and wind raise engineering problems if these become dominant in the energy mix, while the other choices face engineering issues or aren't quite abundant enough to solve the problems themselves.

In the long term—many decades or more in the future—most energy experts seem to be contrasting a successful sun-wind economy with a disastrous exhaustion of fossil fuels in a hard-to-handle greenhouse world. Many plans foresee wiring the sun-wind economy through a smart electrical grid, with an energy-storage system that might include batteries of plug-in hybrid cars, or hydrogen, or some other liquid fuel perhaps made from recaptured CO_2, or other options—the "winners" among the many possibilities are not yet clear.

The smart, distributed architecture of the Internet has revolutionized our world. It is open to everyone, of all "persuasions" (PC and Mac and Unix box, desktop and laptop and smart phone). Everyone can be a user, or a contributor, or both, on a robust system adaptable to local conditions. A smart grid could do the same for power.

In some places it is already possible for every homeowner or farmer to become not only a user but also a supplier of energy, if and only if they want to. Put some solar cells on your roof, and when the sun shines while you are off at the office, you can be selling power to the electric company to run offices. Come home at night, and you buy power back. You could, if you wished, get some batteries and go off grid, but on the grid you have the ability to make money, so conservation isn't just a cash savings, it generates income. Not sunny where you live? Try a windmill. If your coastal community doesn't like wind turbines blocking the view, but there is a vigorous current offshore, you can make electricity in the flow, or use the waves crashing on your breakwater, or the tides rushing into the harbor, to make electricity. You could even buy batteries that you can connect to the grid, sell the storage to the power company, and use those batteries to run your car too.

On the way to that vision, I believe most people see at least some role for nuclear power, and at least some people envision a major role for carbon capture and sequestration. A few people are very happy to discuss geoengineering, but many questions have been raised that are not close to being answered. Importantly, the available science points to an overall expansion of good jobs if the energy system moves away from reliance on fossil fuels, even though some good fossil fuel–linked jobs undoubtedly will be lost in the transition.

When we look ahead to a sustainable system providing enough energy for ten billion people to live in comfort without changing the climate, it is very, very easy to add up all the difficulties and costs and to conclude that the task is impossible. But the task we face in many ways appears smaller than the ones we have accomplished before, including our success in public sanitation. We know that the world has more than enough energy for us, and we have engineering-ready ways to generate that energy for a cost estimated as approximately 1 percent of the world economy per year.

If we wait until things get bad and then try to make the transition rapidly, a lot of economic disruption is likely. If we start to turn early enough, though, things will be much easier—rather than suddenly closing the coal plant and firing the miners, sustainable replacements can be phased in as the coal plant and the miners reach retirement age.

RECOMMENDATIONS

If I were to stop right here, I would be happier, but my high school writing teachers[15] would probably not be happy. This is the point where, in a proper narrative, I would reveal the grand conclusion, point the path to the future, and lead the grand and glorious charge to a brighter tomorrow.

Fat chance.

My impression is that when we scientists try to recommend policy, we have no more insight than anyone else. Asking me what to do runs the risk that someone might believe me because I have expertise in limited, somewhat-related areas. I am a geologist-turned-glaciologist-and-climatologist who took a lot of classes about Earth and enough physics and chemistry and biology and metallurgical engineering to support my interest in the planet, and who has roamed the planet reading the history of climate written in ice and trying to understand the ice well enough to learn what it does to landscapes and sea levels. Having me choose between cap-and-trade or taxes or business as usual, or nuclear or wind or sun or sequestration or geoengineering, is akin to having a U.S. senator use a universal stage to measure the c-axis fabric of a polar ice core—you are better off getting a real expert than you are pretending that someone else can do the job.

Yet, I do have some insights that may be valuable, and I am sensitive enough to the instructions of my high school teachers that I feel I owe you a little more insight on a few of the issues. If someone put a gun to my head and forced me to make recommendations, they would grow out of the following points.

This is science, not revealed truth, but the science is solid. We scientists are simply not socially skilled enough or politically homogeneous enough to organize ourselves to deliberately hoax the world.[16] We might have a little me-too-ism at times, and we do watch the sources of our funding, but these are simply not strong enough for us to mistake black for white, or even red for blue. The basis of global-warming science is unavoidable physics. And once science gives a consistent, long-term, assessed answer, the wisest path is to start from there in deciding how to deal with the world, considering the uncertainties but not throwing away the science because of them. This does not tell you what to do, but any other starting worldview requires discarding hard-won knowledge that is highly likely to be right and useful. Unless you are confident that the U.S. Air Force doesn't know what kind of sensor to put on a heat-seeking missile, you are better off starting from the idea that our CO_2 interacts with radiation, and from there global warming is highly, highly likely to be real. This is not a political statement, no matter what some commentators on the fringe may be telling you; it is physics.[17]

Delaying is not free. If we could keep greenhouse gases and particles at modern levels, the world would continue to warm and the oceans would continue to rise for a while. We have unintentionally geoengineered already, mostly by putting up sun-blocking particles, but these have large health effects and may be cleaned up more in the future, allowing more warming because the CO_2 stays up while the particles fall down. Furthermore, the longer we delay, the less fossil fuel we have in reserve in case the alternatives don't work as well as we hope. And the economically optimal path started to reduce CO_2 emissions a while ago—we have already committed ourselves to a less-than-optimal path by delaying, and the penalty rises with time. Costly delay does not necessarily require immediate action—sometimes your travel plans are so uncertain that you wait to make an airline reservation even though the price is likely to rise—but the best science says that the longer we wait, the harder and more costly the response.

Fossil fuels will run out. We could drill and mine for a while longer, but eventually alternatives are required. To supply ten billion people, or even nine billion, with the good things enjoyed by most of the developed world now, will require *lots* of energy. Trends in the supply and cost of fossil fuels do not provide much encouragement that more drilling and mining for fossil fuels will supply that energy for everyone.

We need alternatives to fossil fuels, and lots of them. My economics professors told us, "There are no needs, only wants." But to reach a stable society, some things look more like needs than wants. The world's population is projected to rise to between nine and ten billion people by the middle of the century, then approximately stabilize.[18] However, there are many assumptions built into the calculations of this uncertain number. Oversimplifying a lot, it appears to me that this projected stabilization of the population requires that the great majority of people must have 1) a reasonable expectation that the children they do have will grow up healthy, and 2) interesting choices about what to do with their lives, which may or may not include having children. These in turn will be very, very difficult to achieve unless people have food, water, medical care, and energy resources. The history of humanity includes a lot of collapsing empires—we have the capability of outgrowing our

Figure 24.1 Ten billion smiling people? We know we can do it.

resources and then facing disaster. We also have the capability of meeting the challenges and succeeding. Powering people sustainably will not come close to solving all of the potential problems facing us, but I believe that we are highly likely to fail in solving those problems unless we provide enough clean, sustainable power.

We haven't been trying very hard. After a brief boom in response to the oil embargo and price shocks of the 1970s, funding for energy research was halved and held low for two decades, with almost two-thirds of that research devoted to fossil fuels, fission, and fusion. Despite recent increases, the money spent on energy research remains a very small fraction of public budgets, or even public research budgets. Yet our ability to generate affordable energy from wind, sun, and other renewables has been improving rapidly. Past performance is not a guarantee of future success, but people who like to bet on past winners are taking note.

Betting on the future can pay off. When my wife and I left our summer of oil-company employment in Dallas, several of our new friends suggested that we could return to the oil company for a bright future. We liked the people, respected their skills and dedication, and enjoyed our jobs. Our decision to pursue an academic career was not controlled by the politics of potential employers in the oil patch. Satisfying people's demand for oil is important, but we decided that we really wanted to help satisfy the curiosity of students. Almost three decades later, we are happier than ever with our decision.

Suppose society were to say to our students, "Your country, your

Figure 24.2 When faced with danger, Greenland's musk oxen often respond initially by running away.

Figure 24.3 After running, musk oxen usually turn as a group to vanquish their foe.

world, wants the ability to sustainably generate enough energy to keep ten billion people happy and healthy. We'll help you with your education and your start-up companies, if you commit to helping us solve the problems." I see a lot of students who would love to have a bigger purpose. They don't want to impress me to get a better grade; they want to learn to make a difference beyond the ivied halls. Many of them are active members of religious or service organizations, running charities or volunteering their time.[19] But we fail to engage many of them, who resort to playing video games or drinking instead. If a little more of that energy could be channeled into providing our energy, we would have a very high chance of succeeding.

POSTSCRIPT

I was born in 1957. The odds are good that I won't live long enough to see ten billion smiling people on the planet—we have a lot of work to do to get there, and with perhaps one billion people hungry already, the task will not be easy. But the odds are quite good that our daughters will live long enough to know whether humanity is really on track to a sustainable future for everyone.

Of the many challenges we face, I believe that the biggest is to get along with each other. Breaking things is so much easier than building them, and we have so much to build that we can't get where we want to go unless we go together.

Figure 24.4 The glowing lava of a Kilauea, Hawaii, night shines over the shoulder of Karen (left) and Janet Alley. Solid science shows clearly that delay makes the problem of global warming bigger and harder to solve. Karen and Janet's future depends on our decisions.

Of the physical challenges facing us, though, I believe that developing a sustainable energy supply is the biggest.[20] The physical challenges of supplying enough water, preserving biodiversity, maintaining productivity of soils, and much more are immense, and not to be taken lightly. But energy use gives us the ability to meet new challenges in these and other areas as they arrive. Our successes so far have rested very heavily on energy use, but our current system is completely unsustainable. Everything we know says that we can solve this—if we really try.

Looking back across the millions of years of Earth's history, and marveling at the beauty and power of the dynamic, planet-wide, sustainable resources waiting for our use, I am optimistic. The Operators' Manual tells us how to power ten billion sustainably smiling people—if we use it.

NOTES

For the information in this book, I have relied primarily on the refereed scientific literature and the scientific assessments of that literature by the appropriate major bodies such as the U.S. National Academy of Sciences. I also used certain highly reliable non-refereed sources, such as governmental statistical compilations. I tried to supply enough references so that you can check on me if you wish. The refereed literature often is not freely available to everyone online, although you usually can arrange to read all of it at a major library, especially at a research university. To make access and reading easier, I also have included pointers to some materials from freely available web-based sources written for popular consumption, such as educational resources from museums or government agencies. However, experience shows that web addresses change quickly—for reliable sources, the museum and the referenced materials usually are still there, but the web site has been redesigned and the address changed. If you attempt to use one of the web addresses given here and it doesn't work, you might try going to the web site of the organization and then searching for the specific information—I find that this usually succeeds.

1: Prepare to Come About

1. Robert F. Kennedy used a variant of this quotation in his June 6, 1966, "Day of Affirmation" speech to the National Union of South African Students at the University of Cape Town: "There is a Chinese curse which says 'May he live in interesting times.' Like it or not, we live in interesting times." The ultimate origin of the saying is obscure, and whether it really is a Chinese curse is disputed; see, for example, http://www.bbc.co.uk/dna/h2g2/A807374 (accessed July 1, 2010).
2. I know of no official assessed estimate of the carrying capacity of the world

for hunter-gatherers, and an authoritative treatment might produce an estimate that differs a lot from mine. However, much scientific research points to the hardships of becoming agriculturalists (see, e.g., Diamond, J., 1987, "The worst mistake in the history of the human race," *Discover* (May): 64–66, and Starling, A. and J. T. Stock, 2007, "Dental indicators of health and stress in early Egyptian and Nubian agriculturalists: A difficult transition and gradual recovery," *American Journal of Physical Anthropology* 134: 28–38). People all over the world made this transition from hunter-gatherer to an apparently less healthy lifestyle, when the world population was one million or slightly more (e.g., Historical Estimates of World Population, http://www.census.gov/ipc/www/worldhis.html [accessed Nov. 16, 2009]). If agriculture was so difficult, but so many people adopted it, a reasonable guess is that the people had a pretty good reason, and feeding themselves is a good reason. Pending better science, this may be a useful estimate.

3. Myers, R. A. and B. Worm, 2003, "Rapid worldwide depletion of predatory fish communities," *Nature* 423: 280–83.

4. Foley, J. A., C. Monfreda, N. Ramankutty and D. Zaks, 2007, "Our share of the planetary pie," *Proceedings of the National Academy of Sciences of the United States of America* 104: 12,585–86.

5. Postel, S. L., G. C. Daily and P. R. Ehrlich, 1996, "Human appropriation of renewable fresh water," *Science* 271: 785–88.

6. Montgomery, D. R., 2007, *Dirt: The Erosion of Civilizations* (University of California Press, Berkeley).

7. For a summary of that history, see Diamond, J., 2005, *Collapse: How Societies Choose to Fail or Succeed* (Viking, New York).

8. The United Nations identified fossil-fuel use, and agriculture and food consumption, as being especially unsustainable aspects of our lives, but a lot of the agriculture is linked to fossil fuels; Hertwich, E., E. van der Voet, S. Suh, et al., UNEP (United Nations Environment Programme), 2010, *Assessing the Environmental Impacts of Consumption and Production: Priority Products and Materials*, a report of the Working Group on the Environmental Impacts of Products and Materials to the International Panel for Sustainable Resource Management.

9. U.S. Department of Energy, Energy Information Administration, "International Energy Annual 2006," http://www.eia.doe.gov/iea/overview.html (accessed Sept. 21, 2009). Primary energy is the bought-and-sold energy use.

10. Nixon quote from http://www.dtp.nci.nih.gov/timeline/noflash/milestones/M4_Nixon.htm (accessed Sept. 21, 2009).

11. British Wreck Commissioner's Court, May, 1912, *On a Formal Investigation Ordered by the Board of Trade into the Loss of the S.S. Titanic, Day 3 Testimony of Robert Hichens*. A transcription of the Wreck Commissioner's Court report is online at http://www.titanicinquiry.org/BOTInq/BOTInq03Hichens01.php.

12. Source unknown. Many versions of this saying exist.

13. Hooke, R. L., 2000, "On the history of humans as geomorphic agents," *Geology* 28: 843–46.

14. Discussed and referenced in my 2000 book, *The Two-Mile Time Machine: Ice Cores, Abrupt Climate Change, and Our Future* (Princeton University Press, Princeton, NJ). Also see Krachler, M., J. C. Zheng, D. Fisher and W. Shotyk, 2009, "Global atmospheric As and Bi contamination preserved in 3000 year old Arctic ice," *Global Biogeochemical Cycles* 23: GB2011; and McConnell, J. R. and R. Edwards, 2008, "Coal burning leaves toxic heavy metal legacy in the Arctic," *Proceedings of the National Academy of Sciences of the United States of America* 105: 12,140–44.

2: Burning to Learn

1. "For the mind requires not like an earthen vessel to be filled up; convenient fuel and aliment only will inflame it with a desire of knowledge and ardent love of truth. Now, as it would be with a man who, going to his neighbor's to borrow fire and finding there a great and bright fire, should sit down to warm himself and forget to go home; so is it with the one who comes to another to learn, if he does not think himself obliged to kindle his own fire within and inflame his own mind, but continues sitting by his master as if he were enchanted, delighted by hearing." *Plutarch's Morals*, translated from the Greek by several hands, corrected and revised by William W. Goodwin, with an introduction by Ralph Waldo Emerson, vol. 1, 1874 (Little, Brown, Boston), p. 463.

2. See Aiello, L. C. and J. C. K. Wells, 2002, "Energetics and the evolution of the genus Homo," *Annual Review of Anthropology* 31: 323–38. Also see Gibbons, A., 2007, "News focus: Food for thought," *Science* 316: 1558–60. These numbers are updated and expanded by Carmody, R. N. and R. W. Wrangham, 2009, "The energetic significance of cooking," *Journal of Human Evolution* 57: 379–91.

3. National Research Council Committee on the Earth System Context for Hominin Evolution, 2010, *Understanding Climate's Influence on Human Evolution* (National Academies Press, Washington, DC). Also see the references in note 2.

4. See note 2.

5. See note 2.

6. Pimentel, D., 1974, "Energy use in world food production," *Environmental Biology Report 74-1*, Cornell University. Also Pimentel, D. and M. Pimentel, 2007, *Food, Energy and Society*, 3rd ed. (CRC Press, Boca Raton, FL), pp. 253–54.

7. Using the numbers from Pimentel and Pimentel, 2007, pp. 253–54, in the previous note, the break-even point would be about 17% efficiency.

8. This example is from the outstanding book by Robert L. Park, 2000, *Voodoo Science: The Road from Foolishness to Fraud* (Oxford University Press, Oxford, UK).

9. You may remember measuring calories in chemistry class, as the energy to warm 1 gram of water by 1°C. You then may have converted that number to joules,

with just over 4 joules/calorie. If so, 1000 of the calories you measured equals 1 of the calories the FDA is discussing; the FDA and others say that 1 calorie is the energy needed to warm 1 kg of water by 1°C. Someone got tired of measuring foods in kilocalories and decided that 1 kilocalorie should be called one calorie; this 1000-calorie calorie is sometimes but not always capitalized as Calorie.

10. Smith, C., 1998, *The Science of Energy: A Cultural History of Energy Physics in Victorian Britain* (Heinemann, London); also, Smith, C., 2004, "Joule, James Prescott (1818–1889)," in *Oxford Dictionary of National Biography* (Oxford University Press, Oxford, UK).

11. This is part of the inscription, by Lord Brougham, on a large white statue of James Watt in St. Paul's Chapel, Westminster Abbey; see http://www.westminster-abbey.org/our-history/people/james-watt (accessed Sept. 9, 2009).

12. The equivalent of 2000 calories/day is about 97 watts (W). But many Americans eat more than 2000 calories/day, especially if they are serious athletes or planning to be fatter in the future. In round numbers, if you leave a 100 W lightbulb running for a day, it will have used up about 2000 calories in that day, or just over 8 million joules. In the United States, the electric company that sells you the power converts to still another set of units. If you use 1000 W for one hour, you have used 1000 watt-hours, or 1 kilowatt-hour, or 1 kWh. Thus, a 100 W lightbulb running for 10 hours, or ten 100 W lightbulbs running for 1 hour, will use 1 kWh. If you leave a 100 W lightbulb turned on for 24 hours, you and the bulb will each have used 2.4 kWh.

13. For Lance Armstrong's energy output of almost 500 W that actually went into moving the bicycle up a mountain stage, see Kolata, G., "Super, sure, but not more than human," *New York Times*, July 24, 2005, calculated using a racing weight for Mr. Armstrong of 74 kg or 162 pounds. In the 2005 race, Floyd Landis, former teammate of Mr. Armstrong's, rode with a device that measured his power output through the bicycle—the heat his body put into the air around him was not measured. For the entire race, Mr. Landis averaged 232 W, but he maintained an average of 379 W over 75 minutes in the final time trial, in which he finished sixth (McClusky, M., "Powering through the tour," *Wired*, July 27, 2005, http://www.wired.com/science/discoveries/news/2005/07/68310, accessed Oct. 19, 2009). Neither Mr. Armstrong nor Mr. Landis is probably the best comparison to use for typical humans, so I will stick with 100 W in this book.

14. The more correct value is 11,600 W/person, but I will round off for convenience here.

15. For energy use versus gross domestic product (GDP) by country, see Key World Energy Statistics 2009, International Energy Agency, http://www.iea.org/Textbase/publications/free_new_Desc.asp?PUBS_ID=1199 (accessed Sept. 10, 2009). A very nice figure at http://commons.wikimedia.org/wiki/File:Energy_consumption_versus_GDP.png (accessed Sept. 22, 2010) shows

energy consumption per capita versus GDP per capita, and was made by Frank van Mierlo. For the United States, see Energy Intensity, 1973–2008, Energy Information Administration of the U.S. Government Department of Energy, http://www.eia.doe.gov/emeu/mer/pdf/pages/sec1_16.pdf (accessed Sept. 10, 2009). Total energy consumption for 1973–2008 increased from approximately 75 to 100 quadrillion BTUs, while the GDP, in inflation-corrected 2005 dollars, increased from approximately $5 to $13 trillion. Thus, the energy use per dollar of GDP was cut by just over half, while energy use still went up by one-third.

16. See note 15.

3: Peak Trees and Peak Whale Oil

1. Melville, H., 1851, *Moby-Dick*, chapter 25 postscript.

2. *The Journal of the Pilgrims at Plymouth, in New England, in 1620.* Reprinted from the original volume, compiled by George B. Cheever (John Wiley, New York, 1848), p. 29.

3. An acre of wood per year to heat a house; see Rubertone, P. E., 1985, "Ecological Transformations," in Part II: Changes in the Coastal Wilderness: Historical Land Use Patterns on Outer Cape Cod, 17th–19th Centuries, in McManamon, F. P. (ed.), *Chapters in the Archaeology of Cape Cod, III: The Historic Period and Historic Period Archaeology*, Cultural Resources Management Study Number 13 (Division of Cultural Resources, North Atlantic Regional Office, National Park Service, U.S. Department of the Interior, Washington, DC), p. 75.

4. See p. 78 in ibid.

5. Eighty years before my father-in-law was climbing Doane Rock, Henry David Thoreau walked the Cape, in the mid-1800s. In his wonderful 1865 book *Cape Cod*, he noted that the formerly forested Cape had been reduced to barren heaths and poverty grass, with houses built of lumber imported from Maine, firewood imported or gathered as driftwood, and a region of trees only 4–5 feet tall referred to as "woods" (his chapter 7). In his chapter 2, Thoreau described a broad region of the town of Dennis in which there were almost no trees except for a square of Lombardy poplars around the meeting-house, all dead.

6. Woods Hole Research Center, Land Cover and Population Changes on Cape Cod, based on 1) Cape Cod Commission, 1997, http://www.capecodcommis sion.org/data/capetrends.htm; and 2) Wilkie, Richard W. and Jack Tager, eds., 1991, *Historical Atlas of Massachusetts* (University of Massachusetts Press, Amherst, MA); U.S. Census, http://www.whrc.org/mapping/capecod/population.html (accessed Sept. 22, 2010).

7. Much of the information here is from the excellent book Eggert, G. G., 1999, *Making Iron on the Bald Eagle: Roland Curtin's Ironworks and Workers' Community* (Pennsylvania State University Press, University Park, PA). That story is focused

on the single furnace of the Eagle Iron Works, but many others were roughly comparable. Also see http://www.nps.gov/cato/historyculture/charcoal.htm from the Catoctin Mountain Park of Maryland, part of the U.S. National Park Service, which includes some nice pictures (accessed Sept. 22, 2010).

8. See note 3.

9. *An Account of the New-Invented Pennsylvanian Fire-places*, 1744, printed and sold by B. Franklin; *The Writings of Benjamin Franklin*, vol. 2, 1722–1750, collected and edited with a life and introduction by Albert Henry Smyth (Macmillan, New York, 1905), pp. 271–72.

10. Bartram, John, 1751, *Observations on the Inhabitants, Climate, Soil, Rivers, Productions, Animals and Other Matters Worthy of Notice. Made by Mr. John Bartram, in His Travels from Pensilvania to Onondago, Oswego, and the Lake Ontario, in Canada* (London, 1751). Quoted in Moon, F. F. and H. C. Belyea, 1920, "Forestry for the private owner," *Bulletin of the New York State College of Forestry* no. 13, vol. 20, no. 2 (July): 11, footnote.

11. See, for example, the Pennsylvania State Parks site, "The Life of Dr. Joseph Trimbel Rothrock," http://www.dcnr.state.pa.us/stateparks/history/history rothrock.aspx (accessed Sept. 18, 2009).

12. For additional information on the charismatic macrofauna, try the Carnegie Museum of Natural History Mammals page, http://www.carnegiemnh.org/ mammals/collections/PAmamm/pamammals.html (accessed Sept. 18, 2009). Their mountain-lion page notes that sightings of mountain lions have persisted to the present day in Pennsylvania, but without verification. The last mountain lion killed in Pennsylvania, in 1967, was a released captive from a southern subspecies. The last really Pennsylvanian mountain lion known was killed in Susquehanna County in 1856. The "fisher" described in the text is a large weasel, noted for being able to dine on porcupine without impaling itself on the quills; see the Carnegie Museum Mammal page or the Pennsylvania Game Commission's Fisher page, http://www.pgc.state.pa.us/pgc/cwp/view.asp?a=498&q=163236 (accessed Sept. 18, 2009). For a bit more on the reforestation of Pennsylvania, see "A century of conservation: The story of Pennsylvania's state parks," *Pennsylvania Heritage Magazine* (Pennsylvania Historical and Museum Commission, Winter 1994), http://www.depweb.state.pa.us/heritage/cwp/view.asp?a=3&q=445063 (accessed Sept. 9, 2009). There can be plenty of discussion about the relative roles of overhunting and other human actions in the extinction of mountain lions and other species in Pennsylvania, but habitat loss was hugely important—if there isn't a tree to hide behind, it is hard to do it in the woods like a bear.

13. Williams, M., 2000, "Dark ages and dark areas: Global deforestation in the deep past," *Journal of Historical Geography* 26: 28–46, doi:10.1006/jhge.1999.0189, available online at http://www.idealibrary.com.

14. Doyle, A. C., "The Adventure of Black Peter," from *The Return of Sherlock*

Holmes, in *Strand Magazine* 27, no. 159 (London, March 1904). You can find this in, for example, Doyle, A. C., 1905, *The Return of Sherlock Holmes: Detective Stories* (W. R. Caldwell, New York), p. 169.

15. For a short and friendly introduction to pioneer lighting, try the Oshkosh Public Museum, of Oshkosh, Wisconsin, http://www.oshkoshmuseum.org/exhibits/cabin_life.htm (accessed Sept. 24, 2009).

16. These stories from the archives of the *New York Times* include: "Camphene lamp explosion," *New York Times*, Feb. 2, 1856, http://query.nytimes.com/gst/abstract.html?res=9901E2DD133DE034BC4A53DFB466838D649FDE; "Death from the bursting of a camphene lamp," *New York Times*, May 23, 1853, http://query.nytimes.com/gst/abstract.html?res=9D04E5DB1331E13BBC4B51DFB3668388649FDE; "Dreadful accident from camphene," *New York Times*, June 1, 1854, http://query.nytimes.com/gst/abstract.html?res=9A06E4D7103AE334BC4953DFB066838F649FDE (all accessed Sept. 24, 2009).

17. The tragic story of the skier and the corset is told in "Corset stay kills skier: Whalebone pierces woman's heart when she falls over a ledge," *New York Times*, Feb. 22, 1922, http://query.nytimes.com/gst/abstract.html?res=9805E7D61E30EE3ABC4A51DFB4668389639EDE (accessed Sept. 24, 2009).

18. See http://www.whalingmuseum.org/library/amwhale/am_index.html to learn about the wonderful New Bedford Whaling Museum and its library.

19. The recent treatment is by Bardi, U., 2007, "Energy prices and resource depletion: Lessons from the case of whaling in the nineteenth century," *Energy Sources, Part B: Economics, Planning, and Policy* 2: 297–304. The data were from Starbuck, A., 1878, *History of the American Whale Fishery* (Secaucus, NJ; repr. Castle Books, 1989).

20. See Dolin, E. J., 2007, *Leviathan: The History of Whaling in America* (W. W. Norton, New York); also see notes 18 and 19.

21. International Whaling Commission, "History and Purpose," http://www.iwcoffice.org/commission/iwcmain.htm (accessed Sept. 25, 2009).

22. www.nps.gov/nebe/historyculture/upload/timeline.pdf (accessed Sept. 25, 2009).

23. Bardi (see note 19) indicated production of roughly 450 million gallons—just over 10 million barrels—of oil from right and sperm whales, the main species hunted, during the main U.S. production time in the 1800s. A large supertanker can hold over 3 million barrels, and the U.S. Department of Energy's Energy Information Administration (http://www.eia.doe.gov/basics/quickoil.html [accessed Oct. 10, 2009]) indicates total U.S. imports of crude oil plus petroleum products of almost 13 million barrels/day. I have not found total production numbers for all whaling nations; the United States dominated in the 1800s, whereas Norway and some others became dominant more recently, and the U.S. dominance followed smaller operations by others including a lot of U.S. produc-

tion in the 1700s. But some back-of-the-envelope scaling arguments suggest that the total production of whale oil by humans over history is no more than a day or two of recent total human use of oils from the ground.

24. Several of the results cited here are from the major review paper by Abrams, M. D. and G. J. Nowacki, 2008, "Native Americans as active and passive promoters of mast and fruit trees in the eastern USA," *Holocene* 18: 1123–37. Abrams and Nowacki provide the William Bartram reference. Also see Black, B. A., C. M. Ruffner and M. D. Abrams, 2006, "Native American influences on the forest composition of the Allegheny Plateau, Northwest Pennsylvania," *Canadian Journal of Forest Research* 36: 1266–75. It is likely that the settlers plowed up the acorns because the native people died of Eurasian diseases before returning to reclaim the food caches.

25. For tropical America, see Nevle, R. J. and D. K. Bird, 2008, "Effects of syn-pandemic fire reduction and reforestation in the tropical Americas on atmospheric CO_2 during European conquest," *Palaeogeography, Palaeoclimatology, Palaeoecology* 264: 25–38. A nice treatment for Australia is Miller, G. H., M. L. Fogel, J. W. Magee, et al., 2005, "Ecosystem collapse in Pleistocene Australia and a human role in megafaunal extinction," *Science* 309: 287–90.

26. Swanton, J. R., 1929, *Smithsonian Institution Bureau of American Ethnology Bulletin* 88.

27. Ruddiman, W. F., 2005, *Plows, Plagues and Petroleum: How Humans Took Control of Climate* (Princeton University Press, Princeton, NJ); also see Ruddiman, W. F., Z. T. Guo, X. Zhou, H. B. Wu and Y. Y. Yu, 2008, "Early rice farming and anomalous methane trends," *Quaternary Science Reviews* 27: 1291–95. For discussion on some of the technical points, see Broecker, W. S. and T. F. Stocker, 2006, "The Holocene CO_2 rise: Anthropogenic or natural?" *Eos* 87: 27–29; and Ruddiman, W. F., 2006, "On 'The Holocene CO_2 rise: Anthropogenic or natural?'" *Eos* 87: 352–53. Also see Olofsson, J. and T. Hickler, 2008, "Effects of human land-use on the global carbon cycle during the last 6,000 years," *Vegetation History and Archaeobotany* 17: 605–15; and Kleinen, T., V. Brovkin, W. von Bloh, D. Archer and G. Munhoven, 2010, "Holocene carbon cycle dynamics," *Geophysical Research Letters* 37: L02705. Humans likely did influence the atmosphere, but perhaps not enough to actually head off an ice age, with orbits doing that for us, as I discuss in chapter 11.

28. Mischler, J. A., T. A. Sowers, R. B. Alley, et al., 2009, "Carbon and hydrogen isotopic composition of methane over the last 1000 years," *Global Biogeochemical Cycles* 23: GB4024. We cannot rule out a climatic contribution to this change, although reduced burning appears to have been the main cause.

29. Brown, K. S., C. W. Marean, A. I. R. Herries, et al., 2009, "Fire as an engineering tool of early modern humans," *Science* 325: 859–62.

4: Fossil Fuelish

1. L'Estrange, Roger, 1724, *Fables of Aesop and Other Eminent Mythologists*, 7th ed. (D. Brown, London), p. 48.

2. Among many possible sources, you can look for the energy use of televisions at the U.S. government site http://www.energystar.gov (accessed May 27, 2010); or see Ostendorp, P., S. Foster and C. Calwell, 2005, "Televisions: Active Mode Energy Use and Opportunities for Energy Savings," Natural Resources Defense Council issue paper, at http://www.nrdc.org/air/energy/energyeff/tv.pdf, under funding from the U.S. Environmental Protection Agency. Many TVs use about 100 W when operating, but some use close to 400 W. Recall that the rate at which energy is used is called "power," and watt is the proper unit for power, so an average human's average power is 100 W. For the output of elite athletes, see chapter 2, note 13.

3. There are lots of ways to calculate energy use. The Energy Information Agency of the U.S. Department of Energy gave World Primary Energy Use—the part that is sold to others—as averaging about 15.7 terawatts (TW) in 2006, or 15.7 $\times$ 10^{12} W, or 15,700,000,000,000 W, or enough to keep one heck of a lot of light-bulbs burning.

    Some experts report energy usage in "quads," where 1 quad is 1 quadrillion British Thermal Units (BTUs), or roughly 1.06 exajoule, or 1.06 $\times$ 10^{18} joule (J). Use of 1 quad/year is the same as use of 1 quad in 31.6 million seconds, or use of 1.06 $\times$ 10^{18}/31.6 $\times$ 10^6 = 3.35 $\times$ 10^{10} W. Then the 15.7 TW of primary energy in the world in 2006 is the same as 469 quad/year, which is the way that the Energy Information Agency reports usage.

    You will also see people reporting energy usage in kilowatt-hours (kWh), which is the amount of energy used inside ten people in 1 hour, or one person in 10 hours, or two people in 5 hours, or so on. Because a year has about 8766 hours, our global primary-energy use of 15.7 TW in a year, which is about 2400 W/person, or 2.4 kW/person, running continuously for a year, translates to 21,000 kWh/person/year. In this book, I will try to stick with the use per person.

    The Department of Energy web site at www.eia.doe.gov has a wealth of resources, with more good information at www.iea.org, the web site for the International Energy Agency. The estimates for the traditional biomass are from the International Energy Agency World Energy Outlook 2002, which reported 891 million tons of oil equivalent per person per year for traditional biomass use in the year 2000, with 1 ton of oil equivalent containing about 42 giga-joules. The International Energy Agency did not project rapid changes in this traditional-biomass use, so I have adopted that number here. International Energy Agency, 2002, "Energy and Poverty," in *World Energy Outlook 2002*, www.iea.org/textbase/nppdf/free/2000/weo2002.pdf (accessed Nov. 16, 2009). Chapter 13 is the most relevant.

4.   U.S. Department of Energy, Energy Information Administration, "International Energy Annual 2006," http://www.eia.doe.gov/iea/overview.html (accessed Sept. 21, 2009).

5.   U.S. Department of Energy, Energy Information Administration, http://www .eia.doe.gov/iea/wecbtu.html (accessed Nov. 17, 2009).

6.   Ibid.

7.   Davis, S. J. and K. Caldeira, 2010, "Consumption-based accounting of CO_2 emissions," *Proceedings of the National Academy of Sciences of the United States of America* 107: 5687–92.

8.   Key numbers are available from the Energy Information Administration of the U.S. Department of Energy, among other sources. The United States used about 40 quads of petroleum in 2006, 22 quads of natural gas, and 22.5 quads of coal. CO_2 emissions are about 64.9 million metric tons/quad for oil, 94.7 million metric tons/quad for coal, and 53.3 million metric tons/quad for natural gas (you would divide by 3.67 to get carbon emissions, which I will discuss soon). Oil is roughly 1.7×10^8 barrels/quad, or, at 7.3 barrels per metric ton, oil is 2.3×10^7 metric tons/quad. Coal is roughly 4.5×10^7 metric tons/quad, and natural gas 2.0×10^7 metric tons/quad. The U.S. population for 2006 was about 298.6 million, according to the U.S. Census Bureau, http://www.census.gov/popest/ states/NST-ann-est.html (accessed Nov. 17, 2009). With a metric ton being 2204 pounds, this yields 17,460 pounds of fossil fuel and 44,000 pounds of CO_2/ person/year in the United States.

Many references report the amount of carbon involved, rather than the CO_2 or the fossil fuels. You will often see that one year of global burning of fossil fuels released about 8.0 billion tons of carbon, abbreviated as Gt C, where the G stands for "giga," one billion is 1,000,000,000, and these are metric tons of 2204 pounds each; this would be about 8.8 billion tons of carbon if you think of a ton as weighing 2000 pounds. Fossil fuels are often about 75% carbon by mass, although this varies notably depending on the type. A little useful information on coal is in Hong, B. D. and E. R. Slatick, 1994, "Carbon dioxide emission factors for coal," *EIA Quarterly Coal Report January–April 1994*, DOE/EIA-0121(94/Q1) (Washington, DC, Aug. 1994): 1–8. Additional CO_2 emissions from deforestation, forest, and peatland degradation are estimated as 1.5 Gt C/year; van der Werf, G. R., D. C. Morton, R. S. DeFries, et al., 2009, "CO_2 emissions from forest loss," *Nature Geoscience* 2: 737–38, doi:10.1038/ngeo671. This is an update of earlier numbers from the Intergovernmental Panel on Climate Change (IPCC); see table 7.1 in Denman, K. L., G. Brasseur, A. Chidthaisong, P. Ciais, P. M. Cox, R. E. Dickinson, D. Hauglustaine, C. Heinze, E. Holland, D. Jacob, U. Lohmann, S. Ramachandran, P. L. da Silva Dias, S. C. Wofsy and X. Zhang, 2007,

"Couplings between Changes in the Climate System and Biogeochemistry," in Solomon, S., D. Qin, M. Manning, Z. Chen, M. Marquis, K. B. Averyt, M. Tignor and H. L. Miller (eds.), *Climate Change 2007: The Physical Science Basis. Contribution of Working Group I to the Fourth Assessment Report of the Intergovernmental Panel on Climate Change* (Cambridge University Press, New York).

9. Ogden, C. L., C. D. Fryar, M. D. Carroll and K. M. Flegal, 2004, "Body weight, height, and body mass index, United States 1960–2002," *Advance Data from Vital and Health Statistics* (U.S. Department of Health and Human Services, Centers for Disease Control and Prevention, National Center for Health Statistics), no. 347, Oct. 27, 2004.

10. See note 8.

11. U.S. Environmental Protection Agency, http://www.epa.gov/reg3wcmd/solid wastesummary.htm (accessed Nov. 20, 2009). "Municipal solid waste . . . includes durable goods, non-durable goods, containers and packaging, food wastes and yard trimmings, and miscellaneous inorganic wastes. . . . The per capita generation rate in 1999 was 4.6 pounds per person per day. . . . Twenty-eight percent of solid waste, or 64 million tons, is recovered and recycled or composted, 15 percent, or 34 million tons, is burned at combustion facilities, and the remaining 57 percent, or 132 million tons, is disposed of in landfills." The calculations appear to be for English tons, of 2000 pounds each. The total waste production in the United States is vastly higher than this, but depends a lot on how you define "waste"—if you strip-mine coal, and dump the rock into the valley, does that count as solid waste? If yes, the rock weighs a lot. For example, Lottermoser, B., 2003, *Mine Wastes: Characterization, Treatment and Environmental Impacts* (Springer, Berlin), gave a very crude estimate of 20 Gt (metric, or 22 Gt English) of mine waste per year globally, comparable to the CO_2 produced.

12. See note 8.

13. The chemistry of photosynthesis is, in English: carbon dioxide plus water plus light energy produces plants plus oxygen. In very simplified chemistry, this is often written as $CO_2 + H_2O$ + sunlight $\rightarrow CH_2O + O_2$. CH_2O is the chemical formula of formaldehyde, which is not the usual product of plants. High school biology texts often multiply everything in the previous equation by 6, except that they say 6 times sunlight is still sunlight, to obtain $6CO_2 + 6H_2O$ + sunlight $\rightarrow C_6H_{12}O_6 + 6O_2$. Here $C_6H_{12}O_6$ is glucose, which is closer to what plants make. For calculations involving the global cycling of various biologically important chemicals, CH_2O works fine as long as everyone recognizes that we aren't really claiming that plants are formaldehyde factories.

14. A fluid moves very slowly near the wall of a pipe or the surface of a grain, so bigger pipes carry a lot more water. Roughly, the flux of water in laminar flow (slow enough that there is no turbulence) increases with the fourth power of the radius of a pipe (Poiseuille's law). That means that if you decided to bring water

down the hill from your spring to your shower, and you had a choice of one big pipe, or of two smaller pipes each with half the inside cross-sectional area of the big pipe, and you kept the pipes full and ran them along the same path, the big pipe would deliver water four times as rapidly. Spaces between very tiny grains are really small, and water just doesn't go through easily, whereas spaces between big grains can transmit much water.

15. See, for example, Falkowski, P. G. and Y. Isozaki, 2008, "The story of O_2," *Science* 322: 540–42.

16. You will find various versions of the chemical reaction for burning "plant"—CH_2O—using sulfate—SO_4^{-2}—or other chemicals. Here's one really simple version for sulfate, in which the sulfate is written as sulfuric acid for convenience. In an ocean, you would find the sulfuric acid separated into the sulfate and H^+, with the H^+ also reacting with carbonate and borate and such, so the reality is a tad more complex, but I hope that you can see how this works:

$$H_2SO_4 + 2CH_2O \rightarrow H_2S + 2CO_2 + 2H_2O + energy$$

If you prefer to see the reaction using the ions dissolved in ocean water, it would look like this:

$$SO_4^{-2} + 2CH_2O \rightarrow H_2S + 2HCO_3^- + energy$$

The net result here is similar to burning with oxygen, breaking down the plant to release energy and CO_2 (the CO_2 is in the HCO_3^- in the second version). In the salt marshes where we like to go kayaking on Cape Cod, and in many other watery environments, a lot of organic material grows, sinks, and is buried rapidly. In these places, stirring up the mud may release a whiff of hydrogen sulfide—the rotten-egg smell of H_2S. You can see black material in the mud, formed when the H_2S reacts with iron to make iron pyrite, FeS_2.

17. I have often relied on the wonderful book by my friends and colleagues at Penn State, Kump, L. R., J. F. Kasting and R. G. Crane, 2009, *The Earth System*, 3rd ed. (Prentice-Hall, Upper Saddle River, NJ). In their figure 8-12, they estimate that the "burning" by oxygen, either in animals or by bacteria in the soils or upper parts of sediments, breaks down a lot more plant material than does the burning without oxygen, by maybe 100-fold.

18. Alcohol groups are often written as –COH, bonded to other things at the dash. Carboxyl groups, –COOH, are also favorites for early use by bacteria. Oddly enough, the parts of the plant material taken off first by the bacteria may not be the most energy-rich ones, but they are chemically the most valuable or else the easiest ones to get; this increases the energy density—the energy per unit mass—of the remaining plant material. See Freeman, K. H., 2001, "Isotopic biogeochemistry of marine organic carbon," in Valley, J. W. and D. Cole (eds.), *Reviews in Mineralogy and Geochemistry* 43: 579–605, doi:10.2138/gsrmg.43.1.579.

19. A good textbook is probably the best place to start if you really want the background, although quicker, friendlier sources are also available. For a comprehensive textbook on petroleum, see Hunt, J. M., 1996, *Petroleum Geochemistry and Geology* (W. H. Freeman, New York). Much more briefly, the U.S. Geological Survey provides Organic Origins of Petroleum, http://energy.er.usgs.gov/gg/research/petroleum_origins.html. Many aspects of coal are covered at least briefly in the text Thomas, L., 2002, *Coal Geology* (John Wiley, New York).

20. See, for example, Soeder, D. J. and Kappel, W. M., 2009, "Water resources and natural gas production from the Marcellus Shale," *U.S. Geological Survey Fact Sheet 2009–3032.*

21. The hypothesis that evolution of the tree-strengthening compound lignin in trees produced material that was difficult for decomposers, and that evolution of efficient decomposers took a long while during which much coal formed, is interesting but surely not proved.

22. Bring the clathrate up slowly, and it melts on the way, allowing the methane to escape. See, for example, Kvenvolden, K. A., 2003, "Gaia's breath—Methane and the future of natural gas," *Sound Waves* (U.S. Geological Survey), http://soundwaves.usgs.gov/2003/06/outreach2.html (accessed Nov. 12, 2009).

23. Denardo, B., L. Pringle, C. DeGrace and M. McGuire, 2001, "When do bubbles cause a floating body to sink?" *American Journal of Physics* 69: 1064–72; May, D. A. and J. J. Monaghan, 2003, "Can a single bubble sink a ship?" *American Journal of Physics* 71: 842–49.

24. Kennett, J. P., K. G. Cannariato, I. L. Hendy and R. J. Behl, 2003, *Methane Hydrates in Quaternary Climate Change: The Clathrate Gun Hypothesis* (American Geophysical Union, Washington, DC).

25. For the first such really confident demonstration, see Severinghaus, J. P., T. Sowers, E. J. Brook, R. B. Alley and M. L. Bender, 1998, "Timing of abrupt climate change at the end of the Younger Dryas interval from thermally fractionated gases in polar ice," *Nature* 391: 141–46. Also see Bock, M., J. Schmitt, L. Möller, et al., 2010, "Hydrogen isotopes preclude marine hydrate CH_4 emissions at the onset of Dansgaard-Oeschger events," *Science* 328: 1686–89.

26. To increase the store of free gas beneath the clathrate layer, the pressure must rise to displace more water. But the pressure cannot rise too high before it will wedge open cracks through the clathrate and let gas out. Also, when the methane and water combine to make clathrate, they reject most of the salt in the water, making salty brines that are harder to freeze, until clathrate, salty water, and free gas can coexist, allowing the free gas to escape in some places by bubbling through the salty water. Liu, X. L. and P. B. Flemings, 2006, "Passing gas through the hydrate stability zone at southern Hydrate Ridge, offshore Oregon," *Earth and Planetary Science Letters* 241: 211–26.

27. Committee on Assessment of the Department of Energy's Methane Hydrate

Research and Development Program: Evaluating Methane Hydrate as a Future Energy Resource, Board on Earth Sciences and Resources, Division on Earth and Life Studies, National Research Council, 2010, *Realizing the Energy Potential of Methane Hydrate for the United States* (National Academies Press, Washington, DC).

28. Buffett, B. and D. Archer, 2004, "Global inventory of methane clathrate: sensitivity to changes in the deep ocean," *Earth and Planetary Science Letters* 227: 185–99; also Kump et al. in note 17. (Note that the fossil fuels are only the tiniest bit—maybe 0.05%—of the total amount of organic carbon spread through the world's sedimentary rocks, too thinly to be worth our while.)

29. See the paper by B. Buffett and D. Archer from the previous note, and Archer, D. and B. Buffett, 2005, "Time-dependent response of the global ocean clathrate reservoir to climatic and anthropogenic forcing," *Geochemistry Geophysics Geosystems* 6: Q03002, doi:10.1029/2004GC000854.

30. Hubbert, M. K., 1956, "Nuclear Energy and the Fossil Fuels," in *Drilling and Production Practice* (American Petroleum Institute, New York).

31. For example, Deffeyes, K. S., 2002, *Hubbert's Peak: The Impending World Oil Shortage* (Princeton University Press, Princeton, NJ).

32. See, for example, the U.S. Geological Survey World Petroleum Assessment 2000, by the USGS World Energy Assessment Team, U.S. Geological Survey Digital Data Series DDS-60, http://pubs.usgs.gov/dds/dds-060/index.html#TOP (accessed May 29, 2010).

33. Net imports were running roughly 11 million barrels/day in the spring of 2010, http://www.eia.doe.gov/dnav/pet/pet_move_wkly_dc_NUS-Z00_mbblpd_w .htm (accessed May 29, 2010). The oil reserves of the Arctic National Wildlife Refuge were estimated in the U.S. Geological Survey Open-File Report 98-34, and described in *Arctic National Wildlife Refuge, 1002 Area, Petroleum Assessment, 1998, Including Economic Analysis*, USGS Fact Sheet FS-028-01, Apr. 2001.

34. Lackner, K. S. and J. D. Sachs, 2005, "A robust strategy for sustainable energy," *Brookings Papers on Economic Activity* 2: 215–69, http://www.jstor.org/ stable/3805122 (accessed Nov. 23, 2009).

35. The atmosphere doesn't care much about the hydrogen in the fossil fuel because it makes water that rains out of the air in just over a week, and while the sulfur in fossil fuels is a big deal for acid rain, the sulfur also falls out quickly and is fairly rare to begin with. So, we talk about carbon. (In chapters 17 and 21, I will consider the importance of the traces of uranium and mercury in coal.)

36. As discussed in note 8, fossil fuel is often about 75% carbon although with notable variations, and CO_2 weighs about 3.67 times as much as the carbon in it.

37. Numbers can be found in Denman et al. in note 8. Also see Kump et al. in note 17. Plants actually use CO_2 with more than 100 Gt C each year, but the plants then burn some of it to keep themselves alive.

38. See Kump et al. in note 17, their figure 8-3.

39. Ibid.

5: Abraham Lincoln or Your Brother-in-Law?

Chapter title. In case anyone is worried, I have four brothers-in-law, and all are friends and absolutely first-rate people. So, I'm not really picking on brothers-in-law!

1. *Discoveries and Inventions: A Lecture by Abraham Lincoln Delivered in 1860* (John Howell, San Francisco, 1915). Available online at various sites, including http://teachingamericanhistory.org/library/index.asp?document=2507 (accessed Nov. 15, 2009).

2. For background, see the extensive educational materials at the Monitor National Marine Sanctuary, http://monitor.noaa.gov/, and especially follow the link to the Monitor Center at the Mariner's Museum of Newport News, Virginia, http://www.mariner.org (both accessed on May 29, 2010). Much additional information is available on this fascinating engagement, including in Tucker, S. C., 2006, *Blue and Gray Navies: The Civil War Afloat* (Naval Institute Press, Annapolis, MD).

3. The invention was never produced, and may not have been especially practical, but represents a commitment to engineering and the manual arts that has not been matched by many presidents. See "Abraham Lincoln's Patent Model: Improvement for Buoying Vessels over Shoals," from the Smithsonian Institution National Museum of American History, http://americanhistory.si.edu/collections/object.cfm?key=35&objkey=19 (accessed Nov. 21, 2009).

4. The web site of the National Academy of Sciences, www.nas.edu, has a good if short history. We don't really know how engaged Lincoln was in founding of the Academy, but he surely signed the act quickly, his administration used the Academy in many ways, and he was a known enthusiast for science and engineering. Even before the Academy was founded, his administration was getting scientific advice from a government agency—the Permanent Commission of the Navy Department—that evaluated ideas sent in by members of the public interested in helping the Union war effort. The Permanent Commission was a body "to which all subjects of a scientific character on which the Government may require information may be referred" (see p. 51 of the Centennial History of the National Academy of Sciences, at the web site above).

5. True, F. W., "A history of the first half-century of the National Academy of Sciences 1863-1913," pp. 213–17, http://www7.nationalacademies.org/archives/iron_ships.html (accessed Oct. 13, 2009).

   The role of the National Academy of Sciences grew with changing circumstances in the twentieth century. Facing the prospect of entry into World War I and the need for scientific advice, the National Research Council was founded at the request of President Woodrow Wilson to mobilize additional expertise to aid the war effort.

Science, engineering, and medicine have always had huge overlaps, but as the fields have grown, the single academy of sciences was broadened into three bodies, through founding of the Institute of Medicine and the National Academy of Engineering, now acting together under the National Research Council. When the U.S. government needs advice on a scientific issue, the National Research Council convenes a committee of experts spanning the main ideas in the relevant field, who agree to work on their own time to aid the public while disclosing any conflicts of interest that might be disqualifying. Presidents Eisenhower (1956) and George H. W. Bush (1993) issued executive orders affirming and broadening the role of the National Research Council, and thus the National Academy of Sciences.

6. Note that even historical sciences can make predictions about data sets not yet collected, so "historical science" is *not* an oxymoron! See, for example, Alley, R. B., 2007, "Wally was right: Predictive ability of the North Atlantic 'conveyor belt' hypothesis for abrupt climate change," *Annual Review of Earth and Planetary Sciences* 35: 241–72. Also note that it has been a long time since a big old idea was truly overthrown rather than being subsumed. For example, Isaac Newton's physics worked far better than Aristotle's ideas in helping us build things that don't fall down or that otherwise do what we want—Newtonian physics comes very close to making truly correct predictions about these things. But Newton's ideas are really bad when applied to tiny things (the size of an atom) or huge things (the size of a galaxy) or things going really fast (nearly the speed of light). The failures of Newton's physics have led to quantum mechanics and to Albert Einstein's relativity. Still, when we want to calculate what buildings and airplanes will do, we usually rely on Newton, because the answers from Newton and from these more correct versions of nature are indistinguishable in a whole lot of cases, and calculations are tremendously easier using Newton. We did not throw away Newton's physics, but found that Newton had supplied a useful approximation of large parts of something better and more inclusive.

7. I might be a bad scientist anyway, but the mere fact of a student outdoing me does not demonstrate my shortcomings.

8. An excellent starting point for investigating these issues is the book Geschwind, C. H., 2001, *California Earthquakes: Science, Risk and the Politics of Hazard Mitigation* (John Hopkins University Press, Baltimore). Also see Williams, J. C., 2000, "Faulty construction: Earthquakes and the culture of prevention in California," *Geojournal* 51: 255–58; and Strupp, C., 2006, "Dealing with Disaster: The San Francisco Earthquake of 1906" (presented at the Symposium "San Francisco Earthquake 1906: Urban Reconstruction, Insurance, and Implications for the Future," Institute of European Studies, University of California at Berkeley,

March 22). It is a thoroughly revised and updated version of an article originally published in German as "'Nothing destroyed that cannot speedily be rebuilt': San Francisco und das Erdbeben von 1906," in Ranft, A. and S. Selzer (eds.), *Städte aus Trümmern: Katastrophenbewältigung zwischen Antike und Moderne* (Vandenhoeck and Ruprecht, Göttingen, 2004), pp. 132–71.

9. The discussion in this paragraph of the text is especially based on Lear, L. J., 1993, "Rachael Carson's 'Silent Spring,'" *Environmental History Review* 17, no. 2: 23–48. This article provides an interesting, accessible, and documented summary of this seminal event in environmental history. There is plenty of additional scholarship available for the interested reader, including Dunlap, T. R., 1981, *DDT: Scientists, Citizens, and Public Policy* (Princeton University Press, Princeton, NJ). Note that in Carson's case, the initial warnings were not from the established groups assessing science, but from someone who was acting as an advocate; although based on science (which was still emerging), her book was designed to achieve change, not simply to assess and inform on the scientific issues. Environmental groups took up the cause, and some of them seem to have been quite energetic and probably occasionally hyperbolic.

10. Baum, R. M., editor-in-chief, "FROM THE EDITOR: Rachael Carson," Editor's Page, *Chemical and Engineering News* 85, no. 23 (June 4, 2007): 5. That editorial is accompanied by the following note: "For an in-depth examination of the current use of DDT in battling malaria, see Naomi Lubick's article 'DDT's Resurrection' in Environmental Science & Technology [2007, 41, 6323]." This is an interesting piece.

11. See, for example, Oreskes, N. and E. M. Conway, 2010, *Merchants of Doubt: How a Handful of Scientists Obscured the Truth on Issues from Tobacco Smoke to Global Warming* (Bloomsbury Press, New York).

12. I have borrowed liberally from Park, R. L., 2000, *Voodoo Science: The Road from Foolishness to Fraud* (Oxford University Press, Oxford, UK), and from references therein.

13. Ibid.

14. National Academy of Sciences, United States Committee for the Global Atmospheric Research Program, 1975, *Understanding Climate Change: A Program for Action* (National Academies Press, Washington, DC).

15. National Academy of Sciences, 1979, *Carbon Dioxide and Climate: A Scientific Assessment* (National Academies Press, Washington, DC); this is the "Charney" report, chaired by Jule G. Charney. The authors, and the people they thank for advice and unpublished results, are a remarkable "who's who" of a large part of climate science.

16. Committee on the Science of Climate Change, National Research Council, 2001, *Climate Change Science: An Analysis of Some Key Questions* (National Academies Press, Washington, DC), www.nap.edu. The "skeptical" scientist mentioned is

Professor Richard S. Lindzen, the Alfred P. Sloan Professor of Meteorology in the Department of Earth, Atmospheric and Planetary Sciences at the Massachusetts Institute of Technology. He commented further on the climate change report in a June 11, 2001, commentary in the *Wall Street Journal*, "Scientists' report doesn't support the Kyoto Treaty," http://interactive.wsj.com/articles/SB992205567633857892.htm.

6: Red, White, and Blue-Green

1. Lowell, J. R. and C. E. Norton (eds.), 1864, "The President's Policy," in *The North American Review*, vol. 98 (Crosby and Nichols, Boston), p. 235.

2. The popular usage "That's just a theory" is sometimes a put-down implying idle speculation. I use "theory" in the scientific way, as a well-accepted, useful, big idea.

3. Harries, J. E., H. E. Brindley, P. J. Sagoo and R. J. Bantges, 2001, "Increases in greenhouse forcing inferred from the outgoing longwave radiation spectra of the Earth in 1970 and 1997," *Nature* 410: 355–57. Also see Griggs, J. A. and J. E. Harries, 2007, "Comparison of spectrally resolved outgoing longwave radiation over the tropical Pacific between 1970 and 2003 using IRIS, IMG, and AIRS," *Journal of Climate* 20: 3982–4001.

4. Thoreau, H. D. to Harrison Blake, May 20, 1860, in Sanborn, F. B. (ed.), *The Writings of Henry David Thoreau*, vol. 6, *Familiar Letters* (Houghton Mifflin, Boston, 1906), p. 360.

5. See "Venus," in Spencer Weart's "The Discovery of Global Warming" (American Institute of Physics), http://www.aip.org/history/climate/Venus.htm (accessed May 31, 2010).

6. In case it bothers you that shorter wavelengths have higher energy, try this: While you are holding one end of a short rope or a jump rope, get a friend to hold the other end. Then move your hand up and down to make waves. If you move your hand slowly, all of the rope will go up and then down, to create a very long wave. Move your hand faster, and you can get two waves between you and your friend, with the center of the rope hardly moving while the part of the rope closer to you moves one way and the part farther away moves the other. Move still faster, and you can get more waves along the length—and you will find your hand getting tired quickly. You must be more energetic to get the shorter waves.

   If the idea of everything always radiating bothers you, note that bodies at the same temperature simply trade energy with no net effect; bodies at different temperatures drift toward a common temperature as the warmer body radiates more to the colder body rather than vice versa.

7. Recall how much warming occurs over a few hours from the cold just before dawn to the heat of mid-afternoon. Then, consider this back-of-the-envelope

calculation. To raise the temperature of 1 kg of dry air at sea level at constant pressure requires approximately 10^3 J/K, where "J" represents joules and "K" represents degrees Kelvin, which is the same as "degrees Celsius" in this usage. The weight of the atmosphere sitting over 1 square meter of Earth at sea level is about 10^4 kg/m^2. Thus, to warm the air over 1 square meter by 1 degree Kelvin takes about 10^7 J, giving 10^7 J/m^2/K. The incoming sunlight supplies about 240 W/m^2, and multiplying by the 3.16×10^7 seconds in a year yields 7.6×10^9 J/m^2/year. Finally, dividing 7.6×10^9 J/m^2/year by 10^7 J/m^2/K yields 760 K/year, or multiplying by 1.8 to convert to degrees Fahrenheit, 1400°F/year. That's a lot of warming. In turn, we can have high confidence that outgoing energy is very close to incoming energy, because the atmosphere is certainly not warming by 1400°F/year!

However, because Earth is warming from the human-caused increase in CO_2, as described in the next chapters, incoming energy is slightly greater than outgoing now. Decadally averaged at the turn of the millennium, the human-caused rise in greenhouse gases was causing Earth to receive approximately 0.75 W/m^2 more from the sun than was going back to space (see Hansen, J., L. Nazarenko, R. Ruedy, et al., 2005, "Earth's energy imbalance: Confirmation and implications," *Science* 308: 1431–35). Only approximately 0.04 W/m^2 of this was going into warming the air, with the great majority going into the ocean. This recent energy imbalance is actually quite large; the imbalance needed to melt the great ice sheets of the ice age over 10,000 years was only about 0.1 W/m^2, following the calculations of Hansen and colleagues in the reference just cited. Overall, while incoming and outgoing energy are not perfectly equal, and the planet warms when incoming energy is larger but cools when outgoing energy is larger, the two are not hugely different.

8.   If you were lying on the sun deck of your spaceship in orbit right next to Mars, being twice as far from the sun as you would be if you were at Venus would mean that you would take about four times as long to get a sunburn. Only the tiniest bit of the sunlight reaching the orbit of Venus is blocked by Venus or Earth or dust or comets before reaching Mars. But the sunlight is spreading in all directions. Think of the sun suspended at the center of a balloon. Initially the balloon is blown up just big enough to touch the orbit of Venus, with the sunlight illuminating the inside of the balloon. Next, inflate the balloon, making it stretch in all directions until it touches the orbit of Mars. The distance from the sun to the surface of the balloon is r, the energy leaving the sun is S, and the surface area illuminated is $4\pi r^2$; the brightness of the sunshine is then $S/4\pi r^2$. This is the energy available at the top of the atmosphere of a planet. This amounts to about 592 W/m^2 at Mars, 1370 W/m^2 at Earth, and 2460 W/m^2 at Venus. Remember that you personally internally generate about 100 W by burning food, and 1 m^2 is just over 3 feet by 3 feet, or the size of a small tabletop, so sunlight is equal to

the energy from 6 of you on a tabletop at the orbit of Mars, 14 at Earth, and 25 at Venus.

9. Take the incoming radiation at the top of the atmosphere, divide by 4 because the energy must be spread around the spherical planet, and multiply by $(1 - A)$, where A is the reflected part, or albedo, so $1 - A$ is the absorbed part, and you express albedo as a decimal so that an albedo of 30% means $A = 0.3$. The energy to warm the surface is then 115 W/m² for Mars, 240 W/m² for Earth, and 132 W/m² for Venus. This is about the internal energy of one person per tabletop on Mars and Venus, and 2 on Earth. So much of the sun's energy is reflected from Venus that it has much less energy available to heat its surface than Earth does!

10. The rate at which energy is radiated away is equal to sT^4, where $s = 5.67 \times 10^{-8}$, W/m²/K⁴ is the Stefan-Boltzmann constant, and T is the absolute temperature in degrees Kelvin (K). Setting this equal to the absorbed energy at the surface, assuming incoming energy equals outgoing energy, and solving for T gives the equilibrium radiative temperature.

In the main text, I stated that a twofold increase in absolute temperature gives a sixteenfold increase in emitted radiation, so let's do that first. If you let the energy radiated be e, then at some temperature T_1, the emitted energy $e_1 = sT_1^4$. At temperature $T_2 = 2T_1$, the energy radiated is $e_2 = sT_2^4 = s(2T_1)^4 = 16sT_1^4 = 16e_1$. Thus, doubling the absolute temperature increases the outgoing radiation sixteenfold.

The text also states that a 1% change in absolute temperature causes a 4% change in outgoing radiation. This isn't absolutely exact; more properly, an infinitesmal relative change in temperature causes a four-times-larger relative change in outgoing radiation, but 1% is small enough for this to be a good approximation. To check, you could just plug in a value for temperature—say, 255°K for equilibrium radiation for the modern Earth—and calculate the outgoing radiation using the Stefan-Boltzmann law given above, finding 240.16 W/m². Repeat with a temperature 1% higher (257.55°K), and you will find 249.92 W/m², an increase of $((249.92 - 240.16)/240.16)*100\% = (9.76/240.16)*100\% = 4.06\%$, pretty close to 4%.

If you want to do it right, note that the percentage increase is either dT/T for temperature or de/e for energy, multiplied by 100%, with dT indicating a small change in T, and de a small change in e. If you happen to know calculus, you can take the derivative of the Stefan-Boltzmann law with respect to temperature, $de/dT = 4sT^3$. Next note that $sT^3 = sT^4/T = e/T$ from the Stefan-Boltzmann law, so $de/dT = 4sT^3 = 4e/T$. Rearrange, and you have $de/e = 4dT/T$, which says that a small relative change in absolute temperature gives a four-times-bigger relative change in energy emitted, or a 1% change in absolute temperature gives about a 4% change in emitted energy.

If you don't know calculus, you might write the Stefan-Boltzmann law for some temperature as $e = sT^4$, and for a slightly higher temperature as $e + de =$

$s (T + dT)^4$, with de and dT again indicating small changes to e and T, respectively. If you expand $(T + dT)^4$, you end up with $(T + dT)^4 = T^4 + 4T^3 dT +$ (additional terms that have higher powers of dT). But, if dT is small, $(dT)^2$ is much smaller, and $(dT)^3$ and $(dT)^4$ are truly tiny, so those "additional terms" just don't matter much. Thus, to good approximation, $e + de = s(T^4 + 4T^3 dT)$. If you subtract the original Stefan-Boltzmann law from this, subtracting e from the left and the equivalent sT^4 from the right, you are left with $de = 4sT^3 dT$. If you then divide by e on the left and by the equivalent sT^4 on the right, you get $de/e = 4dT/T$, the same as we got using the calculus, and again showing that small relative changes in absolute temperature cause four-times-larger relative changes in radiated energy. Whether you do it by examples, or calculus, or expansion and approximation, you get the same answer—we are fortunate that radiation physics keeps temperature changes small.

11. Some of our friends in meteorology *hate* it when the warming effect of gases in the atmosphere is called the "greenhouse effect." The glass in a greenhouse does affect radiation in the same way as "greenhouse" gases, letting visible light enter but blocking the exit of infrared radiation (Silverstein, S. D., 1976, "Effect of infrared transparency on the heat transfer through windows: A clarification of the greenhouse effect," *Science* 193: 229–31). But this contributes to only part of the warming in the greenhouse. The greenhouse is also warmed because the glass stops convection—when the sun heats the ground, which heats the air near the ground, outside a greenhouse the hot air rises in a convection cell that takes the heat away, but inside the greenhouse this motion is blocked by the glass. Unfortunately, use of the term "greenhouse effect" for the effects of certain gases on radiation is so widely known that we couldn't change the terminology if we wanted to, and we haven't found a very good alternative.

12. Very-long-wavelength radar waves can get through, but Venus doesn't send out much energy at those wavelengths.

13. Take some volume of air, say, inside an imaginary balloon. Warm the air a little so that the balloon floats upward. As the balloon rises, there is less air above it, so the air pressure that is squeezing it drops, and the balloon expands. But the expansion requires that it push some surrounding air out of the way. The energy to do this work comes from the internal energy—the heat—in the initial air volume. Therefore, provided that there are no other major sources of energy, the air cools as you go upward in a convection cell. Regions that are frequently affected by convection will follow this rule, even if at a given moment a convection cell is not active.

14. European Space Agency, "SCIAMACHY satellite instrument," http://www.esa .int/esaEO/SEMZHVM5NDF_index_0.html (accessed Oct. 12, 2009).

15. These same wavelengths are absorbed in energy coming down from the sun. But at these wavelengths, Earth supplies more energy to the atmosphere than the

sun does. The total energy coming down from the sun equals the total energy going out from Earth, but the sun sends most of its energy in the shorter, visible wavelengths, whereas Earth sends most of its energy in the longer, infrared wavelengths that interact with greenhouse gases.

16. See note 3.

17. In terms of outgoing radiation now blocked in Earth's atmosphere, water vapor is more important than CO_2. But if most of the CO_2 were taken out of the atmosphere, the resulting cooling would condense a lot of the water vapor and might drop the planet into an ice-covered "snowball" state (see Pierrehumbert, R. T., G. Brogniez and R. Roca, 2007, "On the Relative Humidity of the Atmosphere," in Schneider, T. and A. Sobel (eds.), *The Global Circulation of the Atmosphere* (Princeton University Press, Princeton, NJ), p. 150; also Voigt, A. and J. Marotzke, 2009, "The transition from the present-day climate to a modern Snowball Earth," *Climate Dynamics*, doi 10.1007/s00382-009-0633-5). Thus, you can make an argument that CO_2 is really more important than water vapor as a greenhouse gas, because CO_2 should be credited with its direct effects and with the effects of the water vapor that is in the air because of the warmth from the CO_2.

18. The time between collisions in the atmosphere near Earth's surface is something like 0.0000000001 second. The Glossary of Meteorology from the American Meteorological Society gives the mean free path—the distance between collisions—in air as being on the order of 0.1 micron. See http://amsglossary .allenpress.com/glossary/search?id=mean-free-path1 (accessed Sept. 29, 2009). Mean molecular speed is about 500 microns/second.

19. See Spencer Weart's "The Discovery of Global Warming" (American Institute of Physics), http://www.aip.org/history/climate/.

20. See note 13.

21. The calculation really should be done for each wavelength individually; because you can see the surface from space in some wavelengths but you see high up in the atmosphere in other wavelengths, there is no single height above which radiation escapes to space, but a lot of different heights for different wavelengths. To understand this properly, I recommend Pierrehumbert, R. T., 2010, *Principles of Planetary Climate* (Cambridge University Press, New York). For somewhat lighter fare, the interested reader might consult Weart's "The Discovery of Global Warming," cited in note 19, and especially http://www.aip.org/history/ climate/co2.htm and http://www.aip.org/history/climate/Radmath.htm. The post at the RealClimate web site by Spencer Weart and Ray Pierrehumbert, June 2007, "A Saturated Gassy Argument," http://www.realclimate.org/index.php/ archives/2007/06/a-saturated-gassy-argument/, and the follow-up "Part II: What Ångström Didn't Know," with the same authors but in reversed order, http://www.realclimate.org/index.php/archives/2007/06/a-saturated-gassy- argument-part-ii/, are quite enlightening as well. For a classic treatment, see

Goody, R. M. and G. D. Robinson, 1951, "Radiation in the troposphere and lower stratosphere," *Reviews of Modern Meteorology* 77: 131–87.

22. CO_2 has various absorbtion bands as well, including those for molecules with isotopically "heavy" oxygen or carbon, and "hot" bands that originate in the first excited vibrational state rather than in the ground state. Isotopically substituted molecules and excited molecules are rarer than "normal" CO_2, so these bands are further from "saturation." Bottom line: Earth's atmosphere is not close to saturation from CO_2. Also see note 21.

23. See note 21.

24. One-tenth of an inch in 30 feet is the same as 0.1 inch in 360 inches, or 0.00028, or 280 parts per million (ppm). Recent data on CO_2 are available from the U.S. National Oceanographic and Atmospheric Administration (NOAA) through its Earth System Research Laboratory Global Monitoring Division. "Recent Monthly Mean CO_2 at Mauna Loa," http://www.esrl.noaa.gov/gmd/ccgg/trends/ (accessed Nov. 2, 2009). For refereed scientific literature, the NOAA refers you to Keeling, C. D., R. B. Bacastow, A. E. Bainbridge, et al., 1976, "Atmospheric carbon dioxide variations at Mauna Loa Observatory, Hawaii," *Tellus* 28: 538–51; and Thoning, K. W., P. P. Tans and W. D. Komhyr, 1989, "Atmospheric carbon dioxide at Mauna Loa Observatory: 2. Analysis of the NOAA GMCC data, 1974–1985," *Journal of Geophysical Research* 94: 8549–65.

25. The safe exposure levels for carbon monoxide are from the National Institute for Occupational Safety and Health of the Centers for Disease Control and Prevention, http://www.cdc.gov/niosh/idlh/630080.html (accessed Nov. 2, 2009). I quoted the TWA number, a *time-weighted average* concentration for up to a ten-hour workday during a forty-hour workweek. The Immediately Dangerous to Life and Health (IDLH) level is given as 1500 ppm, and roughly three times that is the lowest concentration reported to kill people quickly (references given for five minutes and half an hour).

26. The value I give is the lethal concentration for half of a population (LC50), and is from http://www.osha.gov/SLTC/healthguidelines/hydrogencyanide/recognition.html (accessed Nov. 2, 2009), which cites Hathaway, G. J., N. H. Proctor, J. P. Hughes and M. L. Fischman, 1991, *Proctor and Hughes' Chemical Hazards of the Workplace*, 3rd ed. (Van Nostrand Reinhold, New York).

7: Canting the Kayak

1. See, for example, Emerson, R. W., 1867, *Poems by Ralph Waldo Emerson* (Houghton Mifflin, Boston), p. 221.

2. Scott, R. F., 1905, *The Voyage of the 'Discovery,'* vol. 2 (repr. Charles Scribner's Sons, New York, 1907), pp. 54–55.

3. See, for example, the Vancouver (British Columbia, Canada) Maritime Museum,

http://www.vancouvermaritimemuseum.com/modules/vmmuseum/treasures
/?artifactid=77 (accessed Oct. 4, 2009).

4. If you were to put some dry air in a sealed bottle with a lot of water, and hold the
    temperature constant for a long time, the amount of water vapor in the bottled
    air would increase for a while but eventually stabilize at a constant level. This is
    called the "equilibrium carrying capacity," or "saturation vapor pressure," and
    you can say that the air is "full" of water, or in balance, or at equilibrium. The
    saturation vapor pressure can be reported in various ways, often as kilograms
    of water per cubic meter of air, or some other mass per volume measure. This
    value increases with temperature, by roughly 7% per degree Celsius or 4% per
    degree Fahrenheit, rounding off just a bit, although this number changes a little
    with temperature, and whether the equilibrium is with water or with ice. Note
    that the rate of hair drying depends on kinetics as well as equilibrium—how
    fast it approaches equilibrium as well as the concentration at equilibrium—so
    this simple scaling won't tell you precisely how much faster a warmer hair dryer
    works.

5. Certain wavelengths are absorbed by both CO_2 and water vapor, so exactly how
    you divide the credit affects the answer. At very high elevations and in other
    really dry places, CO_2 is more important than H_2O. See, for example, Tren-
    berth, K. E., P. D. Jones, P. Ambenje, R. Bojariu, D. Easterling, A. Klein Tank,
    D. Parker, F. Rahimzadeh, J. A. Renwick, M. Rusticucci, B. Soden and P. Zhai,
    2007, "Observations: Surface and Atmospheric Climate Change," in Solomon,
    S., D. Qin, M. Manning, Z. Chen, M. Marquis, K. B. Averyt, M. Tignor and
    H. L. Miller (eds.), *Climate Change 2007: The Physical Science Basis. Contribution
    of Working Group I to the Fourth Assessment Report of the Intergovernmental Panel on
    Climate Change* (Cambridge University Press, New York). Also see note chapter
    6, note 17.

6. As described in the important paper by Pierrehumbert, R. T., G. Brogniez and
    R. Roca, 2007, "On the Relative Humidity of the Atmosphere," in Schneider, T.
    and A. Sobel (eds.), *The Global Circulation of the Atmosphere* (Princeton Univer-
    sity Press, Princeton, NJ); also see Chung, E.-S., D. Yeomans and B. J. Soden,
    2010, "An assessment of climate feedback processes using satellite observations
    of clear-sky OLR," *Geophysical Research Letters* 37: L02702; and the Trenberth ref-
    erence from note 5; I have oversimplified a lot here. The most important issue
    turns out to be how temperature change affects water-vapor change in really dry
    places where a little vapor can have a lot of impact. The air gets to dry places
    from the wet ocean by going through relatively cold places where rain falls, such
    as the high elevations of clouds. Because both cold places and warm places tend
    to warm as the CO_2 level is raised, the water-vapor content of dry places and wet
    places goes up together with warming. So the simple explanation proves to be
    right, although it really should be described the way Raymond Pierrehumbert

and colleagues presented it, and not in my simplified way. Observationally, the relative humidity—the mass of water per volume of air in some sample divided by the equilibrium carrying capacity (as described in note 4) and multiplied by 100—doesn't change much as temperature changes, so the absolute humidity goes up with increasing temperature. If air somewhere is three-fourths "full" of water—75% relative humidity—and the temperature goes up a bit, the air will typically have picked up more water vapor from the ocean and retained extra water vapor through subsequent events, and so will still be about three-fourths "full" of water with 75% relative humidity. The small changes in relative humidity are not assumed; they are observed. Note that it is possible to have relative humidity above 100%, although much more commonly relative humidity is below 100%.

7. See, for example, Lindzen, R. S., M. D. Chou and A. Y. Hou, 2001, "Does the earth have an adaptive infrared iris?" *Bulletin of the American Meteorological Society* 82: 417–32; Hartmann, D. L. and M. L. Michelsen, 2002, "No evidence for iris," *Bulletin of the American Meteorological Society* 83: 249–54; Del Genio, A. D., 2002, "Atmospheric science: The dust settles on water vapor feedback," *Science* 296: 665–66; Sherwood, S. C., R. Roca, T. M. Weckwerth and N. G. Andronova, 2010, "Tropospheric water vapor, convection and climate," *Reviews of Geophysics* 48: RG2001. Also see Murphy, D. M., 2010, "Constraining climate sensitivity with linear fits to outgoing radiation," *Geophysical Research Letters* 37: L09704, doi:10.1029/2010GL042911.

8. Whether you use 1°C or 1°F here doesn't matter. The ice-albedo feedback would amplify a 1°F warming to 1.1°F, and a 1°C warming to 1.1°C.

9. This example, including the numbers used, comes from Hansen, J., A. Lacis, D. Rind, et al., 1984, "Climate Sensitivity: Analysis of Feedback Mechanisms," in Hansen, J. E. and T. Takahashi (eds.), *Climate Processes and Climate Sensitivity*, Geophysical Monograph 29 (American Geophysical Union, Washington, DC), pp. 130–63. A lot of us learned about feedbacks from this classic paper. In this example, the feedback factor for snow and ice is $f_s = 1.1$, amplifying an initial 1° change to 1.1°, and the feedback factor for water vapor is $f_w = 1.6$. The combined $f_{s+w} = f_s f_w / (f_s + f_w - f_s f_w) = 1.9$.

10. Ibid.

11. Sturm, M., C. Racine and K. Tape, 2001, "Climate change: Increasing shrub abundance in the Arctic," *Nature* 411: 546–47.

12. To understand how expansion of deserts contributes a bit of global cooling, consider that much of the energy from sunshine hitting a forest goes into evaporating water from the leaves. This keeps the forest cool. But energy is used to knock the water molecules loose from their neighbors during evaporation, and this energy is released as condensation forms clouds, heating the air there. This warming aloft eventually translates into warming at the surface—see, for exam-

ple, Peixoto, J. P. and A. H. Oort, 1992, *Physics of Climate* (American Institute of Physics, New York), or any other good textbook on atmospheric physics. Changing the albedo does affect the planet's temperature. Note, though, that the clouds really need to be factored in, as I discuss next.

13. Ramanathan, V., R. D. Cess, E. F. Harrison, et al., 1989, "Cloud-radiative forcing and climate: Results from the Earth Radiation Budget Experiment," *Science* 243: 57–63; and Kim, D. Y. and V. Ramanathan, 2008, "Solar radiation budget and radiative forcing due to aerosols and clouds," *Journal of Geophysical Research* 113(D): D02203.

14. See Webb, M. J., C. A. Senior, D. M. H. Sexton, et al., 2006, "On the contribution of local feedback mechanisms to the range of climate sensitivity in two GCM ensembles," *Climate Dynamics* 27: 17–38. The IPCC reports also have much useful information on cloud feedbacks.

15. The IPCC says of climate sensitivity that "it is *likely* [>66% probability] to be in the range 2°C to 4.5°C with a best estimate of about 3°C, and is *very unlikely* [<10%] to be less than 1.5°C. Values substantially higher than 4.5°C cannot be excluded," p. SPM-12, in IPCC, 2007, "Summary for Policymakers," in Solomon, S., D. Qin, M. Manning, Z. Chen, M. Marquis, K. B. Averyt, M. Tignor and H. L. Miller (eds.), *Climate Change 2007: The Physical Science Basis. Contribution of Working Group I to the Fourth Assessment Report of the Intergovernmental Panel on Climate Change* (Cambridge University Press, New York). Also, note that the feedbacks discussed in this chapter act over short times, a human lifetime or less. Somewhat longer times bring in other feedbacks, such as cooling causing growth of glaciers and ice sheets that flow across places where snow is not accumulating on its own. This is a slower amplifier, so the climate response to CO_2 may be larger over longer times, a topic we will revisit when I discuss the ice ages.

16. The actual warming depends not only on the level of CO_2 in the air, but also on how long that level is maintained—if you could magically double CO_2 for a very short time—seconds or minutes—and then drop it again, you wouldn't get nearly the full warming effect. In fact, because the ocean takes so long to warm up, warming would continue for decades, and even very slightly for centuries, after the atmospheric CO_2 was stabilized at some level. So there is a bit more to estimating temperature increase from CO_2 than simply taking so many degrees per doubling, and multiplying by the number of doublings. I will discuss this in the next chapter.

8: Why Accountants and Physicists Care about the Past

1. Credited to Artemus Ward (pen name of Charles Ferrar Browne, 1834–1867). See p. 503 of the piece by John Nichol in Moulton, C. W. (ed.), 1910, *Literary*

Criticism of English and American Authors, vol. 6, *1855–1874* (Henry Maltman, New York).

2. Kennedy, John F., 1961, State of the Union address, National Archives, http://www.archives.gov/press/press-releases/2009/nr09-51-images.html (accessed Sept. 25, 2010).

3. See Caillon, N., J. P. Severinghaus, J. Jouzel, et al., 2003, "Timing of atmospheric CO_2 and Antarctic temperature changes across termination III," *Science* 299: 1728–31; also Loulergue, L., F. Parrenin, T. Blunier, et al., 2007, "New constraints on the gas age-ice age difference along the EPICA ice cores, 0-50 kyr," *Climate of the Past* 3: 527–40. There is still some chance that the CO_2 does not lag temperature, but the preferred interpretation has a lag of a few centuries.

4. Extensive material on development and testing of climate models is given in Randall, D. A., R. A. Wood, S. Bony, R. Colman, T. Fichefet, J. Fyfe, V. Kattsov, A. Pitman, J. Shukla, J. Srinivasan, R. J. Stouffer, A. Sumi and K. E. Taylor, 2007, "Climate Models and Their Evaluation," in Solomon, S., D. Qin, M. Manning, Z. Chen, M. Marquis, K. B. Averyt, M. Tignor and H. L. Miller (eds.), *Climate Change 2007: The Physical Science Basis. Contribution of Working Group I to the Fourth Assessment Report of the Intergovernmental Panel on Climate Change* (Cambridge University Press, New York). Much of the following text is based on that chapter, or material referenced therein. I have never built an atmosphere-ocean general circulation model, but I have built a lot of other models of processes, especially related to glaciers and ice sheets, dating back to my master's thesis in 1983 and the refereed paper that grew out of it (Alley, R. B. and I. M. Whillans, 1984, "Response of the East Antarctic ice sheet to sea-level rise," *Journal of Geophysical Research* 89[C]: 6487–93). I thus have working knowledge of the process of model-building, and the strengths and weaknesses of such modeling. For additional information on the accuracy of climate models, and their improvement over time, see Reichler, T. and J. Kim, 2008, "How well do coupled models simulate today's climate?" *Bulletin of the American Meteorological Society* 89: 303–11.

5. I focus on roughness here not because it is the most important parameter, but because it does matter, and it is especially easy for me to explain. To learn more, see pp. 630–32 in Hansen, J., G. L. Russell, D. Rind, et al., 1983, "Efficient three-dimensional global models for climate studies: Models I and II," *Monthly Weather Review* 111: 609–83.

6. The climate is often taken as the average conditions over the two or three decades just after the mid-1970s, when good satellite data became available and enhanced our ability to map important quantities. The parameter-tuning exercise to match this climate is a wise approach. Very often, a situation arises in which something, say, evaporation rate, is a function of several other things, such as soil moisture and wind speed and soil roughness. If you conduct a field study and measure

each of these individually, you may find that calculating the evaporation rate from the physics of the system and from the central estimates of wind speed, soil roughness, and so on, produces an evaporation rate that is a bit different from the measured value. Many possibilities exist—the measured value of evaporation is off, or the physics are incomplete, or one or more of the central values is not optimal. Suppose, for the sake of argument, that you have measured the evaporation rate several ways, and it is very well known. The physical model is also good. But the roughness has a wide uncertainty range because it is hard to measure, and tweaking the roughness from the central value to a slightly higher one well within the uncertainty range allows you to calculate the evaporation rate accurately. You would be justified in making this adjustment. You would also remain just a little worried. So you would do a few tests to see what happens if you tweak other values instead of this one, and make sure it isn't hugely important. You would also tell your field-science colleagues that this is a number that you need to know better. In addition, you would know that there are enough such decisions involved in building a big model that other teams building independent models are likely to make different adjustments, so you would compare your output with theirs to see how you are doing. You would also watch in the future to see whether your evaporation rates are matching data as new results become available.

7. See chapter 6, note 17.

8. See note 4.

9. Rahmstorf, S., A. Cazenave, J. A. Church, et al., 2007, "Recent climate observations compared to projections," *Science* 316: 709; Le Treut, H., R. Somerville, U. Cubasch, Y. Ding, C. Mauritzen, A. Mokssit, T. Peterson and M. Prather, 2007, "Historical Overview of Climate Change," in Solomon, S., D. Qin, M. Manning, Z. Chen, M. Marquis, K. B. Averyt, M. Tignor and H. L. Miller (eds.), *Climate Change 2007: The Physical Science Basis, Contribution of Working Group I to the Fourth Assessment Report of the Intergovernmental Panel on Climate Change* (Cambridge University Press, New York); and Hansen, J., K. Sato, R. Ruedy, et al., 2006, "Global temperature change," *Proceedings of the National Academy of Sciences of the United States of America* 103: 14,288–93, doi:10.1073/pnas.0606291103. Hansen's 1988 temperature projections have been misrepresented in some parts of the blogosphere. He provided "high," "medium," and "low" warming scenarios. The actual causes of climate change have come closest to his "medium," and the climate change is well within the error bars of his "medium," so most of us consider that this is a successful model test. However, a plot widely disseminated for a while had the "low" and "medium" removed, comparing what actually happened only to the "high" curve. Needless to say, this made the model look bad, which was probably the intent. To get a little of the flavor of how polarized this can become, read the discussion that follows the blog post by Gavin Schmidt at http://www .realclimate.org/index.php/archives/2007/05/hansens-1988-projections/.

10. To learn more about attribution of climate change—did we humans do it, or did something in nature?—see Hegerl, G. C., F. W. Zwiers, P. Braconnot, N. P. Gillett, Y. Luo, J. A. Marengo Orsini, N. Nicholls, J. E. Penner and P. A. Stott, 2007, "Understanding and Attributing Climate Change," in Solomon, S., D. Qin, M. Manning, Z. Chen, M. Marquis, K. B. Averyt, M. Tignor and H. L. Miller (eds.), *Climate Change 2007: The Physical Science Basis, Contribution of Working Group I to the Fourth Assessment Report of the Intergovernmental Panel on Climate Change* (Cambridge University Press, New York); also Santer, B. D., C. Mears, F. J. Wentz, et al., 2007, "Identification of human-induced changes in atmospheric moisture content," *Proceedings of the National Academy of Sciences of the United States of America* 104: 15,248–53; and Santer, B. D., K. E. Taylor, P. J. Glecler, et al., 2009, "Incorporating model quality information in climate change detection and attribution studies," *Proceedings of the National Academy of Sciences of the United States of America* 106: 14,778–83.

11. See the Le Treut et al. reference, cited in note 9. Also see Lorenz, E. N., 1963, "Deterministic nonperiodic flow," *Journal of the Atmospheric Sciences* 20: 130–41. The great Ed Lorenz gave us insights to chaos, and understanding it is hugely important, but it is also widely misunderstood.

9: The Moving Finger Writes

Chapter title. The titles of chapters 9 and 10 come from the following:

> *"The Moving Finger writes: and, having writ,*
> *Moves on: nor all thy Piety nor Wit*
> *Shall lure it back to cancel half a Line,*
> *Nor all thy Tears wash out a Word of it."*

The Rubáiyát of Omar Khayyám, 1889, 5th ed., trans. Edward FitzGerald.

1. Shakespeare, W., 1610–11?, *The Tempest,* 1.2.

2. I am presenting a very short, incomplete summary of a very fascinating field. I wrote a book that covered more aspects of this (Alley, R. B., 2000, *The Two-Mile Time Machine: Ice Cores, Abrupt Climate Change, and Our Future* [Princeton University Press, Princeton, NJ]). There are several good textbooks, including Cronin, T. M., 1999, *Principles of Paleoclimatology* (Columbia University Press, New York); and Bradley, R. S., 1999, *Paleoclimatology—Reconstructing Climate of the Quaternary* (Academic Press, San Diego).

3. Prahl, F. G. and S. G. Wakeham, 1987, "Calibration of unsaturation patterns in long-chain ketone compositions for palaeotemperature assessment," *Nature* 330: 367–69. By analogy, lard is stiff because it has few double bonds ("saturated" fat); vegetable oil is fluid because it has lots of double bonds ("unsaturated" fat). Observations in the ocean and in the laboratory show that the algae change the

ratio of single to double bonds with temperature; that they do this to control brittleness/flexibility of the cell wall is a hypothesis.

4. Additional background information is provided in Alley, R. B., J. J. Fitzpatrick, J. Brigham-Grette, et al., 2009, "Paleoclimate Concepts," in *Past Climate Variability and Change in the Arctic and at High Latitudes,* report by the U.S. Climate Change Science Program and Subcommittee on Global Change Research (U.S. Geological Survey, Reston, VA), pp. 22–76. Sometimes when indicators don't agree, checking shows an error in something. Other times, we learn about additional parts of the climate; see, for example, Denton, G. H., R. B. Alley, G. C. Comer and W. S. Broecker, 2005, "The role of seasonality in abrupt climate change," *Quaternary Science Reviews* 24: 1159–82.

5. In addition to the references in notes 2 and 4, see Alley, R. B., C. A. Shuman, D. A. Meese, et al., 1997, "Visual-stratigraphic dating of the GISP2 ice core: Basis, reproducibility, and application," *Journal of Geophysical Research* 102(C12): 26,367–81. Also see Broecker, W. S., 2002, *The Glacial World according to Wally* (Eldigio Press, New York).

6. See, for example, Jansen, E., J. Overpeck, K. R. Briffa, J.-C. Duplessy, F. Joos, V. Masson-Delmotte, D. Olago, B. Otto-Bliesner, W. R. Peltier, S. Rahmstorf, R. Ramesh, D. Raynaud, D. Rind, O. Solomina, R. Villalba and D. Zhang, 2007, "Palaeoclimate," in Solomon, S., D. Qin, M. Manning, Z. Chen, M. Marquis, K. B. Averyt, M. Tignor and H. L. Miller (eds.), *Climate Change 2007: The Physical Science Basis. Contribution of Working Group I to the Fourth Assessment Report of the Intergovernmental Panel on Climate Change* (Cambridge University Press, New York).

7. Paleomagnetics—"paleomagic," as we sometimes call it—is an almost magical branch of the Earth sciences. Heat a magnet too much, and the frenetic vibrations of its atoms destroy the magnetic properties. Cool it back down, and the magnetization will return, oriented with the magnetic fields around it. After it is cool, turning the magnet will turn its magnetization to point in a direction different from the field around it. Earth has a strong magnetic field, which is always more or less aligned with Earth's rotation, so when a lava flow cools on Earth, the little magnets in the rocks point north-south. Today, the N arrow points north; at certain times in the past, the field flipped ends and the N arrow would have pointed south. But only during the short times when the field was flipping would the arrow have been confused. Note also that Earth's magnetic field is nearly horizontal near the equator but nearly vertical at the poles; explorers near one pole wind a brass wire around the opposite end of a compass needle to keep it nearly horizontal rather than having its tip so steeply that it rubs the top or bottom of the case and won't turn. The angle between a nearly horizontal feature such as a lava flow and the frozen-in "compass needles" of that flow thus tells whether the lava flow was near the poles or the equator when it cooled. See the references in note 2 for more detail.

8. See, for example, Finkel, R. C. and K. Nishiizumi, 1997, "Beryllium 10 concentrations in the Greenland Ice Sheet Project 2 ice core from 3-40 ka," *Journal of Geophysical Research* 102C: 26,699–706; and Muscheler, R., R. Beer, P. W. Kubik and H.-A. Synal, 2005, "Geomagnetic field intensity during the last 60,000 years based on Be-10 and Cl-36 from the Summit ice cores and C-14," *Quaternary Science Reviews* 24: 1849–60. The concentration of beryllium-10 in an ice core or sediment core is not identically the concentration in the air; see Alley, R. B., R. C. Finkel, K. Nishiizumi, et al., 1995, "Changes in continental and sea-salt atmospheric loadings in central Greenland during the most recent deglaciation," *Journal of Glaciology* 41: 503–14.

9. See section 6.6, especially 6.6.3.1, in Jansen et al. cited in note 6. As discussed in the references there, I have oversimplified a bit what the solar scientists do in estimating the brightness of the sun in the past.

10. Carbon-14 and other isotopes are used along with beryllium-10, although carbon-14 brings its own complications because it is involved in Earth's carbon cycle. Note that a change in cosmic-ray flux from outside the solar system would also change production of beryllium-10 and other cosmogenic isotopes, but that thus far we lack any strong evidence of changes in the supply of cosmic rays from deep space.

11. Winckler, G. and H. Fischer, 2006, "30,000 years of cosmic dust in Antarctica," *Science* 313: 491. Also, Winckler, G., R. F. Anderson, M. Stute and P. Schlosser, 2004, "Does interplanetary dust control 100 kyr glacial cycles?" *Quaternary Science Reviews* 23: 1873–78; and Gabrielli, P., J. M. C. Plane, C. F. Boutron, et al., 2006, "A climatic control on the accretion of meteoric and super-chondritic iridium-platinum to the Antarctic ice cap," *Earth and Planetary Science Letters* 250: 459–69.

12. See the Scripps CO_2 page at http://scrippsco2.ucsd.edu/home/index.php (accessed Nov. 8, 2009). Also see Keeling, C. D., 1960, "The concentration and isotopic abundances of carbon dioxide in the atmosphere," *Tellus* 12: 200–203; and Keeling, C. D., 1978, "The Influence of Mauna Loa Observatory on the Development of Atmospheric CO_2 Research," in Miller, J. (ed.), *Mauna Loa Observatory: A 20th Anniversary Report* (National Oceanic and Atmospheric Administration, Environmental Research Laboratories, Boulder, CO), pp. 36–54. Mauna Loa is an active volcano, but not a lot of CO_2 comes out, and Keeling moved safely upwind of the vents into the sea breezes. Before he started at Mauna Loa, he made measurements in other places. Among other things, he learned about variability in atmospheric concentration of CO_2 associated with day-night and summer-winter cycles, with CO_2 dropping as growing trees take it in during daytime in spring and summer, and CO_2 rising at night and in winter from respiration. He also started measurements at the South Pole, about as far from trees as you can get. Sometimes, people on the edge of the blogosphere dredge up

a few scattered pre-Keeling measurements from people who knew less about separating the local effects of trees or fossil-fuel burning from the wider signal, and these blogospherians try to use these inferior results to somehow discredit Keeling and claim that there is no rise in CO_2. This is just silly. For Keeling's manometric and other techniques, see Keeling, C. D., 1958, "The concentration and isotopic abundances of atmospheric carbon dioxide in rural areas," *Geochimica et Cosmochimica Acta* 13: 322–35.

13. See, for example, Raynaud, D., J. M. Barnola, J. Chappellaz, et al., 2000, "The ice record of greenhouse gases: A view in the context of future changes," *Quaternary Science Reviews* 19: 9–17; and Luthi, D., M. Le Floch, B. Bereiter, et al., 2008, "High-resolution carbon dioxide concentration record 650,000–800,000 years before present," *Nature* 453: 379–82. My book, cited in note 2, also may be helpful. The heavier molecules in the air are slightly concentrated in the bubbles by gravity, and small differences between the composition of trapped air and the composition of the free atmosphere may be introduced by thermal diffusion in response to any temperature gradients across the layer of snow turning to ice— called "firn"—but corrections for these are straightforward using nitrogen and argon isotopic ratios; see Severinghaus, J. P., T. Sowers, E. J. Brook, R. B. Alley and M. L. Bender, 1998, "Timing of abrupt climate change at the end of the Younger Dryas interval from thermally fractionated gases in polar ice," *Nature* 391: 141–46. In very impure or warm ice the CO_2 records are notably degraded; see, for example, Campen, R. K., T. Sowers and R. B. Alley, 2003, "Evidence of microbial consortia metabolizing within a low-latitude mountain glacier," *Geology* 31: 231–34. CO_2 records from Greenland ice cores usually differ slightly from those from Antarctica, with the deviations in Greenland correlated to the chemistry of the ice, because there are just enough more impurities in Greenland than in Antarctica to cause a little chemistry involving CO_2; as a consequence, the CO_2 records discussed in assessments are all from Antarctica, because the Greenland data aren't quite good enough.

Ice-core bubbles are trapped a few tens of meters down, with air moving through spaces in the firn above. The time for air to diffuse to the trapping depth is much less than the time for snow to be buried to that depth; and the time for burial to the trapping depth at different sites is different. If some characteristic of the ice were controlling CO_2 concentrations in the Antarctic records, then when major climatic changes occurred, the CO_2, methane, and other gases trapped in bubbles would change in the same ice as the changes in impurities and ice isotopic ratios (e.g., Monnin, E., A. Indermühle, A. Dällenbach, et al., 2001, "Atmospheric CO_2 concentrations over the last glacial termination," *Science* 291: 112–14); instead, the gas and ice records show changes in samples of the same age but different depths, with the depth offset different in different cores owing to different snowfall rates and temperatures, further evidence that the atmospheric composition of the past is being recorded, and not some artifact of contamination from the ice.

The ice cores are generally sampled discretely for CO_2—a sample here, and then one there—for technical reasons, even though some other things in ice cores are analyzed continuously. But it is very unlikely that any large changes in CO_2 have been missed between samples. Not all of the bubbles in a sample closed at exactly the same time, so any discrete sample is characterized both by a "midpoint age" when about half of its bubbles had closed, and an age span. The gaps between midpoint ages of samples that have been analyzed so far are not large—an average of 570 years for samples from 650,000 to 800,000 years ago (Luthi, D., M. Le Floch, B. Bereiter, et al., 2008, "High-resolution carbon dioxide concentration record 650,000–800,000 years before present," *Nature* 453: 379–82), and only 180 years between samples from 22,000 to 9,000 years ago as described in the Monnin reference in the previous paragraph. The age span means that in many cases the samples do overlap a little, with no large gaps at any age. Any large change in atmospheric CO_2 that was trapped in at least a few bubbles would influence the average value for a sample, so the observed smoothness of the records shows that there haven't been large, short-lived CO_2 changes. This is reinforced because the discrete samples from different ice cores will have slightly different midpoint ages, further filling in the time gaps. We can thus be confident that all of the big changes in CO_2 have been detected over the ice-core record. Our physical understanding of the carbon cycle also shows that very large and very short-lived changes cannot occur; we will see this in chapter 10 when I discuss the hundreds of thousands of years required to draw down the CO_2 from the Paleocene-Eocene Thermal Maximum (PETM).

14. Beerling, D. J., 1999, "Stomatal Density and Index: Theory and Application," in Jones, T. P. and N. P. Rowe (eds.), *Fossil Plants and Spores: Modern Techniques* (Geological Society, London); also Royer, D. L., R. A. Berner and D. J. Beerling, 2001, "Phanerozoic atmospheric CO_2 change: Evaluating geochemical and paleobiological approaches," *Earth-Science Reviews* 54: 349–92; Royer, D. L., 2008, "Linkages between CO_2, climate, and evolution in deep time," *Proceedings of the National Academy of Sciences of the United States of America* 105: 407–8; and Kurschner, W. M., Z. Kvacek and D. L. Dilcher, 2008, "The impact of Miocene atmospheric carbon dioxide fluctuations on climate and the evolution of terrestrial ecosystems," *Proceedings of the National Academy of the United States of America* 105: 449–53.

15. Almost all carbon is 6-proton/6-neutron carbon-12, but a little is 7-neutron carbon-13, and a tiny bit is radioactive 8-neutron carbon-14. All are chemically carbon, but the carbon-12 diffuses a little faster, reacts a little more easily, and so is preferentially incorporated into growing plants. When CO_2 is rare, the plant has to use what is available, including more of the carbon-13–bearing CO_2. Hence, the carbon-12 : carbon-13 ratio of a plant is a measure of CO_2 avail-

ability. This can be applied in marine settings (see, e.g., Freeman, K. H. and J. M. Hayes, 1992, "Fractionation of carbon isotopes by phytoplankton and estimates of ancient CO_2 levels," *Global Biogeochemical Cycles* 6: 185–98; and Pagani, M., M. A. Arthur and K. H. Freeman, 1999, "Miocene evolution of atmospheric carbon dioxide," *Paleoceanography* 14: 273–92), and to special plants growing in fresh water called "bryophytes" (Fletcher, B. J., D. J. Beerling, S. J. Brentnall and D. L. Royer, 2005, "Fossil bryophytes as recorders of ancient CO_2 levels: Experimental evidence and a Cretaceous case study," *Global Biogeochemical Cycles* 19, doi:10.1029/2005GB002495), and to carbonates formed in soils from carbon in organic matter that had lived in plants above (Cerling, T. E., 1992, "Further comments on using carbon isotopes in palaeosols to estimate the CO_2 content of the paleo-atmosphere," *Journal of the Geological Society of London* 149: 673–75; and Cerling, T. E., 1992, "Use of carbon isotopes in paleosols as an indicator of the PCO_2 of the paleoatmosphere," *Global Biogeochemical Cycles* 6: 307–14).

CO_2 also can be estimated by looking at the history of boron in the ocean. The boron there occurs primarily as either $B(OH)_3$ or $B(OH)_4^-$. If the ocean becomes more acidic, the H^+ of the acid grabs an OH^- off the $B(OH)_4^-$, so increasing acidity shifts more of the boron to $B(OH)_3$. The heavy isotope boron-11 is enriched in $B(OH)_3$ compared to $B(OH)_4^-$, which concentrates the light boron-10. The total boron content of the ocean, and its isotopic composition, change only slowly in the ocean (over many millions of years or longer). With the average boron isotopic composition remaining fixed, while the $B(OH)_3$ is always isotopically heavier than the $B(OH)_4^-$, a paleo-pH meter for acidity is available. When the acidity is high, most of the boron is $B(OH)_3$, so its isotopic composition almost matches the mean composition in the ocean, with the $B(OH)_4^-$ notably lighter. Low acidity shifts most of the boron to the $B(OH)_4^-$ form, so its isotopic composition must nearly match the boron isotopic composition of the mean ocean, with the little bit of $B(OH)_3$ isotopically heavier. The charged form $B(OH)_4^-$ goes into shells in place of the charged carbonate. See Honisch, B., N. G. Hemming, D. Archer, M. Siddall and J. F. McManus, 2009, "Atmospheric carbon dioxide concentration across the mid-Pleistocene transition," *Science* 324: 1551–54; also Tripati, A. K., C. D. Roberts and R. A. Eagle, 2009, "Coupling of CO_2 and ice sheet stability over major climate transitions of the last 20 million years," *Science* 326: 1394–97.

16. The models that estimate atmospheric CO_2 over time are very powerful Earth-system biochemical-cycling models; Berner, R. A., A. C. Lasaga and R. M. Garrels, 1983, "The carbonate-silicate geochemical cycle and its effect on atmospheric carbon dioxide over the past 100 millions years," *American Journal of Science* 283: 641–83; Berner, R. A., 1991, "A model for atmospheric CO_2 over Phanerozoic time," *American Journal of Science* 291: 339–76; Berner, R. A., 2004, *The Phanerozoic Carbon Cycle: CO_2 and O_2* (Oxford University Press, Oxford, UK); Berner, R. A.,

2006, "GEOCARBSULF: A combined model for Phanerozoic atmospheric O_2 and CO_2," *Geochimica et Cosmochimica Acta* 70: 5653–64; and Royer, D. L., R. A. Berner, I. P. Montanez, N. J. Tabor and D. J. Beerling, 2004, "CO_2 as a primary driver of Phanerozoic climate change," *GSA Today* 14: 4–10.

17. For a little more background, see Alley, R. B., 2009, "The Biggest Control Knob: Carbon Dioxide in Earth's Climate History" (presented at A23A, Bjerknes Lecture, AGU, San Francisco, December), http://www.agu.org/meetings/fm09/lectures/videos.php.

18. You may also see this big rock from space called an asteroid, or bolide, or impactor. "Asteroid" is probably used most commonly, but I will stick with "meteorite" here.

19. There exists an immense literature on the meteorite that killed the dinosaurs, but the best starting point is probably the major review article by Schulte, P., L. Alegret, I. Arenillas, et al., 2010, "The Chicxulub asteroid impact and mass extinction at the Cretaceous-Paleogene boundary," *Science* 327: 1214–18.

20. For example, see Koeberl, C., K. A. Farley, B. Peucker-Ehrenbrink and M. A. Sephton, 2004, "Geochemistry of the end-Permian extinction event in Austria and Italy: No evidence for an extraterrestrial component," *Geology* 32: 1053–56.

21. Firestone, R. B., A. West, J. P. Kennett, et al., 2007, "Evidence for an extraterrestrial impact 12,900 years ago that contributed to the megafaunal extinctions and the Younger Dryas cooling," *Proceedings of the National Academy of Sciences of the United States of America* 104: 16,016–21; and Kennett, D. J., J. P. Kennett, A. West, et al., 2009, "Shock-synthesized hexagonal diamonds in Younger Dryas boundary sediments," *Proceedings of the National Academy of Sciences of the United States of America* 106: 12,623–28.

22. Marlon, J. R., P. J. Bartlein, M. K. Walsh, et al., 2009, "Wildfire responses to abrupt climate change in North America," *Proceedings of the National Academy of Sciences of the United States of America* 106: 2519–24; Buchanan, B., M. Collard and K. Edinborough, 2008, "Paleoindian demography and the extraterrestrial impact hypothesis," *Proceedings of the National Academy of Sciences of the United States of America* 105: 11,651–54; Surovell, T. A., V. T. Holliday, J. A. M. Gingerich, et al., 2009, "An independent evaluation of the Younger Dryas extraterrestrial impact hypothesis," *Proceedings of the National Academy of Sciences of the United States of America* 106: 18,155–58; Kerr, R., 2007, "Mammoth-killer impact gets mixed reception from Earth scientists," *Science* 316: 1264–65; Kerr, R., 2008, "Experts find no evidence for a mammoth-killer impact," *Science* 319: 1331–32; Kerr, R., 2009, "Did the mammoth slayer leave a diamond calling card?" *Science* 323: 26; Haynes, C. V., Jr., J. Boerner, K. Domanik, et al., 2010, "The Murray Springs Clovis site, Pleistocene extinction, and the question of extraterrestrial impact," *Proceedings of National Academy of Sciences of the United States of America* 107: 4010–15;

Haynes, C. V., Jr., J. Boerner, K. Domanik, et al., 2010, "Reply to Firestone et al.: No confirmation of impact at the lower Younger Dryas boundary at Murray Springs, AZ, E106," *Proceedings of National Academy of Sciences of the United States of America* 107: E106, published ahead of print date of June 8, 2010; and Scott, A. C., N. Pinter, M. E. Collinson, et al., 2010, "Fungus, not comet or catastrophe, accounts for carbonaceous spherules in the Younger Dryas 'impact layer,'" *Geophysical Research Letters* 37: L14302, doi: 10.1029/2010GL043345.

23. Continuous sedimentary records are available from the sea floor and would reveal a dinosaur-killer–sized event with high fidelity, but old, cold sea floor tends to sink back into the mantle, so except for a little region of the Mediterranean, the oldest sea floor is less than 200 million years old; Muller, R. D., M. Sdrolias, C. Gaina and W. R. Roest, 2008, "Age, spreading rates and spreading symmetry of the world's ocean crust," *Geochemistry Geophysics Geosystems* 9: Q04006, doi:10.1029/2007GC001743. Shelly fossils became common a bit over 500 million years ago, helping make a clearer and more complete fossil record, so even without sea floor, it is likely that we would know if a dinosaur-killer–sized event had occurred less than about 500 million years ago.

24. Sleep, N. H., K. J. Zahnle, J. F. Kasting and H. J. Morowitz, 1989, "Annihilation of ecosystems by large asteroid impacts on the early Earth," *Nature* 342: 139–42.

25. See note 11.

26. With thanks to James F. (Jim) Kasting for instruction on how to calculate this. The "Mars-sized" object that formed the moon might have been as much as twice the mass of Mars, making it 20% the mass of Earth. It can't have been going too fast, though, or it would have blasted material too far away to be captured to form the moon. If the speed is taken as 10% more than Earth's, and moving in the same direction as Earth, the change in angular momentum of Earth assuming retention of all the mass blasted out would be the mass of the impactor, 0.2 times Earth's mass, multiplied by the 0.1 excess in velocity, or 0.02 = 2%. Putting all of this angular-momentum change into a change in the distance of Earth from the sun would give a 2% change, which is an upper limit. For the meteorite that killed the dinosaurs at the end of the Cretaceous, the mass was about 1.6×10^{-10} that of Earth—huge if it hit you on the head, but miniscule compared to the planet. It was probably coming rather rapidly, roughly 20 km/second, compared to the 30 km/second orbital speed of Earth. In this case, the change in angular momentum would be about 1×10^{-10}, or 1 part in 10 billion. As we are about 1.5×10^{10} cm from the sun, that translates into something like 1.5 cm as the maximum change in Earth's distance from the sun, or roughly half an inch.

27. Meehl, G. A., J. M. Arblaster, K. Matthes, F. Sassi and H. van Loon, 2009,

"Amplifying the Pacific climate system response to a small 11-year solar cycle," *Science* 325: 1114–18.

28. See note 6.

29. See, for example, Bond, G., B. Kromer, J. Beer, et al., 2001, "Persistent solar influence on north Atlantic climate during the Holocene," *Science* 294: 2130–36; Denton, G. H. and W. Karlen, 1973, "Holocene climatic variations—Their pattern and possible cause," *Quaternary Research* 3: 155–205; and Wang, Y. J., H. Cheng, R. L. Edwards, et al., 2005, "The Holocene Asian monsoon: Links to solar changes and North Atlantic climate," *Science* 308: 854–57. However, also see Haam, E. and P. Huybers, 2010, "A test for the presence of covariance between time-uncertain series of data with application to the Dongge Cave speleothem and atmospheric radiocarbon records," *Paleoceanography* 25: PA2209, doi:10.1029/2008PA001713.

30. See note 27.

31. Mann, M. E., Z. H. Zhang, S. Rutherford, et al., 2009, "Global signatures and dynamical origins of the Little Ice Age and Medieval Climate Anomaly," *Science* 326: 1256–60.

32. Coworkers and I made another attempt to find an amplifier of solar forcing in the paleoclimatic record. I really enjoyed this, learning about the physics of stochastic resonance. However, our mechanism seems to require bigger ice sheets around the North Atlantic than we have today, and I fear that our mechanism wouldn't work even then; see Alley, R. B., S. Anandakrishnan and P. Jung, 2001, "Stochastic resonance in the North Atlantic," *Paleoceanography* 16: 190–98; and Alley, R. B., 2007, "Wally was right: Predictive ability of the North Atlantic 'conveyor belt' hypothesis for abrupt climate change," *Annual Review of Earth and Planetary Sciences* 35: 241–72.

33. For a discussion, see Damon, P. E. and P. Laut, 2004, "Pattern of strange errors plagues solar activity and terrestrial climate data," *Eos, Transactions, American Geophysical Union* 85: 370, 374.

34. Svensmark, H., T. Bondo and J. Svensmark, 2009, "Cosmic ray decreases affect atmospheric aerosols and clouds," *Geophysical Research Letters* 36: L15101, doi:10.1029/2009GL038429.

35. Calogovic, J., C. Albert, F. Arnold, et al., 2010, "Sudden cosmic ray decreases: No change of global cloud cover," *Geophysical Research Letters* 37: L03802, doi:10.1029/2009GL041327; Laken, B., A. Wolfendale and D. Kniveton, 2009, "Cosmic ray decreases and changes in the liquid water cloud fraction over the oceans," *Geophysical Research Letters* 36: L23803, doi:10.1029/2009GL040961; and Kulmala, M., I. Riipinen, T. Nieminen, et al., 2010, "Atmospheric data over a solar cycle: No connection between galactic cosmic rays and new particle formation," *Atmospheric Chemistry and Physics* 10: 1885–98.

36. Pierce, J. R. and P. J. Adams, 2009, "Can cosmic rays affect cloud condensation nuclei by altering new particle formation rates?" *Geophysical Research Letters* 36: L09820, doi: 10.1029/2009GL037946. For additional insight on cosmic rays and climate, or some of the more fanciful treatments of solar variability and climate, there are several useful sources: Pittock, B., 2009, "Can solar variations explain variations in the Earth's climate? An editorial comment," *Climatic Change* 96: 483–87, and the several interesting papers by Pittock referenced therein; Duffy, P. B., B. D. Santer and T. M. L. Wigley, 2009, "Solar variability does not explain late-20th-century warming," *Physics Today* 62: 48–49; Forster, P., V. Ramaswamy, P. Artaxo, T. Berntsen, R. Betts, D. W. Fahey, J. Haywood, J. Lean, D. C. Lowe, G. Myhre, J. Nganga, R. Prinn, G. Raga, M. Schulz and R. Van Dorland, 2007, "Changes in Atmospheric Constituents and in Radiative Forcing," in Solomon, S., D. Qin, M. Manning, Z. Chen, M. Marquis, K. B. Averyt, M. Tignor and H. L. Miller (eds.), *Climate Change 2007: The Physical Science Basis. Contribution of Working Group I to the Fourth Assessment Report of the Intergovernmental Panel on Climate Change* (Cambridge University Press, New York); and Lockwood, M. and C. Frohlich, 2007, "Recent oppositely directed trends in solar climate forcings and the global mean surface air temperature," *Proceedings of the Royal Society of London Series A* 463: 2447–60. Also of interest is section 6.11.2.2 of Ramaswamy, V., O. Boucher, J. Haigh, D. Hauglustaine, J. Haywood, G. Myhre, T. Nakajima, G. Y. Shi and S. Solomon, 2001, "Radiative Forcing of Climate Change," in Houghton, J. T., Y. Ding, D. J. Griggs, M. Noguer, P. J. van der Linden, X. Dai, K. Maskell and C. A. Johnson (eds.), *Climate Change 2001: The Scientific Basis. Contribution of Working Group I to the Third Assessment Report of the Intergovernmental Panel on Climate Change* (Cambridge University Press, New York), http://www .grida.no/publications/other/ipcc_tar/, and in particular http://www.ipcc.ch/ ipccreports/tar/wg1/246.htm. There is no trend in cosmic rays over the instrumental record, as shown by, for example, Richardson, I. G., E. W. Cliver and H. V. Cane, "Long-term trends in interplanetary magnetic field strength and solar wind structure during the twentieth century," *Journal of Geophysical Research* 107(A): article 1304, doi:10.1029/2001JA000507.

37. See note 8.

38. See note 6.

39. See note 6.

40. See Huybers, P. and C. Langmuir, 2009, "Feedback between deglaciation, volcanism, and atmospheric CO_2," *Earth and Planetary Science Letters* 286: 479–91. Higher pressure tends to reduce melting in the mantle, so the weight of ice growing on a volcano tends to suppress eruptions. Lots of volcanoes had more ice cover during ice ages than now, so this would have reduced eruptions as the ice grew and favored eruptions as the ice melted. Moving water from the ocean to the glaciers on land would have favored undersea eruptions while the ice was

growing and reduced on-land eruptions, but the bigger changes in load, and in volcanic eruptions, were on land. In addition, the flexing of the land as glaciers and sea level changed may have helped crack rocks and allow melted rock to reach the surface and erupt. Huybers and Langmuir looked at the theory of volcanic eruptions and ice ages, and at the history of eruptions, and suggested that the effect of ice ages on CO_2 release may have been bigger than the effect on sun-blocking sulfuric acid, and that changes in volcanic CO_2 may have contributed to the total change in atmospheric CO_2 from the ice age to today, as described in chapter 11, providing part of the positive feedback. As I am writing this, the Huybers and Langmuir results are quite new and haven't been digested by the community, but they are fascinating.

If the question is "Do volcanoes warm the climate, or cool it?" one perfectly accurate answer is "yes." An increase in volcanic activity would be expected to cool over short times of centuries by putting up particles that block the sun, and warm over much longer times as CO_2 accumulates in the atmosphere while the particles fall down. A possible analogy—if you start eating more, you will gain weight. But if the first thing you eat is a bunch of hot peppers, which send you to the bathroom in a hurry, you may lose weight over the short term. The time frame you are considering really does change the answer in some cases. If you think this is really cool, either you're a scientist or you could have been one. If it drives you crazy, or you wish that scientists would just give you straight answers rather than this mumbo-jumbo, you almost certainly didn't get this far in the end notes of this book, and you're missing out on what we scientists think is fun.

41. Prothero, D. R. and R. H. Dott Jr., 2010, *Evolution of the Earth*, 8th ed. (McGraw-Hill, New York); or Kump, L. R., J. F. Kasting and R. G. Crane, 2009, *The Earth System*, 3rd ed. (Prentice-Hall, Upper Saddle River, NJ).

42. Barron, E. J., P. J. Fawcett, W. H. Peterson, D. Pollard and S. L. Thompson, 1995, "A simulation of mid-Cretaceous climate," *Paleoceanography* 10: 953–62; also Donnadieu, Y., R. Pierrehumbert, R. Jacob and F. Fluteau, 2006, "Modelling the primary control of paleogeography on Cretaceous climate," *Earth and Planetary Science Letters* 248: 426–37; also see Kump et al. in note 41.

10: And Having Writ, Moves On

1. Twain, M., 1883, chapter 22 in *Life on the Mississippi* (Chatto and Wincus, Picadilly).

2. See, for example, Bice, K. L., D. Birgel, P. A. Meyers, et al., 2006, "A multiple proxy and model study of Cretaceous upper ocean temperatures and atmospheric CO_2 concentrations," *Paleoceanography* 21: PA2002; Royer, D. L., 2010, "Fossil soils constrain ancient climate sensitivity," *Proceedings of the National Academy of Sciences of the United States of America* 107: 517–18; and Breecker, D. O., Z. D. Sharp

and L. D. McFadden, 2010, "Atmospheric CO_2 concentrations during ancient greenhouse climates were similar to those predicted for 2100 A.D.," *Proceedings of the National Academy of Sciences of the United States of America* 107: 576–80. The uncertainties in tropical temperature are probably a few degrees. The estimated CO_2 concentrations were roughly three to four times preindustrial values, but with notable uncertainties—two times and five times are probably consistent with the data too. Much hotter temperatures and much higher CO_2 levels then than now are virtually unavoidable.

3. Kasting, J. F., 1993, "Earth's early atmosphere," *Science* 259: 920–26. We don't have direct observations, or even useful proxies, for the brightness of the sun that far back in time, but the increase in brightness from the aging of the sun leads to about a 1% change over 100 million years.

4. Donnadieu, Y., R. Pierrehumbert, R. Jacob and F. Fluteau, 2006, "Modelling the primary control of paleogeography on Cretaceous climate," *Earth and Planetary Science Letters* 248: 426–37.

5. See note 2.

6. Prothero, D. R. and R. H. Dott Jr., 2010, *Evolution of the Earth*, 8th ed. (McGraw-Hill, New York).

7. Royer, D. L., R. A. Berner, I. P. Montanez, N. J. Tabor and D. J. Beerling, 2004, "CO_2 as a primary driver of Phanerozoic climate change," *GSA Today* 14: 4–10; Royer, D. L., 2006, "CO_2-forced climate thresholds during the Phanerozoic," *Geochimica et Cosmochimica Acta* 70: 5665–75; and Came, R. E., J. M. Eiler, J. Veizer, et al., 2007, "Coupling of surface temperatures and atmospheric CO_2 concentrations during the Palaeozoic era," *Nature* 449: 198–203. CO_2 levels probably were a bit higher than recently, offsetting the slightly dimmer sun, as I discuss later in this chapter.

8. Royer, D. L, R. A. Berner and J. Park, 2007, "Climate sensitivity constrained by CO_2 concentrations over the past 420 million years," *Nature* 446: 530–32.

9. See, for example, Kump, L. R., M. A. Arthur, M. E. Patzkowsky, et al., 1999, "A weathering hypothesis for glaciation at high atmospheric pCO_2 during the Late Ordovician," *Palaeogeography, Palaeoclimatology, Palaeoecology* 152: 173–87, together with Royer, D. L., 2006, "CO_2-forced climate thresholds during the Phanerozoic," *Geochimica et Cosmochimica Acta* 70: 5665–75. Also see Royer, D. L., 2008, "Linkages between CO_2, climate, and evolution in deep time," *Proceedings of the National Academy of Sciences of the United States of America* 105: 407–8; and Kurschner, W. M., Z. Kvacek and D. L. Dilcher, 2008, "The impact of Miocene atmospheric carbon dioxide fluctuations on climate and the evolution of terrestrial ecosystems," *Proceedings of the National Academy of Sciences of the United States of America* 105: 449–53. Also, Tripati, A. K., C. D. Roberts and R. A. Eagle, 2009, "Coupling of CO_2 and ice sheet stability over major climate transitions of the last 20 million years," *Science* 326: 1394–97; Pagani, M., Z. Liu, J. LaRiviere and A. C. Ravelo, 2010,

"High Earth-system climate sensitivity determined from Pliocene carbon dioxide concentrations," *Nature Geoscience* 3: 27–30; Young, S. A., M. R. Saltzman, W. I. Ausich, A. Desrochers and D. Kaljo, 2010, "Did changes in atmospheric CO_2 coincide with latest Ordovician glacial-interglacial cycles?" *Palaeogeography, Palaeoclimatology, Palaeoecology* 296: 376–88, doi:10.1016/j.palaeo.2010.02.033; and Ruddiman, W. F., 2010, "Climate: A paleoclimatic enigma?" *Science* 328: 838–39.

10. Royer, D. L., 2006, "CO_2-forced climate thresholds during the Phanerozoic," *Geochimica et Cosmochimica Acta* 70: 5665–75.

11. See Alley, R. B., 2009, "The Biggest Control Knob: Carbon Dioxide in Earth's Climate History" (presented at A23A, Bjerknes Lecture, AGU, San Francisco, December), http://www.agu.org/meetings/fm09/lectures/videos.php.

12. Prentice, I. C., G. D. Farquhar, M. J. R. Fasham, M. L. Goulden, M. Heimann, V. J. Jaramillo, H. S. Kheshgi, C. Le Quéré, R. J. Scholes and D. W. R. Wallace, 2001, "The Carbon Cycle and Atmospheric Carbon Dioxide," in Houghton, J. T., Y. Ding, D. J. Griggs, M. Noguer, P. J. van der Linden, X. Dai, K. Maskell and C. A. Johnson (eds.), *Climate Change 2001: The Scientific Basis. Contribution of Working Group I to the Third Assessment Report of the Intergovernmental Panel on Climate Change* (Cambridge University Press, New York), http://www.grida.no/publications/other/ipcc_tar/. Note also that some CO_2 is released by metamorphism before the rocks melt, but that this generally is included in estimates of CO_2 release from Earth.

13. The headstones made of granite usually change more slowly than any others in the graveyard, but all of them are changing. Even in the granite, cracks may be visible, with a bit of rust on some of the darker mineral grains; microscopic examination will reveal other changes.

14. There is much chemical complexity in a rock formed from stuff thrown out of a volcano, or a granite solidifying beneath a volcano, but the rock's formula is sometimes oversimplified to $CaSiO_3$, ignoring the sand, rust, clay, and salt for seawater. The breakdown of this simplified "rock" to make things that dissolve in water is as follows:

$$CaSiO_3 + 3H_2O + 2CO_2 \rightarrow Ca^{+2} + H_4SiO_4 + 2HCO_3^-$$

with streams dissolving everything on the right side of the equation to be carried to the ocean. There, corals and clams and coccolithophores build their shells from calcium carbonate, whereas sponges and diatoms and radiolaria build their shells from silica, so the next step is

$$Ca^{+2} + H_4SiO_4 + 2HCO_3^- \rightarrow CaCO_3 + SiO_2 + 3H_2O + CO_2$$

You end up with the same amount of water as before, but one CO_2 has been used to make a clam shell. Just adding these rock-weathering and shell-growth equations together, and crossing out all the H_2O and the one CO_2 that appear on both sides, gives

$$CaSiO_3 + CO_2 \rightarrow CaCO_3 + SiO_2$$

In words, volcanoes release rock and CO_2, which recombine to make shells. Over millions of years or longer, these are then subducted (dragged down by sinking tectonic plates) to feed volcanoes, or are cooked in the hearts of mountain belts to make metamorphic rocks and feed hot springs, with basically the same effect as volcanoes. Note that if there weren't shell-making critters in the ocean, the shell-making chemicals would build up and eventually be deposited to make rocks in some places without help from living things.

15. A lot of the shells, or their inorganically precipitated equivalents, sit for a while in rocks before being heated to release their CO_2. Earth has enough carbon to give us a Venusian atmosphere with surface temperatures hot enough to melt lead. We have avoided such a fate because so much of our carbon is in rocks rather than in the air. All of the carbon in the CO_2 in the air weighs less than 800 gigatons (800 Gt C), compared to perhaps as much as 6000 Gt C in fossil fuels, 10,000,000 Gt C of formerly living things that would make fossil fuels if they were much more concentrated, and 40,000,000 Gt C in carbonate rocks, most of which started off as shells in the sea (Kump, L. R., J. F. Kasting and R. G. Crane, 2009, *The Earth System*, 3rd ed. [Prentice-Hall, Upper Saddle River, NJ]; see their figure 8-3). Note also that I am using "shell" loosely, including the carbonate produced by coralline algae, for example. The carbon in these carbonate rocks cannot be burned to get energy because it is already in the oxidized/burned CO_2 form.

16. Over about a millennium or so, the CO_2 in the ocean and the CO_2 in the atmosphere reach a balance, and there is a whole lot more CO_2 dissolved in the ocean (mainly as bicarbonate, HCO_3^-) than in the atmosphere. If extra CO_2 is released to the air over millennia or longer, this new CO_2 has time to come into balance with the CO_2 in the ocean, so almost all of the total CO_2 will end up in the ocean. Hence, the answer to the question of how long it takes for a change in rock-weathering or volcanic CO_2 to affect the atmospheric composition is more or less the total amount of CO_2 in the ocean plus that in atmosphere, divided by the volcanic flux, or crudely half a million years. See Kump et al. in note 15, and Walker, J. C. G., P. B. Hays and J. F. Kasting, 1981, "A negative feedback mechanism for the long-term stabilization of Earth's surface temperature," *Journal of Geophysical Research* 86: 9776–82.

17. See Sagan, C. and G. Mullen, 1972, "Earth and Mars: Evolution of atmospheres and surface temperatures," *Science* 177: 52–56; Walker, J. C. G., P. B. Hays and J. F. Kasting, 1981, "A negative feedback mechanism for the long-term stabilization of Earth's surface temperature," *Journal of Geophysical Research* 86: 9776–82; Halevy, I., R. T. Pierrehumbert and D. P. Schrag, 2009, "Radiative transfer in

CO_2-rich paleoatmospheres," *Journal of Geophysical Research—Atmospheres* 114: D18112; Chyba, C. F., 2010, "Countering the early faint sun," *Science* 328: 1238–39; Valley, J. W., W. H. Peck, E. M. King and S. A. Wilde, 2002, "A cool early Earth," *Geology* 30: 351–54; and Kasting, J. F., 2010, "Early Earth: Faint young sun redux," *Nature* 464: 687–89.

18. An excellent place to start reading about snowball Earth is Hoffman, P. F. and D. P. Schrag, 2002, "The snowball Earth hypothesis: Testing the limits of global change," *Terra Nova* 14: 129–55. Use of iridium anomalies to constrain the duration of the snowball is described by Bodiselitsch, B., C. Koeberl, S. Master and W. U. Reimold, 2005, "Estimating duration and intensity of Neoproterozoic snowball glaciations from Ir anomalies," *Science* 308: 239–42. Use of boron and calcium isotopes to find evidence of acidification of the ocean as the snowball melted, and then of enhanced weathering contributing to the cap carbonate, is given by Kasemann, S. A., C. J. Hawkesworth, A. R. Prave, A. E. Fallick and P. N. Pearson, 2005, "Boron and calcium isotope composition in Neoproterozoic carbonate rocks from Namibia: Evidence for extreme environmental change," *Earth and Planetary Science Letters* 231: 73–86. To learn about the ice physics on a snowball, see Pollard, D. and J. F. Kasting, 2005, "Snowball Earth: A thin-ice solution with flowing sea glaciers," *Journal of Geophysical Research—Oceans* 110(C7): C07010. Note that this is a rapidly evolving scientific field, and the story is not yet completely told; see, for example, Le Hir, G., Y. Donnadieu, Y. Godderis, et al., 2009, "The snowball Earth aftermath: Exploring the limits of continental weathering processes," *Earth and Planetary Science Letters* 277: 453–63.

19. Raymo, M. E. and W. F. Ruddiman, 1992, "Tectonic forcing of late Cenozoic climate," *Nature* 359: 117–22; and Gibbs, M. T., G. J. S. Bluth, P. J. Fawcett and L. R. Kump, 1999, "Global chemical erosion over the last 250 my: Variations due to changes in paleogeography, paleoclimate, and paleogeology," *Science* 1999: 611–51.

20. See Berner, R. A., 2003, "The long-term carbon cycle, fossil fuels and atmospheric composition," *Nature* 426: 323–26; and Falkowski, P. G. and Y. Isozaki, 2008, "The story of O_2," *Science* 322: 540–42.

21. Yes, I know that the birds are sufficiently closely related to dinosaurs that some people argue that the dinosaurs didn't die, they still fly. But those dinosaurs that could flatten your house by wagging their tails indeed are gone, along with a lot of smaller ones. Also, following note 18 in chapter 9, the term "meteorite" can be used for big or small things that hit Earth, but "asteroid" is restricted to big things, so many people prefer to refer to the dinosaur-killer as an asteroid.

22. The early geologists picked certain points in geologic history to change names. We now can see that most of these choices were wise, because they were marked by major events. The early geologists knew what they were doing.

23. Zachos, J. C., M. W. Wara, S. Bohaty, et al., 2003, "A transient rise in tropical

sea surface temperature during the Paleocene-Eocene Thermal Maximum," *Science* 302: 1551–54; and Kennett, J. P. and L. D. Stott, 1991, "Abrupt deep-sea warming, paleoceanographic changes and benthic extinctions at the end of the Paleocene," *Nature* 353: 225–28.

24. Zachos, J. C., U. Rohl, S. A. Schellenberg, et al., 2005, "Rapid acidification of the ocean during the Paleocene-Eocene Thermal Maximum," *Science* 308: 1611–15.

25. Pagani, M., K. Caldeira, D. Archer and J. C. Zachos, 2006, "An ancient carbon mystery," *Science* 314: 1556–57; Zachos, J. C., S. M. Bohaty, C. M. John, et al., 2007, "The Palaeocene-Eocene carbon isotope excursion: Constraints from individual shell planktonic foraminifer records," *Philosophical Transactions of the Royal Society A: Mathematical, Physical and Engineering Sciences* 365: 1829–42; and Gibbs, S. J., H. M. Stoll, P. R. Bown and T. J. Bralower, 2010, "Ocean acidification and surface water carbonate production across the Paleocene-Eocene thermal maximum," *Earth and Planetary Science Letters* 295: 583–92.

26. The extinction was prominent in bottom-dwelling foraminifera; Thomas, E. and N. J. Shackleton, 1996, "The Paleocene-Eocene benthic foraminiferal extinction and stable isotope anomalies," in Knox, R. W. O. B., R. M. Corfield and R. E. Dunay (eds.), *Correlation of the Early Paleogene in Northwest Europe*, vol. 101 (Geological Society of London), pp. 401–41.

27. Smith, J. J., S. T. Hasiotis, M. J. Kraus and D. T. Woody, 2009, "Transient dwarfism of soil fauna during the Paleocene-Eocene Thermal Maximum," *Proceedings of the National Academy of Sciences of the United States of America* 106: 17,655–60; Woodburne, M. O., G. F. Gunnell and R. K. Stucky, 2009, "Climate directly influences Eocene mammal faunal dynamics in North America," *Proceedings of the National Academy of Sciences of the United States of America* 106: 13,399–403; and Alroy, J., P. L. Koch and J. C. Zachos, 2000, "Global climate change and North American mammalian evolution," in Erwin, D. H. and S. L. Wing (eds.), *Paleobiology* 26, no. 4, Supplement: 259–88.

28. Currano, E. D., P. Wilf, S. L. Wing, et al., "Sharply increased insect herbivory during the Paleocene-Eocene thermal maximum," *Proceedings of the National Academy of Sciences of the United States of America* 105: 1960–64.

29. Rohl, U., T. Westerhold, T. J. Bralower and J. C. Zachos, 2007, "On the duration of the Paleocene-Eocene thermal maximum (PETM)," *Geochemistry Geophysics Geosystems* 8: Q12002. The CO_2 for the PETM started out as living or formerly living organic material, not inorganic CO_2 from volcanoes. The event was mostly completed in just a couple hundred thousand years, slow on human time scales but faster than expected for the rock-weathering/CO_2 thermostat acting alone. However, we would expect that if a lot of methane came out of sea-floor clathrates to help cause the event, natural leakage of methane from below the clathrates would have been greatly reduced after the event because there wouldn't have been much methane remaining to leak. Then, consumption of CO_2 to form

plants to form methane would not have been balanced by leakage of methane to be oxidized to CO_2, so formation of new clathrates would have lowered atmospheric CO_2. Hence, the "fast" (couple-hundred-thousand-year) end of the event is consistent with our understanding of the relevant processes (Dickens, G. R., 2003, "Rethinking the global carbon cycle with a large, dynamic and microbially mediated gas hydrate capacitor," *Earth and Planetary Science Letters* 213: 169–83).

30. As described in chapter 9, note 15, living things preferentially are composed of isotopically light carbon. Records of carbon isotopes in sediments from land and sea across the PETM show that it was marked by very light values, so the extra carbon came from living or formerly living things and not from volcanoes. Carbon isotopic compositions across the event are mentioned in many of the papers cited above and below, including the paper by Zachos et al. in note 25.

31. Panchuk, K., A. Ridgwell and L. R. Kump, 2008, "Sedimentary response to Paleocene-Eocene Thermal Maximum carbon release: A model-data comparison," *Geology* 36: 315–18; and Higgins, J. A. and D. P. Schrag, 2006, "Beyond methane: Towards a theory for the Paleocene-Eocene Thermal Maximum," *Earth and Planetary Science Letters* 245: 523–37.

32. Dickens, G. R., J. R. O'Neil, D. K. Rea and R. M. Owen, 1995, "Dissociation of oceanic methane hydrate as a cause of the carbon isotope excursion at the end of the Paleocene," *Paleoceanography* 10: 965–71; and Buffett, B. and D. Archer, 2004, "Global inventory of methane clathrate: Sensitivity to changes in the deep ocean," *Earth and Planetary Science Letters* 227: 185–99.

33. Bralower, T. J., D. J. Thomas, J. C. Zachos, et al., 1997, "High-resolution records of the late Paleocene thermal maximum and circum-Caribbean volcanism: Is there a causal link?" *Geology* 25: 963–66; Svensen, H., S. Planke, A. Malthe-Sorenssen, et al., 2004, "Release of methane from a volcanic basin as a mechanism for initial Eocene global warming," *Nature* 429: 542–45; Kurtz, A. C., L. R. Kump, M. A. Arthur, J. C. Zachos and A. Paytan, 2003, "Early Cenozoic decoupling of the global carbon and sulfur cycles," *Paleoceanography* 18: article 1090; and Higgins, J. A. and D. P. Schrag, 2006, "Beyond methane: Towards a theory for the Paleocene-Eocene Thermal Maximum," *Earth and Planetary Science Letters* 245: 523–37. The article by Higgins and Schrag is a good starting point.

34. Nicolo, M. J., G. R. Dickens, C. J. Hollis and J. C. Zachos, 2007, "Multiple early Eocene hyperthermals: Their sedimentary expression on the New Zealand continental margin and in the deep sea," *Geology* 35: 699–702.

35. Ridgwell, A. and D. N. Schmidt, 2010, "Past constraints on the vulnerability of marine calcifiers to massive carbon dioxide release," *Nature Geoscience* 3: 196–200.

36. One of the most interesting geologic events is the great dying of 251 million years ago, and it is linked to warmth and deposition of mud-rocks rich in organic material—black shales. Perhaps 95% of the known species on the planet became extinct at that time. A species can survive if even a very few individuals hang

on, so virtually all of the things living then must have died. A rush of science in the last few years is rapidly nailing down what happened in this end-Permian extinction. It coincides in time with the greatest volcanic outpouring for more than half a billion years. It appears that the resulting warmth from the volcanic CO_2, and the sulfur and nutrients from enhanced weathering of the new volcanic rocks, were instrumental in using up the oxygen in much of the ocean water, not just in the sediments. When many dead plants are available without oxygen, bacteria use sulfate from the water for burning and release hydrogen sulfide. This rotten-egg-stink chemical is highly toxic to many living things in even rather low concentrations (U.S. Centers for Disease Control and Prevention, National Institute for Occupational Safety and Health IDLH [Immediate Danger to Life or Health] level is 100 parts per million in the air; http://www.cdc.gov/niosh/npg/npgd0337.html [accessed Nov. 12, 2009]).

Green sulfur bacteria, *Chlorobiaceae*, use hydrogen sulfide instead of water for photosynthesis, producing sulfur instead of oxygen, and these bacteria make distinctive organic molecules. These molecules are found in sediments deposited during the end-Permian extinction, pointing to a vast change in ocean chemistry that introduced widespread hydrogen sulfide to the shallow, sunlit zone of the ocean. In turn, this likely released large quantities of hydrogen sulfide to the air from an anoxic and sulfidic ocean (Grice, K., C. Q. Cao, G. D. Love, et al., 2005, "Photic zone euxinia during the Permian-Triassic superanoxic event," *Science* 307: 706–9; also see Hotinski, R. M., K. L. Bice, L. R. Kump, R. G. Najjar and M. A. Arthur, 2001, "Ocean stagnation and end-Permian anoxia," *Geology* 29: 7–10; and Kump, L. R., A. Pavlov and M. A. Arthur, 2005, "Massive release of hydrogen sulfide to the surface ocean and atmosphere during intervals of oceanic anoxia," *Geology* 33: 397–400).

Waters with levels of hydrogen sulfide that are toxic to many species are called "euxinic." Ironically, this name comes from the Black Sea, where deep salty waters in a basin are largely trapped by a fresher lens floating on top so that oxygen is not carried down, producing one of the rare euxinic environments on Earth today. The irony arises because the Black Sea's name in Greek, *Euxineos Pontos*, means "hospitable sea," named for the uppermost, oxygenated layer and not for the deeper, euxinic layers that were apparently unknown to the Greeks. Unlike the Black Sea, the end-Permian ocean seems to have extended the deep euxinia upward to poison the air. At least two other major extinctions, one before and one after the end-Permian extinction, also are associated with times of warmth, euxinia, and black shales (Meyer, K. M. and L. R. Kump, 2008, "Oceanic euxinia in Earth history: Causes and consequences," *Annual Reviews of Earth and Planetary Science* 36: 251–88). Excess CO_2 that built up in the anoxic ocean may have contributed to extinction events (Knoll A. H., R. K. Barnbach,

J. L. Payne, S. Pruss and W. W. Fischer, 2007, "Paleophysiology and end-Permian mass extinction," *Earth and Planetary Science Letters* 256: 295–313).

Thus far, there is no credible suggestion that human-induced warming will even approach triggering such an event in our future. Still, if you want an absolutely worst-case nightmare scenario of what excess warming and associated fertilization might do, belching poison gases from an overheated, anoxic, euxinic ocean to kill almost every living thing on the planet is a powerfully bad dream indeed!

For additional insights on CO_2 and warmth, see, for example, Kauffman, E. G., 1995, "Global change leading to biodiversity crisis in a greenhouse world: The Cenomanian-Turonian (Cretaceous) mass extinction," in *Effects of Past Global Change on Life* (National Academies Press, Washington, DC), pp. 47–71; Pancost, R. D., N. Crawford, S. Magness, et al., 2004, "Further evidence for the development of photic-zone euxinic conditions during Mesozoic oceanic anoxic events," *Journal of the Geological Society of London* 161: 353–64; Jenkyns, H. C., 2010, "Geochemistry of oceanic anoxic events," *Geochemistry Geophysics Geosystems* 22: Q03004; and Erba, E., C. Bottini, H. J. Weissert and C. E. Keller, 2010, "Calcareous nannoplankton response to surface-water acidification around Oceanic Anoxic Event 1a," *Science* 329: 428–32.

11: The Great Ice That Covers the Land

1. Agassiz, L., 1886, "Ice-Period in America," in *Geological Sketches*, 2nd series (Houghton Mifflin, Boston), p. 99.

2. Zachos, J., M. Pagani, L. Sloan, E. Thomas and K. Billups, 2001, "Trends, rhythms, and aberrations in global climate 65 Ma to present," *Science* 292: 686–93; Climate Change Science Program (Alley, R. B., J. Brigham-Grette, G. H. Miller, L. Polyak and J. W. C. White [coordinating lead authors]), 2009, *Past Climate Variability and Change in the Arctic and at High Latitude,* report by the U.S. Climate Change Science Program and Subcommittee on Global Change Research (U.S. Geological Survey, Reston, VA).

3. DeConto, R. M. and D. Pollard, "Rapid Cenozoic glaciation of Antarctica induced by declining atmospheric CO_2," *Nature* 421: 245–49.

4. Lemke, P., J. Ren, R. B. Alley, I. Allison, J. Carrasco, G. Flato, Y. Fujii, G. Kaser, P. Mote, R. H. Thomas and T. Zhang, 2007, "Observations: Changes in Snow, Ice and Frozen Ground," in Solomon, S., D. Qin, M. Manning, Z. Chen, M. Marquis, K. B. Averyt, M. Tignor and H. L. Miller (eds.), *Climate Change 2007: The Physical Science Basis. Contribution of Working Group I to the Fourth Assessment Report of the Intergovernmental Panel on Climate Change* (Cambridge University Press, New York).

5. Jansen, E., J. Overpeck, K. R. Briffa, J.-C. Duplessy, F. Joos, V. Masson-Delmotte,

D. Olago, B. Otto-Bliesner, W. R. Peltier, S. Rahmstorf, R. Ramesh, D. Raynaud, D. Rind, O. Solomina, R. Villalba and D. Zhang, 2007, "Palaeoclimate," in Solomon, S., D. Qin, M. Manning, Z. Chen, M. Marquis, K. B. Averyt, M. Tignor and H. L. Miller (eds.), *Climate Change 2007: The Physical Science Basis. Contribution of Working Group I to the Fourth Assessment Report of the Intergovernmental Panel on Climate Change.*(Cambridge University Press, New York).

6.   Milankovitch, M., 1941, *Canon of Insolation and the Ice Age Problem* (Belgrade); Imbrie, J. and K. P. Imbrie, 1979, *Ice Ages: Solving the Mystery* (Harvard University Press, Cambridge, MA); Broecker, W. S., 2002, *The Glacial World according to Wally* (Eldigio Press, New York); and Alley, R. B., 2000, *The Two-Mile Time Machine: Ice Cores, Abrupt Climate Change, and Our Future* (Princeton University Press, Princeton, NJ).

7.   More accurately, the precession takes a bit over 26,000 years, but it interacts with a pivoting of the whole orbit to cause the changing distribution of sunshine to repeat roughly every 19,000 or 23,000 years.

8.   Note that under very special conditions, it might just be possible to roll a planet over; Williams, D. M., J. F. Kasting and L. A. Frakes, 1998, "Low-latitude glaciation and rapid changes in the Earth's obliquity explained by obliquity-oblateness feedback," *Nature* 396: 453–55. For recent times and behavior, this doesn't matter.

9.   Eccentricity changes primarily from the slight tug of Jupiter's gravity as we pass it in our respective orbit, and varies with a 400,000-year cycle as well as the better-known 100,000-year cycle. Eccentricity is calculated as the difference between the maximum and minimum distances of Earth from the sun, divided by the sum of these distances, and ranges from 0.0034 (virtually circular) to 0.058 (still pretty close to circular). Modern eccentricity is fairly low (0.0167) and decreasing slowly. The precession and obliquity of the child's top change as Earth's gravity tugs on it; for Earth, the precession and obliquity vary as the gravity of the moon and sun act on the slight bulge at the equator caused by the planet's rotation. Today, our obliquity is about 23.4 degrees and roughly halfway through a decrease. Precession has little effect on the total sunshine a place receives for a year (the total number of watts reaching the top of the atmosphere in a year), because more-intense summers caused by closer approach to the sun are also shorter; with a noncircular orbit around an off-center sun, Earth spends less time close to the sun and more time far from the sun during a year. The trade-off between distance and time spent at that distance also means that variations in eccentricity cause the total sunshine received by the planet in a year to change a tiny bit—less than 0.1%. Changes in the obliquity and precession have a vanishingly small influence on the total sunshine received by the planet in a year.

10.   See the references in notes 5 and 6.

11.   Most of the recent ice ages terminated with southern warming when the local

peak-summer sunshine was falling (usually estimated as the number of watts received at a place at the top of the atmosphere on the first day of summer, or during some other chosen days or weeks close to the first day of summer), even though the total-year sunshine (the number of watts received at a place at the top of the atmosphere for the whole year) was rising. However, the southern warming at the end of the ice age about 250,000 years ago occurred when both local peak-summer sunshine and local total-year sunshine were dropping, arguing that the south is being controlled from elsewhere in ice-age cycles. The ice ages seem to end when total-year sunshine is high at both poles and northern midsummer sunshine is rising. Cheng, H., R. L. Edwards, W. S. Broecker, et al., 2009, "Ice age terminations," *Science* 326: 248–52.

12. A good place to start is the source in note 5. Also useful is Hansen, J., M. Sato, P. Kharecha, et al., 2008, "Target atmospheric CO_2: Where should humanity aim?" *Open Atmospheric Science Journal* 2: 217–31. Also, Cuffey, K. M. and E. J. Brook, 2000, "Ice sheets and the ice-core record of climate change," in Jacobson, M. C., R. J. Charlson, H. Rodhe and G. H. Orians (eds.), *Earth System Science: From Biogeochemical Cycles to Global Change*, International Geophysics Series, vol. 72 (Academic Press, London), pp. 459–97. I reviewed and extended some of this in Alley, R. B., 2003, "Paleoclimatic insights into future climate challenges," *Philosophical Transactions of the Royal Society of London Series A: Mathematical, Physical and Engineering Sciences* 361: 1831–49, 10.1098/rsta.2003.1254.

13. The most direct treatment of this issue is probably the one by Hansen et al., cited in note 12. Note that as one goes from millennia to millions of years, the rock-weathering thermostat pulls the CO_2 concentration down.

14. Ibid.

15. See Anderson, R. F., S. Ali, L. I. Bradtmiller, et al., 2009, "Wind-driven upwelling in the Southern Ocean and the deglacial rise in atmospheric CO_2," *Science* 323: 1443–48; also Denton, G. H., R. F. Anderson, J. R. Toggweiler, et al., 2010, "The last glacial termination," *Science* 328: 1652–56. The story is surely not this simple; for more on the causes of the CO_2 changes over ice-age cycles, start with Jansen et al. as cited in note 5, or Broecker's *Glacial World according to Wally*, cited in note 6. Also very useful is Brovkin, V., A. Ganopolski, D. Archer and S. Rahmstorf, 2007, "Lowering of glacial atmospheric CO_2 in response to changes in oceanic circulation and marine biogeochemistry," *Paleoceanography* 22: PA4202; and Archer, D., A. Winguth, D. Lea and N. Mahowald, "What caused the glacial/interglacial atmospheric pCO(2) cycles?" *Reviews of Geophysics* 38: 159–89. There were probably fewer plants on land during ice ages, so the extra CO_2 didn't go into land plants. Changes in rock-weathering, fossil fuels, and other things involving rocks are primarily too slow to really matter. Just as cooling a glass of soda pop causes it to hold more of its CO_2 "fizz," the colder ocean would have held more CO_2, but the ocean also was saltier during the ice age because fresh water evaporated

to grow the ice sheets, and salting your pop or your ocean drives off CO_2, almost offsetting the effect of the cooling.

16. Some friendly sources on zombie worms were prepared by the Monterey Bay Aquarium Research Institute, to accompany release of important scientific studies. These include "Whale carcass yields bone-devouring worms," http://www.mbari.org/news/news_releases/2004/whalefall.html, and "Whale falls: Islands of abundance and diversity in the deep sea," http://www.mbari.org/news/news_releases/2002/dec20_whalefall.html (both accessed Oct. 28, 2009). There is a large and growing technical literature on the subject. One important paper is Rouse, G. W., S. K. Goffredi and R. C. Vrijenhoek, 2004, "*Osedax*—Bone-eating marine worms with dwarf males," *Science* 305: 668–71. Also see Goffredi, S. K., C. K. Paull, K. Fulton-Bennett, L. A. Hurtado and R. C. Vrijenhoek, 2004, "Unusual benthic fauna associated with a whale fall in Monterey Canyon, California," *Deep Sea Research, Part I: Oceanographic Research Papers* 51: 1295–1306; and Baco, A. R. and C. R. Smith, 2003, "High species richness in deep-sea chemoautotrophic whale skeleton communities," *Marine Ecology-Progress Series* 260: 109–14. The chemistry of digesting a fallen whale is complex. Sulfate-reducers produce sulfides, and chemosynthetic bacteria use the sulfides late in the degradation process.

17. See note 15.

18. See note 5.

19. White, J. W. C., R. B. Alley, A. Jennings, et al., 2009, "Past Rates of Climate Change in the Arctic," in *Past Climate Variability and Change in the Arctic and at High Latitudes*, report by the U.S. Climate Change Program and Subcommittee on Global Change Research. (U.S. Geological Survey, Reston, VA), pp. 247–302.

12: Kindergarten Soccer and the Last Century of Climate

1. "Dr. Corrie, in his diary, December 10, 1838, remarks that, passing by Parker's Piece on that day, he 'saw some forty gownsmen playing at football. The novelty and liveliness of the scene was very amusing.'" Gray, A., 1902, *Jesus College* (F. E. Robinson, London), p. 148. Dr. G. E. Corrie was master of Jesus College, Cambridge, UK.

2. White, J. W. C., R. B. Alley, A. Jennings, et al., 2009, "Past Rates of Climate Change in the Arctic," in *Past Climate Variability and Change in the Arctic and at High Latitudes,* report by the U.S. Climate Change Program and Subcommittee on Global Change Research (U.S. Geological Survey, Reston, VA), pp. 247–302.

3. Jansen, E., J. Overpeck, K. R. Briffa, J.-C. Duplessy, F. Joos, V. Masson-Delmotte, D. Olago, B. Otto-Bliesner, W. R. Peltier, S. Rahmstorf, R. Ramesh, D. Raynaud, D. Rind, O. Solomina, R. Villalba and D. Zhang, 2007, "Palaeoclimate," in Solomon, S., D. Qin, M. Manning, Z. Chen, M. Marquis, K. B. Averyt, M. Tignor

and H. L. Miller (eds)., *Climate Change 2007: The Physical Science Basis. Contribution of Working Group I to the Fourth Assessment Report of the Intergovernmental Panel on Climate Change* (Cambridge University Press, New York). Also see Bennike, O., D. Bloshiyanov, J. Dowdeswell, et al. (CAPE Project Members), 2001, "Holocene paleoclimate data from the Arctic: Testing models of global climate change," *Quaternary Science Reviews* 20: 1275–87; also Renssen, H., H. Goosse, T. Fichefet, et al., 2005, "Simulating the Holocene climate evolution at northern high latitudes using a coupled atmosphere-sea ice-ocean-vegetation model," *Climate Dynamics* 24: 23–43; and Kaufman, D. S., T. A. Ager, N. J. Anderson, et al., 2004, "Holocene thermal maximum in the western Arctic (0–180 degrees W)," *Quaternary Science Reviews* 23: 529–60.

4. Many scientists avoid the term "Medieval Warm Period" because even the northern lands most affected were not warm everywhere, and the warmth that did occur was at somewhat different times in different places. However, "Medieval Climate Anomaly," while scientifically more accurate, doesn't have quite the same name recognition. For information on the Medieval Climate Anomaly and the Little Ice Age, see Jansen et al. in note 3, and the report from the National Research Council Committee on Surface Temperature Reconstructions for the Last 2,000 Years, Board on Atmospheric Sciences and Climate, 2006, *Surface Temperature Reconstructions for the Last 2,000 Years* (National Academies Press, Washington, DC). I provided a requested oral presentation to the committee, concerning ice-core records. For a possible effect of the thermohaline circulation, see Broecker, W. S., 2000, "Was a change in thermohaline circulation responsible for the Little Ice Age?" *Proceedings of the National Academy of Sciences of the United States of America* 97: 1339–42. For a possible role of the North Atlantic Oscillation, see Trouet, V., J. Esper, N. E. Graham, et al., 2009, "Persistent positive North Atlantic Oscillation mode dominated the Medieval Climate Anomaly," *Science* 324: 78–80. Also see Fagan, B. M., 2000, *The Little Ice Age: How Climate Made History, 1300–1850* (Basic Books, New York).

5. If we use marine sediments to extend the record, CO_2 levels might have reached just a shade higher than 300 ppm over the last two million years, but modern levels are higher than they have been in two million years or longer, with fairly high confidence; see Honisch, B., N. G. Hemming, D. Archer, M. Siddall and J. F. McManus, 2009, "Atmospheric carbon dioxide concentration across the Mid-Pleistocene Transition," *Science* 324: 1551–54. Notice that whether or not CO_2 levels are really the highest in millions of years is not hugely important in projecting impacts of future CO_2 rise on humans. Being able to say "highest in more than two million years" is a bit of an attention-grabber, though.

6. Denman, K. L., G. Brasseur, A. Chidthaisong, P. Ciais, P. M. Cox, R. E. Dickinson, D. Hauglustaine, C. Heinze, E. Holland, D. Jacob, U. Lohmann, S. Ramachandran, P. L. da Silva Dias, S. C. Wofsy and X. Zhang, 2007, "Couplings

between Changes in the Climate System and Biogeochemistry," in Solomon, S., D. Qin, M. Manning, Z. Chen, M. Marquis, K. B. Averyt, M. Tignor and H. L. Miller (eds.), *Climate Change 2007: The Physical Science Basis. Contribution of Working Group I to the Fourth Assessment Report of the Intergovernmental Panel on Climate Change* (Cambridge University Press, New York). Just over half of our CO_2 has stayed in the air, with the largest part of the remainder going into the ocean, a bit into fertilizing plants, and a little into faster rock-weathering or other places; Rafelski, L. E., S. C. Piper and R. F. Keeling, 2009, "Climate effects on atmospheric carbon dioxide over the last century," *Tellus, Series B—Chemical and Physical Meteorology* 61: 718–31.

7. See Prentice, I. C., G. D. Farquhar, M. J. R. Fasham, M. L. Goulden, M. Heimann, V. J. Jaramillo, H. S. Kheshgi, C. Le Quéré, R. J. Scholes and D. W. R. Wallace, 2001, "The Carbon Cycle and Atmospheric Carbon Dioxide," in Houghton, J. T., Y. Ding, D. J. Griggs, M. Noguer, P. J. van der Linden, X. Dai, K. Maskell and C. A. Johnson (eds.), *Climate Change 2001: The Scientific Basis. Contribution of Working Group I to the Third Assessment Report of the Intergovernmental Panel on Climate Change* (Cambridge University Press, New York), http://www.grida.no/publications/other/ipcc_tar/. A summary of volcanic sources is given in Huybers, P. and C. Langmuir, 2009, "Feedback between deglaciation and volcanic emissions of CO_2," *Earth and Planetary Science Letters* 286: 479–91, which reports that 0.10–0.15 Gt of CO_2 comes from volcanoes each year. To see some of the techniques used to measure emissions from volcanoes, including by my colleagues at Penn State, see Werner, C., S. L. Brantley and K. Boomer, 2000, "CO_2 emissions related to the Yellowstone volcanic system, 2. Statistical sampling, total degassing, and transport mechanisms," *Journal of Geophysical Research* 105(B): 10,831–46; also Werner, C., G. Chiodini, D. Voigt, et al., 2003, "Monitoring volcanic hazard using eddy covariance at Solfatara volcano, Naples, Italy," *Earth and Planetary Science Letters* 210: 561–77.

8. Battle, M., M. L. Bender, P. P. Tans, et al., 2000, "Global carbon sinks and their variability inferred from atmospheric O_2 and delta C-13," *Science* 287: 2467–70. I discussed some of this in chapter 10 with the PETM as well.

9. Suess, H. E., 1955, "Radiocarbon concentration in modern wood," *Science* 122: 415–17. The "Suess effect," as CO_2 from fossil fuels lacking carbon-14 diluted the CO_2 containing carbon-14 in the atmosphere, was much easier to measure before people decided to explode atomic bombs in the air, because the bombs made a lot of carbon-14. For a while after the bombs, detecting our fossil-fuel CO_2 in the air was difficult, but the bomb-produced carbon-14 has been moving out of the air, and it is once again easy to observe the fossil-fuel effect in diluting the atmosphere's carbon-14; see, for example, Turnbull, J., P. Rayner, J. Miller, et al., 2009, "On the use of $^{14}CO_2$ as a tracer for fossil fuel CO_2: Quantifying uncer-

tainties using an atmospheric transport model," *Journal of Geophysical Research* 114: D22302, doi:10.1029/2009JD012308.

10. For changes in oxygen, see Keeling, R. F., 1996, "Global and hemispheric CO_2 sinks deduced from changes in atmospheric O_2 concentration," *Nature* 281: 218–21; also, Bender, M. L., D. T. Ho, M. B. Hendricks, et al., 2005, "Atmospheric O_2/N_2 changes, 1993–2002: Implications for the partitioning of fossil fuel CO_2 sequestration," *Global Biogeochemical Cycles* 19: GB4017.

11. As described by Spencer Weart in "The Discovery of Global Warming" (American Institute of Physics), http://www.aip.org/history/climate/, the greenhouse effect was discovered in 1824 by Joseph Fourier, the ability of CO_2 and other gases to trap heat was first measured by John Tyndall in 1859, the first estimate of the climate sensitivity (see chapter 7) to CO_2 was calculated by Svante Arrhenius in 1896, and research by the U.S. Air Force and others filled in a lot of the foundations by the 1950s.

12. "*Radiative forcing* is a measure of the influence that a factor has in altering the balance of incoming and outgoing energy in the Earth-atmosphere system and is an index of the importance of the factor as a potential climate change mechanism. Positive forcing tends to warm the surface while negative forcing tends to cool it. In this report, radiative forcing values are for 2005 relative to pre-industrial conditions defined at 1750 and are expressed in watts per square metre (W m^{-2})." Footnote 2, p. 2, in IPCC, 2007, "Summary for Policymakers," in Solomon, S., D. Qin, M. Manning, Z. Chen, M. Marquis, K. B. Averyt, M. Tignor and H. L. Miller (eds.), *Climate Change 2007: The Physical Science Basis. Contribution of Working Group I to the Fourth Assessment Report of the Intergovernmental Panel on Climate Change* (Cambridge University Press, New York).

13. Table SPM-2 in ibid. provides additional explanation and uncertainties.

14. Most of the methane comes out the front of the cow, with only a little coming out the back. See, for example, chapter 3 of Energy Information Administration, Office of Integrated Analysis and Forecasting, 1995, *Emissions of Greenhouse Gases in the United States 1987–1994* (U.S. Department of Energy, Washington, DC), http://www.eia.doe.gov/oiaf/1605/archive/95report/contents.html (accessed June 18, 2010).

15. There are many good sources, including section 7.4.2 in Denman et al. in note 6. Chapter 4 in *Emissions of Greenhouse Gases in the United States 1987–1994* in note 14 has a short overview.

16. You may still meet someone who is convinced that the world's scientists are nuts and that we should be trying to head off a coming ice age by burning fossil fuels as fast as possible. However, if we really wanted to heat things up in a hurry, we would be wiser to use other gases such as halocarbons that are much more efficient warmers.

17. Forster, P., V. Ramaswamy, P. Artaxo, T. Berntsen, R. Betts, D. W. Fahey, J. Haywood, J. Lean, D. C. Lowe, G. Myhre, J. Nganga, R. Prinn, G. Raga, M. Schulz and R. Van Dorland, 2007, "Changes in Atmospheric Constituents and in Radiative Forcing," in Solomon, S., D. Qin, M. Manning, Z. Chen, M. Marquis, K. B. Averyt, M. Tignor and H. L. Miller (eds.), *Climate Change 2007:The Physical Science Basis. Contribution of Working Group I to the Fourth Assessment Report of the Intergovernmental Panel on Climate Change* (Cambridge University Press, New York). Some of the larger eruptions can have a small multiyear or even decadal cooling impact. And while increases and decreases in the sun can occur, volcanoes either cool or they don't. The total cooling from volcanoes in a century can be substantial.

18. Ibid.

19. Folland, C. K. and D. E. Parker, 1995, "Correction of instrumental biases in historical sea-surface temperature data," *Quarterly Journal of the Royal Meteorological Society* 121: 319–67. For the citation information, I used the ISI Web of Science, accessed on Nov. 1, 2009. The start and end of World War II caused changes in which nations were making measurements and contributing data; in particular, the return of information collected by British ships to the global database after the war shifted temperatures a little because the British and the United States were measuring temperatures in different ways; see Thompson, D. W. J., J. J. Kennedy, J. M. Wallace and P. D. Jones, 2008, "A large discontinuity in the mid-twentieth century in observed global-mean surface temperature," *Nature* 453: 646–49.

20. See section 3.2.2.2 of Trenberth, K. E., P. D. Jones, P. Ambenje, R. Bojariu, D. Easterling, A. Klein Tank, D. Parker, F. Rahimzadeh, J. A. Renwick, M. Rusticucci, B. Soden and P. Zhai, 2007, "Observations: Surface and Atmospheric Climate Change," in Solomon, S., D. Qin, M. Manning, Z. Chen, M. Marquis, K. B. Averyt, M. Tignor and H. L. Miller (eds.), *Climate Change 2007: The Physical Science Basis. Contribution of Working Group I to the Fourth Assessment Report of the Intergovernmental Panel on Climate Change* (Cambridge University Press, New York). Also see Hansen, J., R. Ruedy, M. Sato, et al., 2001, "A closer look at United States and global surface temperature change," *Journal of Geophysical Research* 106: 23,947–63; Jones, P. D., P. Y. Groisman, M. Coughlan, et al., 1990, "Assessment of urbanization effects in time series of surface air temperature over land," *Nature* 347: 169–72; Karl, T. R., H. F. Diaz and G. Kukla, 1988, "Urbanization: Its detection and effect in the United States climate record," *Journal of Climate* 1: 1099–1123; Parker, D. E., 2004, "Large-scale warming is not urban," *Nature* 432: 290; Parker, D. E., 2006, "A demonstration that large-scale warming is not urban," *Journal of Climate* 19: 2882–95; Peterson, T. C., 2003, "Assessment of urban versus rural in situ surface temperatures in the contiguous United States: No difference found," *Journal of Climate* 16: 2941–59; and Peterson T. C, K. P. Gallo, J. Lawrimore, et al.,

1999, "Global rural temperature trends," *Geophysical Research Letters* 26: 329–32. Problems with the placement of the observing system in the United States may have caused a very slight underestimate of the warming that has occurred, but the effect is small; Menne, M. J., C. N. Williams Jr. and M. A. Palecki, "On the reliability of the U.S. surface temperature record," *Journal of Geophysical Research* 115: D11108, doi:10.1029/2009JD013094.

21. Ibid.

22. Among the key papers are Christy, J. R. and R. T. McNider, 1994, "Satellite greenhouse signal," *Nature* 367: 325; and Christy, J. R. and J. D. Goodridge, 1995, "Precision global temperatures from satellites and urban warming effects of non-satellite data," *Atmospheric Environment* 29: 1957–61.

23. The models do simulate the observed cooling in the upper stratosphere associated with the greenhouse warming at the surface, and successful simulation of this signal is one of many demonstrations of model skill. The satellite data being discussed here were for the troposphere, where models simulated warming with a warming surface.

24. The earlier study by the Panel on Reconciling Temperature Observations, Climate Research Committee, Board on Atmospheric Sciences and Climate, 2000, *Reconciling Observations of Global Temperature Change* (National Academies Press, Washington, DC), is useful. Key updates and interpretations are provided in Karl, T. R., S. J. Hassol, C. D. Miller and W. L. Murray (eds.), 2006, *Temperature Trends in the Lower Atmosphere: Steps for Understanding and Reconciling Differences*, report by the Climate Change Science Program and the Subcommittee on Global Change Research (Washington, DC).

25. Lemke, P., J. Ren, R. B. Alley, I. Allison, J. Carrasco, G. Flato, Y. Fujii, G. Kaser, P. Mote, R. H. Thomas and T. Zhang, 2007, "Observations: Changes in Snow, Ice and Frozen Ground," in Solomon, S., D. Qin, M. Manning, Z. Chen, M. Marquis, K. B. Averyt, M. Tignor and H. L. Miller (eds.), *Climate Change 2007: The Physical Science Basis. Contribution of Working Group I to the Fourth Assessment Report of the Intergovernmental Panel on Climate Change* (Cambridge University Press, New York).

26. Rosenzweig, C., D. Karoly, M. Vicarelli, et al., 2008, "Attributing physical and biological impacts to anthropogenic climate change," *Nature* 453: 353–57, doi:10.1038/nature06937.

27. A little of this post–World War II signal may be an artifact of the changing observational network, as explained in Thompson et al. in note 19.

28. Hegerl, G. C., F. W. Zwiers, P. Braconnot, N. P. Gillett, Y. Luo, J. A. Marengo Orsini, N. Nicholls, J. E. Penner and P. A. Stott, 2007, "Understanding and Attributing Climate Change," in Solomon, S., D. Qin, M. Manning, Z. Chen, M. Marquis, K. B. Averyt, M. Tignor and H. L. Miller (eds.), *Climate Change 2007: The Physical Science Basis, Contribution of Working Group I to the Fourth Assessment*

Report of the Intergovernmental Panel on Climate Change (Cambridge University Press, New York). Note that the experiments I'm summarizing here run up to 2005; the scientists had to complete the runs, get the scientific papers through peer review, and assess the results for the 2007 IPCC report.

29. If we could instantaneously raise the atmospheric level of CO_2 without changing the long-term rate of supply (say, one big CO_2 belch), within a few years the excess in the air would drop by about half as CO_2 shifted from the air to the ocean, but the remaining anomaly in the air would last a lot longer. There is a technicality here that is sometimes badly abused in the "skeptical" blogosphere. CO_2 moves rapidly between air and plants, and between air and ocean. As noted earlier, every breaking wave throws spray with CO_2 into the air, where the water evaporates to release the CO_2, and takes bubbles with CO_2 into the water where the CO_2 can dissolve. If you ask how long a single molecule of CO_2 stays in the air, it is only a few years. But if CO_2 goes into a leaf in the spring and comes back into the air as the leaf dies and rots in the fall, or goes into the ocean in a bubble and comes out in a spray droplet in the next wave, nothing has happened to the amount of CO_2 in the air that is contributing to the greenhouse effect. So people who tell you that the CO_2 we put in the air stays there for only a few years are technically correct for the individual molecules, but completely wrong for the things that really matter—once we raise the concentration of CO_2 in the air, the concentration takes a long time to come back down, even if the individual molecules contributing to that high concentration are exchanged into and out of plants or into and out of the surface ocean.

30. Explosive volcanoes have more or less the same issues. Tropospheric particles are washed out in a couple of weeks, particles in the stratosphere and thus above the rain can last a year or two before falling or being mixed down to the troposphere and then being washed out, whereas the CO_2 from the volcanoes lasts much longer but builds up more slowly. Therefore, a sudden increase in explosive volcanism would cause cooling followed by warming.

31. For example, Rahmstorf, S., A. Cazenave, J. A. Church, et al., 2007, "Recent climate observations compared to projections," *Science* 316: 709; Le Treut, H., R. Somerville, U. Cubasch, Y. Ding, C. Mauritzen, A. Mokssit, T. Peterson and M. Prather, 2007, "Historical overview of climate change," in Solomon, S., D. Qin, M. Manning, Z. Chen, M. Marquis, K. B. Averyt, M. Tignor and H. L. Miller (eds.), *Climate Change 2007: The Physical Science Basis, Contribution of Working Group I to the Fourth Assessment Report of the Intergovernmental Panel on Climate Change* (Cambridge University Press, New York); and Hansen, J., K. Sato, R. Ruedy, et al., 2006, "Global temperature change," *Proceedings of the National Academy of Sciences of the United States of America* 103: 14,288–93, doi:10.1073/pnas.0606291103.

13: But My Brother-in-Law Said . . .

Chapter title. Recall from the first note in chapter 5 that I have great brothers-in-law, and I am not picking on them. But, many people hear loud shouting in politics and the blogosphere, correctly infer that some of the shouting is not providing good information, don't know how to find Abe Lincoln's reliable sources, and instead get information from friends and relatives. However, all too often, the friends and relatives got their information from the shouters.

1.   *"The Moving Finger writes: and, having writ,*
        Moves on: nor all thy Piety nor Wit
        Shall lure it back to cancel half a Line,
        Nor all thy Tears wash out a Word of it."

        The Rubáiyát of Omar Khayyám, 1889, 5th ed., trans. Edward FitzGerald.

2.   Knutti, R. and G. C. Hegerl, 2008, "The equilibrium sensitivity of the Earth's temperature to radiation changes," *Nature Geoscience* 1: 735–43.

3.   Note that the volcanic particles are primarily in the stratosphere above the clouds, so we can estimate their effects much more accurately than for the human particles; the volcanic source of particles to the troposphere is small compared to the human source.

4.   See note 2.

5.   Accessed Nov. 2, 2009.

6.   Box, J. F., 1987, "Guinness, Gosset, Fisher, and small samples," *Statistical Science* 2: 45–52.

7.   Student, 1908a, "The probable error of a mean," *Biometrika* 6: 1–25; and Student, 1908b, "Probable error of a correlation coefficient," *Biometrika* 6: 302–10.

8.   For the underestimated warming by the UK record in comparison to the physically interpolated values, see Simmons, A. J., K. M. Willett, P. D. Jones, P. W. Thorne and D. P. Dee, 2010, "Low-frequency variations in surface atmospheric humidity, temperature, and precipitation: Inferences from reanalyses and monthly gridded observational data sets," *Journal of Geophysical Research* 115: D01110, doi:10.1029/2009JD012442.

9.   Borenstein, S., "Statisticians reject global cooling," Associated Press, Oct. 26, 2009.

10.   The statement was made at a Capitol Hill briefing that included Dr. Richard L. Smith, the distinguished climate scientist Warren Washington, and me. The briefing, "Climate Change: Key Questions and Answers," was sponsored by the American Association for the Advancement of Science (AAAS), the American Chemical Society, the American Geophysical Union, the American Statistical Association, the University Corporation for Atmospheric Research, the Ameri-

can Institute of Biological Sciences, the American Meteorological Society, the American Society of Agronomy, the Crop Science Society of America, the Ecological Society of America, the Geological Society of America, the National Ecological Observatory Network, and the Soil Science Society of America. The statements were reported in Lempinen, E. W., "Researchers at AAAS Capitol Hill Briefing Offer Data to Reaffirm Global Climate Change," June 1, 2010, http://www.aaas.org/news/releases/2010/0601climate_briefing.shtml.

11. Warming is unequivocal; IPCC, 2007, "Summary for Policymakers," in Solomon, S., D. Qin, M. Manning, Z. Chen, M. Marquis, K. B. Averyt, M. Tignor and H. L. Miller (eds.), *Climate Change 2007: The Physical Science Basis. Contribution of Working Group I to the Fourth Assessment Report of the Intergovernmental Panel on Climate Change* (Cambridge University Press, New York), p. SPM-5.

12. During the early 1970s, the small but rapidly growing climate-science community was investigating the cooling effect of particles, together with the observation of little warming or even slight cooling since the 1940s. And the 100,000-year ice-age cycling had just been figured out, with the realization that the slow slide into a new ice age might be due to start soon, but without the subsequent realization that the 400,000-year modulation of the 100,000-year cycle likely means that a new ice age is not due naturally for another 20,000 years or more (Berger, A. and M. F. Loutre, 2002, "An exceptionally long interglacial ahead?" *Science* 297: 1287–88). As a result, some scientists were concerned about cooling, although scientists recognized the unavoidable warming effect of the CO_2 from our rapidly rising fossil-fuel use. Overall, the scientific literature in the 1970s included many more "warming" than "cooling" papers, with the "warming" ones having much more impact, as shown by the number of citations (Peterson, T. C., W. M. Connolley and J. Fleck, 2008, "The myth of the 1970s global cooling scientific consensus," *Bulletin of the American Meteorological Society* 89: 1325–37). The National Academy of Sciences, in 1975 (National Academy of Sciences, United States Committee for the Global Atmospheric Research Program, 1975, *Understanding Climate Change: A Program for Action* [National Academies Press, Washington, DC]), discussed the various issues and suggested that serious research would be needed to provide answers. A statement in an appendix about the ice-age cycling is sometimes quoted out of context to suggest that the Academy predicted cooling, but there is no such prediction in the report. Instead, my reading of the report indicates more focus on warming, consistent with the underlying science of the time. And, by 1979, the National Academy's "Charney report" (National Academy of Sciences, 1979, *Carbon Dioxide and Climate: A Scientific Assessment* [National Academies Press, Washington, DC]) gave a clear view of future warming from continuing CO_2 emissions, which has simply been strengthened by subsequent science. *Newsweek* and a small number of other press outlets in the 1970s did combine the post–World War II cooling effect from particles with the ice-age cycling to make

a very scary cooling story (Gwynne, P., "The cooling world," *Newsweek,* Apr. 28, 1975). But even the popular media had a lot of coverage on warming. For example, the *New York Times* ran two articles in 1975 by the same reporter (Sullivan, W., "Scientists ask why world climate is changing; major cooling may be ahead," *New York Times,* May 21, 1975; and Sullivan, W., "Warming trend seen in climate; two articles counter view that cold period is due," *New York Times,* Aug. 14, 1975), the first suggesting cooling and the second answering and suggesting warming. Overall, the science in the 1970s pointed primarily to warming. And as governments and other sources funded scientists to implement the 1975 recommendations of the National Academy of Sciences and generate useful answers, the science just got stronger and stronger that the long-term concern is warming from CO_2.

13. For references on the UK data set, see Brohan, P., J. J. Kennedy, I. Harris, S. F. B. Tett and P. D. Jones, 2006, "Uncertainty estimates in regional and global observed temperature changes: A new dataset from 1850," *Journal of Geophysical Research* 111: D12106, doi: 10.1029/2005JD006548; Jones, P. D., M. New, D. E. Parker, S. Martin, S. and I. G. Rigor, 1999, "Surface air temperature and its variations over the last 150 years," *Reviews of Geophysics* 37: 173–99; Rayner, N. A., P. Brohan, D. E. Parker, et al., 2006, "Improved analyses of changes and uncertainties in marine temperature measured in situ since the mid-nineteenth century: The HadSST2 dataset," *Journal of Climate* 19: 446–69; and Rayner, N. A., D. E. Parker, E. B. Horton, et al., 2003, "Globally complete analyses of sea surface temperature, sea ice and night marine air temperature, 1871–2000," *Journal of Geophysical Research* 108: 4407, doi:10.1029/2002JD002670. The Hadley Centre web site is http://www.metoffice.gov.uk/climatechange/science/hadleycentre/ and the Climate Research Unit (CRU) site is http://www.cru.uea.ac.uk/.

14. My opinion is that almost every elected public figure who has run for reelection, at least in the United States, understands what it means to have words or votes taken out of context and used by the opposition in an attempt to sway votes, and doesn't like it.

15. See, for example, Borenstein, S., M. Ritter and R. Satter, "Climategate: Science not faked, but not pretty," *US News and World Report,* Dec. 12, 2009, http://www.usnews.com/articles/news/energy/2009/12/12/climategate-science-not-faked-but-not-pretty_print.htm (accessed Jan. 20, 2010).

16. The Science and Technology Committee (P. Willis, chair) of the United Kingdom House of Commons, 2010, "The disclosure of climate data from the Climatic Research Unit at the University of East Anglia," Eighth Report of Session 2009–10, http://www.publications.parliament.uk/pa/cm200910/cmselect/cmsctech/387/38702.htm; Oxburgh, R., H. Davies, K. Emanuel, L. Graumlich, D. Hand, H. Huppert and M. Kelly, 2010, "Report of the International Panel Set Up by the University of East Anglia to Examine the Research of the Cli-

matic Research Unit," submitted to the University 12 April 2010, updated 19 April 2010; http://www.uea.ac.uk/mac/comm/media/press/CRUstatements/ SAP; also see various statements from the University at http://www.uea.ac.uk/ mac/comm/media/press/CRUstatements; and Sir Muir Russell, chair, with G. Boulton, P. Clarke, D. Eyton and J. Norton, 2010, "The Independent Climate Change E-mail Review," University of East Anglia, http://www.cce-review .org/Meetings.php (accessed July 7, 2010). Also, Foley, H. C., A. W. Scaroni and C. A. Yekel, 2010, "RA-10 Inquiry Report: Concerning the Allegations of Research Misconduct against Dr. Michael E. Mann, Department of Meteorology, College of Earth and Mineral Sciences, The Pennsylvania State University," www.research.psu.edu/orp/Findings_Mann_Inquiry.pdf (all accessed June 26–27, 2010); and Assman, S. M., W. Castleman, M. J. Irwin, N. G. Jablonski, F. W. Vondracek, and C. Yekel, 2010, "RA-1O Final Investigation Report Involving Dr. Michael E. Mann, The Pennsylvania State University, June 4, 2010," live .psu.edu/fullimg/userpics/10026/Final_Investigation_Report.pdf (accessed July 3, 2010).

17. See the first source in note 16.

18. Mann, M. E., R. S. Bradley and M. K. Hughes, 1998, "Global-scale temperature patterns and climate forcing over the past six centuries," *Nature* 392: 779–87; and Mann, M. E., R. S. Bradley and M. K. Hughes, 1999, "Northern hemisphere temperatures during the past millennium: Inferences, uncertainties, and limitations," *Geophysical Research Letters* 26: 759–62. When these were written, Mike Mann was at the University of Virginia, but he has since moved to Penn State.

19. Jansen, E., J. Overpeck, K. R. Briffa, J.-C. Duplessy, F. Joos, V. Masson-Delmotte, D. Olago, B. Otto-Bliesner, W. R. Peltier, S. Rahmstorf, R. Ramesh, D. Raynaud, D. Rind, O. Solomina, R. Villalba and D. Zhang, 2007, "Palaeoclimate," in Solomon, S., D. Qin, M. Manning, Z. Chen, M. Marquis, K. B. Averyt, M. Tignor and H. L. Miller (eds.), *Climate Change 2007: The Physical Science Basis. Contribution of Working Group I to the Fourth Assessment Report of the Intergovernmental Panel on Climate Change* (Cambridge University Press, New York).

20. Mann, M. E., Z. H. Zhang, M. K. Hughes, et al., 2008, "Proxy-based reconstructions of hemispheric and global surface temperature variations over the past two millennia," *Proceedings of the National Academy of Sciences of the United States of America* 105: 13,252–57.

21. National Research Council, Committee on Surface Temperature Reconstructions for the Last 2,000 Years, Board on Atmospheric Sciences and Climate, 2006, *Surface Temperature Reconstructions for the Last 2,000 Years* (National Academies Press, Washington, DC).

22. See note 2.

23. Mann, M. E., Z. H. Zhang, S. Rutherford, et al., 2009, "Global signatures and dynamical origins of the Little Ice Age and Medieval Climate Anomaly," *Science*

326: 1256–60. This work suggests that the small size of the climate response from medieval times to Little Ice Age in many places is linked to dynamical changes in El Niño and related factors, which are not powerful enough to offset many decades of human-released CO_2 but which might notably affect changes over shorter times.

24. See note 21, which supports my suspicion.

25. Use of medieval to Little Ice Age changes is further complicated by uncertainty about the causes of the climate changes, so high confidence is unlikely.

26. For responses to these issues: No, some data couldn't be shared, but future sharing will probably be quicker and easier; maybe the tests weren't exactly right, but the answer was pretty good; and, no, there wasn't a statistically significant pause.

27. In general, if someone starts a discussion by stating that he or she will disprove some particular person's science—Darwin's or Einstein's, for example—or that events related to a particular person disprove a large body of science, be very cautious. The premise is not consistent with properly assessed science, and the odds are quite high that the discussion is not going to involve the best science.

14: The Future

1. Whewell, W., 1840, *Philosophy of the Inductive Sciences, Founded upon their History*, vol. 1 (John W. Parker, London), p. 228. In the original, "foretell" was spelled "foretel."

2. See chapter 8. The limits on predictability of weather are often called "chaos."

3. Hawkins, E. and R. Sutton, 2009, "The potential to narrow uncertainty in regional climate predictions," *Bulletin of the American Meteorological Society* 90: 1095–1107; these authors found that internal variability of the climate system dominates uncertainties out to about 10–15 years, model uncertainty dominates for the next decades, and then human choices dominate. Also see Baker, M. B. and G. H. Roe, 2009, "The shape of things to come: Why is climate change so predictable," *Journal of Climate* 22: 4574–89.

4. See, for example, p. 18 in IPCC, 2007, "Summary for Policymakers," in Solomon, S., D. Qin, M. Manning, Z. Chen, M. Marquis, K. B. Averyt, M. Tignor and H. L. Miller (eds.), *Climate Change 2007: The Physical Science Basis. Contribution of Working Group I to the Fourth Assessment Report of the Intergovernmental Panel on Climate Change* (Cambridge University Press, New York).

5. McCarthy, J. J., 2009, "Reflections on: Our planet and its life, origins and futures," *Science* 326: 1646–55; Manning, M. R., J. Edmonds, S. Emori, et al., 2010, "Misrepresentation of the IPCC CO_2 emission scenarios," *Nature Geoscience* 3: 376–77. I was writing this in 2010, so we didn't know where the emissions would end up for that year, with the persistent global economic recession.

6. Rahmstorf, S., A. Cazenave, J. A. Church, et al., 2007, "Recent climate observations compared to projections," *Science* 316: 709.

7. Meehl, G. A., T. F. Stocker, W. D. Collins, P. Friedlingstein, A. T. Gaye, J. M. Gregory, A. Kitoh, R. Knutti, J. M. Murphy, A. Noda, S. C. B. Raper, I. G. Watterson, A. J. Weaver and Z.-C. Zhao, 2007, "Global Climate Projections," in Solomon, S., D. Qin, M. Manning, Z. Chen, M. Marquis, K. B. Averyt, M. Tignor and H. L. Miller (eds.), *Climate Change 2007: The Physical Science Basis. Contribution of Working Group I to the Fourth Assessment Report of the Intergovernmental Panel on Climate Change* (Cambridge University Press, Cambridge, UK, and New York). To understand the shape of the curves, you can read the technical literature.

8. The climate community is a little obsessive about understanding what is going on in the models, and linking that behavior back to the fundamentals of the climate, and for good reason. For examples of trying to understand the model outputs physically, see Held, I. M. and B. J. Soden, 2007, "Robust responses of the hydrological cycle to global warming," *Journal of Climate* 19: 5686–99; and Pierrehumbert, R. T., G, Brogniez and R. Roca, 2007, "On the relative humidity of the atmosphere," in Schneider, T. and A. Sobel (eds.), *The Global Circulation of the Atmosphere* (Princeton University Press, Princeton, NJ).

9. See Committee on Stabilization Targets for Atmospheric Greenhouse Gas Concentrations, National Research Council, 2010, *Climate Stabilization Targets: Emissions, Concentrations, and Impacts over Decades to Millennia* (National Academies Press, Washington, DC).

10. All of the IPCC results are given in Celsius. I converted the data to Fahrenheit and provide temperatures for both scales.

11. See note 4.

12. Sokolov, A. P., P. H. Stone, C. E. Forest, et al., 2009, "Probabilistic forecast for twenty-first-century climate based on uncertainties in emissions (without policy) and climate parameters," *Journal of Climate* 22: 5175–5204.

13. For the lowest-CO_2 scenario, the most-likely warming is 3.2°F, but the center of the "likely" range is 3.6°F (most-likely, 1.8°C; center of range, 2.0°C); for the highest-CO_2 scenario, the most-likely warming is 7.2°F, but the center of the range is 7.9°F (most-likely, 4.0°C; center of range, 4.4°C). See Roe, G. H., 2007, "Why is climate sensitivity so unpredictable?" *Science* 318: 629–32; also Roe, G., 2009, "Feedbacks, timescales, and seeing red," *Annual Review of Earth and Planetary Sciences* 37: 93–115. If you are interested, you can put numbers into the equation in chapter 7, note 9, on feedbacks to see how the interactions among feedbacks work.

14. The issues of CO_2 being released from soils with warming are quite complex and not easily summarized; chapter 10 (see note 7) and other chapters of the IPCC

Fourth Assessment Report are a good place to start. A recent study that raises important issues is Dorrepaal, E., S. Toet, R. S. P. van Lagtestijn, et al., 2009, "Carbon respiration from subsurface peat accelerated by climate warming in the subarctic," *Nature* 460: 616–19.

15. See figure SPM-6 in the "Summary for Policymakers" cited in note 4; also see Urban, N. M. and K. Keller, 2009, "Complementary observational constraints on climate sensitivity," *Geophysical Research Letters* 36: L04708, doi:10.1029/2008GL036457.

16. The behavior of sea ice is easier to understand, and project, in the Arctic than in the Antarctic. Sea ice that is formed by ocean freezing near the Antarctic coastline can be blown rapidly out into the vast surrounding ocean, so the rate of melting is competing with the rate of transport. Transport matters in the Arctic as well, but it is a lot slower than in the Antarctic. Antarctic natural variability in sea ice is very large, but there does appear to have been a bit of expansion recently, probably at least in part because the ozone hole has strengthened the atmospheric circulation around the continent, and this in turn has pushed the sea ice more vigorously away from the continent; see Turner, J., J. C. Comiso, G. J. Marshall, et al., 2009, "Non-annular atmospheric circulation change induced by stratospheric ozone depletion and its role in the recent increase of Antarctic sea ice extent," *Geophysical Research Letters* 36: L08502. Freshening of the Antarctic surface waters also might contribute to sea-ice expansion; back in 1992, model results from Suki Manabe and coworkers indicated that rising CO_2 would cause an expansion of Antarctic sea ice and a contraction of Arctic sea ice (Manabe, S., M. J. Spelman and R. J. Stouffer, 1992, "Transient responses of a coupled ocean-atmosphere model to gradual changes of atmospheric CO_2. Part II: Seasonal response," *Journal of Climate* 5: 105–26; also see Liu, J. and J. A. Curry, 2010, "Accelerated warming of the Southern Ocean and its impacts on the hydrological cycle and sea ice," *Proceedings of the National Academy of the United States of America* 107: 14,987–92). Shrinkage of Arctic sea ice is expected from the full range of models, and is occurring; models have been less accurate and less consistent in expectations for Antarctic sea ice, with the real possibility of growth occurring before shrinkage in response to warming. The blogospherians who insist on comparing Antarctic and Arctic sea ice to assess impacts of global warming sound like they are just being fair, but they surely are not.

17. As I was writing this in January of 2010, there was a minor tempest about the second paragraph of section 10.6.2 in Cruz, R. V., H. Harasawa, M. Lal, S. Wu, Y. Anokhin, B. Punsalmaa, Y. Honda, M. Jafari, C. Li and N. Huu Ninh, 2007, "Asia," in Parry, M. L., O. F. Canziani, J. P. Palutikof, P. J. van der Linden and C. E. Hanson (eds.), *Climate Change 2007: Impacts, Adaptation and Vulnerability. Contribution of Working Group II to the Fourth Assessment Report of the Intergovernmental*

Panel on Climate Change (Cambridge University Press, New York), 469–506. This gave an unrealistically early time for total melting of Himalayan glaciers, based on an inappropriate citation. This mistake should not have happened, but in no way does it change the reality that the glaciers of Earth have been melting and are highly likely to continue melting. A lot of very careful work is summarized by Working Group I in chapter 4 of the IPCC's 2007 report (Lemke, P., J. Ren, R. B. Alley, I. Allison, J. Carrasco, G. Flato, Y. Fujii, G. Kaser, P. Mote, R. H. Thomas and T. Zhang, 2007, "Observations: Changes in Snow, Ice and Frozen Ground," in Solomon, S., D. Qin, M. Manning, Z. Chen, M. Marquis, K. B. Averyt, M. Tignor and H. L. Miller [eds.], *Climate Change 2007: The Physical Science Basis. Contribution of Working Group I to the Fourth Assessment Report of the Intergovernmental Panel on Climate Change* [Cambridge University Press, New York]). This work shows with very high confidence that individual glaciers do many interesting things. However, if you look at many glaciers over many years in a glaciated region (such as Patagonia, or the high mountains of Asia), you will see that there has been shrinkage overall, with warming playing a very important role in causing this change, but with other processes contributing in some places. As we look to the future, glacier shrinkage is projected to accelerate, averaged over appropriately many glaciers and years in a region, until a region starts to run out of glaciers to melt (see note 7). The problems with the Working Group II paragraph were identified by scientists, and pointed out in the scientific literature (see Cogley, J. G., J. S. Kargel, G. Kaser and C. J. Van der Veen, 2010, "Tracking the source of glacier misinformation," *Science* 327: 522). The IPCC is not perfect, but it works. For a recent update on the effects of melting Himalayan glaciers, see Immerzeel, W. W., L. P. H. van Beek and M. F. P. Bierkens, 2010, "Climate change will affect the Asian water towers," *Science* 328: 1382–85.

18. Simply converting the numbers in the table in the "Summary for Policymakers," cited in note 4, gives 0.18–1.90 feet.

19. See Meehl, G. A., C. Tebaldi, G. Walton, D. Easterling and L. McDaniel, 2009, "Relative increase of record high maximum temperatures compared to record low minimum temperatures in the US," *Geophysical Research Letters* 36: L23701.

20. Sheffield, J. and E. F. Wood, 2008, "Projected changes in drought occurrence under future global warming from multi-model, multi-scenario, IPCC AR4 simulations," *Climate Dynamics* 31: 79–105.

21. See Bender, M. A., T. R. Knutson, R. E. Tuleya, et al., 2010, "Modeled impact of anthropogenic warming on the frequency of intense Atlantic hurricanes," *Science* 327: 454–58; and Knutson, T. R., J. L. McBride, J. Chan, et al., 2010, "Tropical cyclones and climate change," *Nature Geoscience* 3: 157–63; also table SPM-2 in "Summary for Policymakers," cited in note 4. For influences on damages from tropical cyclones, see Pielke, R. A., Jr., 2007, "Future economic damage from tropical cyclones: Sensitivities to societal and climate changes," *Philosophical*

Transactions of the Royal Society A: Mathematical, Physical and Engineering Sciences 365: 2717–29.

22. McGranahan, G., D. Balk and B. Anderson, 2007, "The rising tide: Assessing the risks of climate change and human settlements in low elevation coastal zones," *Environment and Urbanization* 19: 17–37.

23. See table 5 of Darwin, R. F. and R. S. J. Tol, 2001, "Estimates of the economic effects of sea level rise," *Environmental and Resource Economics* 19: 113–29.

24. I am not blaming Katrina on global warming. But, suppose that sea level rises in the future because of global warming, and walls are built to protect cities. If a storm then overtops a wall and destroys much of a city, large damages will have occurred because of a less-than-wise response. The overtopping of the levees during Katrina was not a planned event, and the existence of such a failure suggests that the assumption of a wise response to changing conditions may not always be justified.

25. Organisms in the tropics seem to be especially vulnerable; see Tewksbury, J. J., R. B. Huey and C. A. Deutsch, 2008, "Putting the heat on tropical animals," *Science* 320: 296–97.

26. Easterling, W. E., P. K. Aggarwal, P. Batima, K. M. Brander, L. Erda, S. M. Howden, A. Kirilenko, J. Morton, J.-F. Soussana, J. Schmidhuber and F. N. Tubiello, 2007, "Food, Fibre and Forest Products," in Parry, M. L., O. F. Canziani, J. P. Palutikof, P. J. van der Linden and C. E. Hanson (eds.), *Climate Change 2007: Impacts, Adaptation and Vulnerability. Contribution of Working Group II to the Fourth Assessment Report of the Intergovernmental Panel on Climate Change* (Cambridge University Press, New York), pp. 273–313; also see Long, S. P., E. A. Ainsworth, A. D. B. Leakey, J. Nosberger and D. R. Ort, 2006, "Food for thought: Lower-than-expected crop yield stimulation with rising CO_2 concentrations," *Science* 312: 1918–21.

27. This 10% estimate is not given in the IPCC report; it comes from my reading of the results given in the report and elsewhere for 550 ppm CO_2 in the atmosphere, and should be considered "order-of-magnitude"; some situations may lead to a more than 20% increase for doubled CO_2, and others, less than 5%.

28. See, for example, Stacey, D. A. and M. D. E. Fellowes, 2002, "Influence of elevated CO_2 on interspecific interactions at higher trophic levels," *Global Change Biology* 8: 668–78; Taub, D. R., B. Miller and H. Allen, "Effects of elevated CO_2 on the protein concentrations of food crops: A meta-analysis," *Global Change Biology* 14: 565–75; Hogy, P., H. Wieser, P. Kohler, et al., 2009, "Effects of elevated CO_2 on grain yield and quality of wheat: Results from a 3-year free-air CO_2 enrichment experiment," *Plant Biology* 11, Supplement 1: 60–69; Bloom, A. J., M. Burger, J. S. R. Asensio and A. B. Cousins, 2010, "Carbon dioxide enrichment inhibits nitrate assimilation in wheat and *Arabidopsis,*" *Science* 328: 899–903.

29. Once again, the simplest answer turns out to be simplistic. In June of 2010, searching on the web using such terms as "CO_2 is life" or "CO_2 gives life" or

"CO_2: We call it life" yielded many, many sites online repeating the claim that more CO_2 is better for plants, life, food, and other things we value. That surely is part of the story, but not the whole story.

30. Mohan, J. E., L. H. Ziska, W. H. Schlesinger, et al., 2006, "Biomass and toxicity responses of poison ivy (Toxicodendron radicans) to elevated atmospheric CO_2," *Proceedings of the National Academy of Sciences of the United States of America* 103: 9086–89; Schnitzer, S. A., R. A. Londré, J. Klironomos and P. B. Reich, "Biomass and toxicity responses of poison ivy (*Toxicodendron radicans*) to elevated atmospheric CO_2: Comment," *Ecology* 89: 581–85; and Mohan, J. E., L. H. Ziska, R. B. Thomas, et al., 2008, "Biomass and toxicity responses of poison ivy (*Toxicodendron radicans*) to elevated atmospheric CO_2: Reply," *Ecology* 89: 585–87.

31. See section C2 in Parry, M. L., O. F. Canziani, J. P. Palutikof, P. J. van der Linden and C. E. Hanson (eds.), 2007, "Cross-Chapter Case Studies," in Parry, M. L., O. F. Canziani, J. P. Palutikof, P. J. van der Linden and C. E. Hanson (eds.), *Climate Change 2007: Impacts, Adaptation and Vulnerability. Contribution of Working Group II to the Fourth Assessment Report of the Intergovernmental Panel on Climate Change* (Cambridge University Press, New York), pp. 843–68; and box 7.3 in Denman, K. L., G. Brasseur, A. Chidthaisong, P. Ciais, P. M. Cox, R. E. Dickinson, D. Hauglustaine, C. Heinze, E. Holland, D. Jacob, U. Lohmann, S. Ramachandran, P. L. da Silva Dias, S. C. Wofsy and X. Zhang, 2007, "Couplings between Changes in the Climate System and Biogeochemistry," in Solomon, S., D. Qin, M. Manning, Z. Chen, M. Marquis, K. B. Averyt, M. Tignor and H. L. Miller (eds.), *Climate Change 2007: The Physical Science Basis. Contribution of Working Group I to the Fourth Assessment Report of the Intergovernmental Panel on Climate Change* (Cambridge University Press, New York). Also, Fabry, V. J., 2008, "Marine calcifiers in a high-CO_2 ocean," *Science* 320: 1020–22, and the papers in the journal issue headed by Doney, S. C., W. M. Balch, V. J. Fabry and R. A. Feely, 2009, "Ocean acidification: A critical emerging problem for the ocean sciences," *Oceanography* 22: 16–25.

32. For the recent decline, see Boyce, D. G., M. R. Lewis and B. Worm, 2010, "Global phytoplankton decline over the past century," *Nature* 466: 591–96; and Siegel, D. A. and B. A. Franz, 2010, "Oceanography: Century of phytoplankton change," *Nature* 466: 569–71. For the future, see Steinacher, M., F. Joos, T. L. Frölicher, et al., 2010, "Projected 21st century decrease in marine productivity: A multi-model analysis," *Biogeosciences* 7: 979–1005. Surprising negative impacts on fish are also possible; see Munday, P. L., D. L. Dixson, M. I. McCormick, et al., 2010, "Replenishment of fish populations is threatened by ocean acidification," *Proceedings of the National Academy of Sciences of the United States of America* 107, www.pnas.org/cgi/doi/10.1073/pnas.1004519107.

33. See note 26.

34. Battisti, D. S. and R. L. Naylor, 2009, "Historical warnings of future food insecu-

rity with unprecedented seasonal heat," *Science* 323: 240–44; also see Holden, C., 2009, "Higher summer temperatures seen reducing global harvests," *Science* 323: 193; Hockley, N., J. M. Gibbons and G. Edwards-Jones, 2009, "Risks of extreme heat and unpredictability," *Science* 324: 177; and Battisti, D. S. and R. L. Naylor, 2009, "Risks of extreme heat and unpredictability: Response," *Science* 324: 178.

35. Confalonieri, U., B. Menne, R. Akhtar, K. L. Ebi, M. Hauengue, R. S. Kovats, B. Revich and A. Woodward, 2007, "Human Health," in Parry, M. L., O. F. Canziani, J. P. Palutikof, P. J. van der Linden and C. E. Hanson (eds.), *Climate Change 2007: Impacts, Adaptation and Vulnerability. Contribution of Working Group II to the Fourth Assessment Report of the Intergovernmental Panel on Climate Change* (Cambridge University Press, New York), pp. 391–431.

36. Gething, P. W., D. L. Smith, A. P. Patil, et al., 2010, "Climate change and the global malaria recession," *Nature* 465: 342–46.

37. You may wish to refer back to the Pielke paper in note 21. Also see National Research Council, 2002, *Abrupt Climate Change: Inevitable Surprises* (National Academies Press, Washington, DC); and Alley, R. B., J. Marotzke, W. D. Nordhaus, et al., 2003, "Abrupt climate change," *Science* 299: 2005–10.

38. This is the Younger Dryas cold event. It was concentrated in the wintertime. It did not really return Europe to the ice age, and there is no danger that global warming will cause a new ice age. But it did matter. My recent review may be helpful: Alley, R. B., 2007, "Wally was right: Predictive ability of the North Atlantic 'conveyor belt' hypothesis for abrupt climate change," *Annual Review of Earth and Planetary Sciences* 35: 241–72. Also see Denton, G. H., R. B. Alley, G. C. Comer and W. S. Broecker, 2005, "The role of seasonality in abrupt climate change," *Quaternary Science Reviews* 24: 1159–82.

39. Shakun, J. D. and A. E. Carlson, 2010, "A global perspective on Last Glacial Maximum to Holocene climate change," *Quaternary Science Reviews* 29: 1801–16.

40. To learn more about possible economic impacts of such an event, see Keller, K., B. M. Bolker and D. F. Bradford, 2004, "Uncertain climate thresholds and optimal economic growth," *Journal of Environmental Economics and Management* 48: 723–41; also see Keller, K., G. Yohe and M. Schlesinger, 2008, "Managing the risks of climate thresholds: Uncertainties and information needs," *Climatic Change* 91: 5–10; McInerney, D. and K. Keller, 2008, "Economically optimal risk reduction strategies in the face of uncertain climate thresholds," *Climatic Change* 91: 29–41; Weitzman, M. L., 2009, "On modeling and interpreting the economics of catastrophic climate change," *Review of Economics and Statistics* 91: 1–19; and Tol, R. S. J., 2003, "Is the uncertainty about climate change too large for expected cost-benefit analysis?" *Climatic Change* 56: 265–89.

41. See note 4.

42. See, for example, Clark, P. U., A. J. Weaver (coordinating lead authors), E. Brook, E. R. Cook, T. L. Delworth and K. Steffen (chapter lead authors), 2008, *Abrupt*

Climate Change, report by the U.S. Climate Change Science Program and the Subcommittee on Global Change Research (U.S. Geological Survey, Reston, VA); National Research Council, 2002, *Abrupt Climate Change: Inevitable Surprises* (National Academies Press, Washington, DC); and Alley et al. in note 37.

43. White, J. W. C., R. B. Alley, A. Jennings, et al., 2009, "Past Rates of Climate Change in the Arctic," in *Past Climate Variability and Change in the Arctic and at High Latitudes*, report by the U.S. Climate Change Science Program and Subcommittee on Global Change Research (U.S. Geological Survey, Reston, VA), pp. 247–302.

15: Valuing the Future

1. "The vast possibilities of our great future will become realities only if we make ourselves, in a sense, responsible for that future. The planned and orderly development and conservation of our natural resources is the first duty of the United States. It is the only form of insurance that will certainly protect us against the disasters that lack of foresight has in the past repeatedly brought down on nations since passed away." Gifford Pinchot, first chief of the U.S. Forest Service, 1910, in *The Fight for Conservation* (Doubleday, Page, New York), p. 20.

2. You might start with the book Nordhaus, W., 2008, *A Question of Balance: Weighing the Options on Global Warming Policies* (Yale University Press, New Haven, CT), and at the very least, read the first chapter. A clean copy requires purchase, but in January of 2010, an advanced pre-publication copy was available online at http://www.econ.yale.edu/~nordhaus/homepage/Balance_2nd_proofs.pdf. Also see Nordhaus, W. D., 1992, "An optimal transition path for controlling greenhouse gases," *Science* 258: 1315–19. The discussion in letters that followed this 1992 paper is also fascinating; all three letters are under the title "Pondering greenhouse policy" (Schneider, S. H., 1993, *Science* 259: 1381; Dowlatabadi, H. and L. B. Lave, 1993, *Science* 259: 1381–82; Oppenheimer, M., 1993, *Science* 259: 1382–83), and the response from W. H. Nordhaus follows them (1993 *Science* 259: 1383–84). The papers may be a little old by now, but the ideas are still quite fresh. I enjoyed serving on a committee of the National Academy of Sciences with Bill Nordhaus, and learned a lot from him. The numbers I use in the text are from Nordhaus's 2008 book; a recent update changes them slightly, but does not cause any fundamental shift in the main results; see Nordhaus, W. D., 2010, "Economic aspects of global warming in a post-Copenhagen environment," *Proceedings of the National Academy of Sciences of the United States of America* 107: 11,721–26.

3. Here is another way to understand the "efficient" or "optimal" path: if you are trying to reach a target such as keeping warming less than 3°C, the economist solves for the least-cost way to reach the target.

4. In learning what people actually value in their decisions, economists not only

rely on the actual exchanges in the marketplace but also use surveys and expert opinions of people's valuations. However, reliance is placed on what people do, not on what the economists think is right or want to happen.

5. If I give you \$1, and you agree to repay \$1.05 in a year, the interest rate is (1.05 −1.00)/1.00 = 0.05, or 5%/year. The discount rate is (1.05 − 1.00)/1.05 = 0.0476; or 4.76%—the change is compared to the value at the end, not the beginning. The discount rate may be easier to understand if you think about bonds. Suppose a government bond matures with a value of \$1.00 in one year, and sells for \$0.95 today. Then the discount rate is (1.00 − 0.95)/1.00 = 0.05, or 5%, and the interest rate is (1.00 − 0.95)/0.95 = 0.0526, or 5.26%. If the bond were selling for \$0.9524 today, the discount rate would be 4.76% and the interest rate would be 5%. If the interest rate is i, the discount rate is $d = i/(1 + i)$, or $i = d/(1 - d)$, using the decimal form for i and d (so 0.05, not 5%). Given how poorly we know the discount rate, the difference between 5% and 4.76% doesn't matter much.

6. I am giving the Ramsey equation, in which the discount rate is taken as being equal to the real rate of return on capital, r^*. Given a number of assumptions, the discount rate is equal to the growth rate of per capita consumption in the economy, g^*, multiplied by the elasticity of utility of consumption, α, which measures how rapidly the utility of consumption (the "good" we get from the stuff we consume) decreases as our total consumption increases, added to the pure rate of time preference, ρ. Thus, $r^* = \rho + \alpha\, g^*$. See the first reference in note 2.

7. See Nordhaus's *Question of Balance*, p. 10, cited in note 2.

8. Actually, we should have started well before 2005 (see Keller, K., A. Robinson, D. Bradford and M. Oppenheimer, 2007, "The regrets of procrastination in climate policy," *Environmental Research Letters* 2: paper 024004), because the same answer has been coming up for quite a while. But we don't have a time machine.

9. Whether we should stay pretty close to conducting business as usual, or move closer to panic, depends greatly on the discount rate used. To see this, consider a simple example. Suppose a problem exists (it may be related to global warming, although this works for any problem). You know that you could fix the problem now, and you know that the problem will cause \$1 million in damages a century from now if you don't fix it. Should you fix the problem now, or invest the money, let it grow for a century, and give the money to your descendants to compensate them for their \$1 million loss?

    To do the calculation that follows, note that the formula for the current value of \$$M$ in t years with a discount rate of d is $M/(1 + d)^t$. So, with a 4% = 0.04 discount rate, the current value of \$1000 after 100 years is \$19.80, and of \$1,000,000 is \$19,800. But, raise the discount rate to 7% and the current value of \$1000 in 100 years is only \$1.15, whereas at 1% the current value is \$370.

    If the discount rate is 4%, then \$19,800 invested today will be worth \$1 million in a century. So, if you can solve the problem for less than $19,800, you are economically efficient to do so. But if the solution costs $20,000 today and you have that much money, invest it, let it grow to $1.01 million in a century, and your descendants will have $10,000 left over after they get the $1 million for the damages.

But, this assumes that the long-term discount rate is 4%/year. Increase the 4% rate to 7%, and you should spend no more than $1150 now to head off $1 million in damages in a century; decrease to 1%, and you should spend as much as $370,000 now to head off $1 million in damages in a century. You can find a range from 1% or less to perhaps 7% or more in discussions of global warming, with 3–4% probably most common.

10. Halsnæs, K., P. Shukla, D. Ahuja, G. Akumu, R. Beale, J. Edmonds, C. Gollier, A. Grubler, M. Ha Duong, A. Markandya, M. McFarland, E. Nikitina, T. Sugiyama, A. Villavicencio and J. Zou, 2007, "Framing Issues," in Metz, B., O. R. Davidson, P. R. Bosch, R. Dave and L. A. Meyer (eds.), *Climate Change 2007: Mitigation. Contribution of Working Group III to the Fourth Assessment Report of the Intergovernmental Panel on Climate Change* (Cambridge University Press, New York); Stern, N., 2007, *The Economics of Climate Change: The Stern Review* (Cambridge University Press, New York); Nordhaus, W., 2007, "Critical assumptions in the Stern Review on climate change," *Science* 317: 201–2; Schultz, P. A. and J. F. Kasting, 1997, "Optimal reductions in CO_2 emissions," *Energy Policy* 25: 491–500; Mendelsohn, R., 2008, "Is the *Stern Review* an economic analysis?" *Review of Environmental Economics and Policy* 2: 45–60; Sterner, T. and U. M. Persson, 2008, "An even Sterner Review: Introducing relative prices into the discounting debate," *Review of Environmental Economics and Policy* 2: 61–76; Weyant, J. P., 2008, "A critique of the *Stern Review*'s mitigation cost analyses and integrated assessment," *Review of Environmental Economics and Policy Advance* 2: 77–93; and Dietz, S. and N. Stern, 2008, "Why economic analysis supports strong action on climate change: A response to the *Stern Review*'s critics," *Review of Environmental Economics and Policy* 2: 94–113.

11. See Yohe, G. W., R. D. Lasco, Q. K. Ahmad, N. W. Arnell, S. J. Cohen, C. Hope, A. C. Janetos and R. T. Perez, 2007, "Perspectives on Climate Change and Sustainability," in Parry, M. L., O. F. Canziani, J. P. Palutikof, P. J. van der Linden and C. E. Hanson (eds.), *Climate Change 2007: Impacts, Adaptation and Vulnerability. Contribution of Working Group II to the Fourth Assessment Report of the Intergovernmental Panel on Climate Change* (Cambridge University Press, New York), pp. 811–41. Nordhaus obtained $27, or $30 in round numbers, as above.

12. See Nordhaus's *Question of Balance*, p. 4, cited in note 2. Much more research could be done on this topic, explicitly comparing different options for using money rather than aggregating many parts of the economy and possible uses of money into sectors and then conducting the calculations.

13. Nordhaus, W. D., 2006, "Geography and macroeconomics: New data and new findings," *Proceedings of the National Academy of Sciences of the United States of America* 103: 3510–17.

14. Dell, M., B. F. Jones and B. A. Olken, 2009, "The economic impacts of climate change: Temperature and income: Reconciling new cross-sectional and panel estimates," *American Economic Review: Papers and Proceedings* 99: 198–204.

15. See Nordhaus's *Question of Balance*, p. 13, cited in note 2.

16. Ibid.

17. U.S. Department of Defense, 2010, *Quadrennial Defense Review Report*, pp. 84–85, http://www.defense.gov/QDR/.

18. See *The Civil War Diary of General Josiah Gorges*; an excerpt can be found online at the CSS *Virginia* home page, Tyson, M. and M. H. Tyson, 1994–2009, http://cssvirginia.org/vacsn/base/afterbtl.htm (accessed June 24, 2010). One can't help but wonder whether Abraham Lincoln's boat-floating invention would have helped.

19. Anderson, I. H., Jr., 1975, "The 1941 de facto embargo on oil to Japan: A bureaucratic reflex," *Pacific Historical Review* 44: 201–31.

20. Halliday, F., 1991, "The Gulf War and its aftermath: First reflections," *International Affairs* 67: 223–34.

21. President George H. W. Bush, Address to Congress on the Persian Gulf Crisis, Sept. 11, 1990.

22. The Materials Policy Commission was also called the "Paley Commission," after chair William Paley. Paley, W. S., chair, 1952, *President's Materials Policy Commission*, vol. 1, *Foundations for Growth and Security*; vol. 2, *The Outlook for Key Commodities*; vol. 3, *The Outlook for Energy Sources*; vol. 4, *The Promise of Technology*; and vol. 5, *Selected Reports to the Commission* (Government Printing Office, Washington, DC). The National Security Resources Board reinforced the Paley report, adding atomic, wind, and tide energy to the list of alternatives to be explored. See note 23.

23. Strum, H., 1984, "Eisenhower's solar energy policy," *Public Historian* 6: 37–50; Paley, W. S., 1952, *President's Materials Policy Commission, Resources for Freedom*, (Government Printing Office, Washington, DC); and National Security Resources Board, 1952, *The Objectives of United States Materials Resources Policy and Suggested Initial Steps in their Accomplishment* (Washington, DC).

24. Eady, D. S., S. B. Siegel, R. S. Bell and S. H. Dicke, 2009, *Sustain the Mission Project: Casualty Factors for Fuel and Water Resupply Convoys*, Army Environmental Policy Institute, http://www.aepi.army.mil/publications/sustainability/ (accessed June 24, 2010). Some commentators have suggested that the U.S. military efforts in Iraq and Afghanistan would have been more successful if the civilian effort to supply power for air-conditioning and other purposes had been more successful; see Dickey, C., "Power and light: Getting Iraqis and Afghans on the grid could

have helped build peace, but the Americans couldn't quite deliver," *Newsweek*, July 28, 2010, http://www.newsweek.com/2010/07/28/power-and-light.pdf (accessed July 28, 2010).

25. Burke, M. B., E. Miguel, S. Satyanath, J. A. Dykema and D. B. Lobell, 2009, "Warming increases the risk of civil war in Africa," *Proceedings of the National Academy of Sciences of the United States of America* 106: 20,670–74. This pioneering study found a strong relation between increasing temperature and increasing civil war in Africa that is not explained solely by the hot years being dry. Simply extrapolating the linear relation between African civil wars and rising temperature using projected warming produces an increase by more than 50% by the year 2030, with almost 400,000 additional battle deaths if future wars are similar to recent ones. However, a more recent study questions the techniques used and thus the significance of the result (Buhaug, H., 2010, "Climate not to blame for African civil wars," *Proceedings of the National Academy of Sciences of the United States of America* 107: 16,477–82). A slightly earlier study addressing the relatively cold Little Ice Age (Zhang, D. D., P. Brecke, H. F. Lee, Y.-Q. He and J. Zhang, 2007, "Global climate change, war, and population decline in recent human history," *Proceedings of the National Academy of Sciences of the United States of America* 104: 19,214–19) found more wars with anomalously cold climates, likely linked to harvest failures. Possible correlations between temperature and rainfall were not assessed, except to note that dry conditions also correlated with wars in China. Nonetheless, the suggestion is that humanity is prepared for the climate we have, and we have difficulties when that climate changes. And those difficulties can lead to serious security issues, environmental refugees, and wars. See, in addition, Feng, S., A. B. Krueger and M. Oppenheimer, 2010, "Linkages among climate change, crop yields and Mexico-US cross-border migration," *Proceedings of the National Academy of Sciences of the United States of America* 107: 14,257-62. Also see Tol, R. S. J. and S. Wagner, 2010, "Climate change and violent conflict in Europe over the last millennium," *Climatic Change* 99: 65–79, which shows that industrialization has reduced the effect of climate on conflict in Europe.

26. See the Defense Department's 2010 *Quadrennial Defense Review Report*, p. 87, cited in note 17.

27. Ibid. The idea of a military effort proving sufficiently innovative and successful to fundamentally change something as big as the energy system is not far-fetched. The Internet, after all, has its roots deeply in the U.S. Department of Defense through the ARPANET (Advanced Research Projects Agency Network), which is how I knew it when I first logged on while a student at the University of Wisconsin. The main long-haul roads in the United States, the Dwight D. Eisenhower National System of Interstate and Defense Highways, were justified in important ways based on national-security issues. Eisenhower himself was led to think about the defense implications of good roads by his participation in

a 1919 cross-country convoy that took two months; Eisenhower, D. D., 1967, "Through Darkest America with Truck and Tank," in *At Ease: Stories I Tell to Friends* (Doubleday, Garden City, NY). For a short perspective on a military role, see Sarawitz, D., 2010, "Missing weapons: The US defence department should be at the centre of the nation's energy policy," *Nature* 464: 672.

28. See, for example, Nordhaus's *Question of Balance*, p. 21, cited in note 2.

29. Rear Admiral David Titley, the U.S. Navy's senior oceanographer, described how the naval oceanography program provides environmental information to safer and more effective operation of the fleet. He noted that "we are operating in nature's casino; I intend to count the cards." Freeman, B., 2009, "Navy task force assesses changing climate," Special to American Forces Press Service, July 31, 2009, http://www.defense.gov/news/newsarticle.aspx?id=55327.

30. He actually used more-colorful terminology with the same significance.

31. In a ski drag, the pilot flies the plane while resting enough weight on the skis touching the snow surface to have a good chance of breaking any snow bridges across crevasses. The pilot then gains elevation, comes around, and inspects the tracks, landing in the same place if and only if there still are no crevasses visible. The commander almost certainly was planning a ski drag anyway.

32. In fact, I was correct. But, I've never discussed this with the commander since that meeting in Antarctica.

33. See, for example, Keller, K., M. Hall, S. R. Kim, D. F. Bradford and M. Oppenheimer, 2005, "Avoiding dangerous anthropogenic interference with the climate system," *Climatic Change* 73: 227–38; and Weitzman, M. L., 2009, "On modeling and interpreting the economics of catastrophic climate change," *Review of Economic Studies* 91: 1–19.

34. See, for example, Clark, P. U., A. J. Weaver (coordinating lead authors), E. Brook, E. R. Cook, T. L. Delworth and K. Steffen (chapter lead authors), 2008, *Abrupt Climate Change*, report by the U.S. Climate Change Science Program and the Subcommittee on Global Change Research (U.S. Geological Survey, Reston, VA); National Research Council, 2002, *Abrupt Climate Change: Inevitable Surprises* (National Academies Press, Washington, DC); and Alley, R. B., J. Marotzke, W. D. Nordhaus, et al., 2003, "Abrupt climate change," *Science* 299: 2005–10. The increasing possibility of a major discontinuity, tipping point, or whatever you want to call it is highlighted in Smith, J. B., H.-J. Schellnhuber, M. M. Q. Mirza, S. Fankhauser, R. Leemans, L. Erda, L. Ogallo, B. Pittock, R. Richels, C. Rosenzweig, U. Safriel, R. S. J. Tol, J. Weyant and G. Yohe, 2001, "Vulnerability to Climate Change and Reasons for Concern: A Synthesis," in McCarthy, J., O. Canziana, N. Leary, D. Dokken and K. White (eds.), *Climate Change 2001: Impacts, Adaptation, and Vulnerability, IPCC Third Assessment Report* (Cambridge University Press, New York), pp. 913–67; and Smith, J. B., S. H. Schneider, M. Oppenheimer, et al., 2009, "Assessing dangerous climate change through an update

of the Intergovernmental Panel on Climate Change (IPCC) 'reasons for concern,'" *Proceedings of the National Academy of Sciences of the United States of America* 106: 4133–37.

35. Keller, K., B. M. Bolker and D. F. Bradford, 2004, "Uncertain climate thresholds and optimal economic growth," *Journal of Environmental Economics and Management* 48: 723–41; also see Keller, K., G. Yohe and M. Schlesinger, 2008, "Managing the risks of climate thresholds: Uncertainties and information needs," *Climatic Change* 91: 5–10; McInerney, D. and K. Keller, 2008, "Economically optimal risk reduction strategies in the face of uncertain climate thresholds," *Climatic Change* 91: 29–41; Weitzman, M. L., 2009, "On modeling and interpreting the economics of catastrophic climate change," *Review of Economics and Statistics* 91: 1–19; and Tol, R. S. J., 2003, "Is the uncertainty about climate change too large for expected cost-benefit analysis?" *Climatic Change* 56: 265–89.

36. Barker, T., I. Bashmakov, A. Alharthi, M. Amann, L. Cifuentes, J. Drexhage, M. Duan, O. Edenhofer, B. Flannery, M. Grubb, M. Hoogwijk, F. I. Ibitoye, C. J. Jepma, W. A. Pizer and K. Yamaji, 2007, "Mitigation from a Cross-Sectoral Perspective," in Metz, B., O. R. Davidson, P. R. Bosch, R. Dave and L. A. Meyer (eds.), *Climate Change 2007: Mitigation. Contribution of Working Group III to the Fourth Assessment Report of the Intergovernmental Panel on Climate Change* (Cambridge University Press, New York); see section 11.8.2.

37. Sims, R. E. H., R. N. Schock, A. Adegbululgbe, J. Fenhann, I. Konstantinaviciute, W. Moomaw, H. B. Nimir, B. Schlamadinger, J. Torres-Martínez, C. Turner, Y. Uchiyama, S. J. V. Vuori, N. Wamukonya and X. Zhang, 2007, "Energy Supply," in Metz, B., O. R. Davidson, P. R. Bosch, R. Dave and L. A. Meyer (eds.), *Climate Change 2007: Mitigation. Contribution of Working Group III to the Fourth Assessment Report of the Intergovernmental Panel on Climate Change* (Cambridge University Press, New York); see section 4.5.3.

38. Levine, M., D. Ürge-Vorsatz, K. Blok, L. Geng, D. Harvey, S. Lang, G. Levermore, A. Mongameli Mehlwana, S. Mirasgedis, A. Novikova, J. Rilling and H. Yoshino, 2007, "Residential and Commercial Buildings," in Metz, B., O. R. Davidson, P. R. Bosch, R. Dave and L. A. Meyer (eds.), *Climate Change 2007: Mitigation. Contribution of Working Group III to the Fourth Assessment Report of the Intergovernmental Panel on Climate Change* (Cambridge University Press, New York); see section 6.6.4

39. Sathaye, J., A. Najam, C. Cocklin, T. Heller, F. Lecocq, J. Llanes-Regueiro, J. Pan, G. Petschel-Held, S. Rayner, J. Robinson, R. Schaeffer, Y. Sokona, R. Swart and H. Winkler, 2007, "Sustainable Development and Mitigation," in Metz, B., O. R. Davidson, P. R. Bosch, R. Dave and L. A. Meyer (eds.), *Climate Change 2007: Mitigation. Contribution of Working Group III to the Fourth Assessment Report of the Intergovernmental Panel on Climate Change* (Cambridge University Press, New York); see section 12.3.1.

40. Wei, M., S. Patadia and D. M. Kammen, 2010, "Putting renewables and energy efficiency to work: How many jobs can the clean energy industry generate in the US?" *Energy Policy* 38: 919–31.

41. For example, see Branker, K. and J. M. Pearce, 2010, "Financial return for government support of large-scale thin-film solar photovoltaic manufacturing in Canada," *Energy Policy* 38: 4291–4303; Sastresa, E. L., A. A. Uson, I. Z. Bribian and S. Scarpellini, 2010, "Local impact of renewables on employment: Assessment methodology and case study," *Renewable and Sustainable Energy Reviews* 14: 679–90; Moreno, B. and A. J. Lopez, 2008, "The effect of renewable energy on employment. The case of Asturias (Spain)," *Renewable and Sustainable Energy Reviews* 12: 732–51; and Stoddard, L., J. Abiecunas and R. O'Connell, 2006, *Economic, Energy, and Environmental Benefits of Concentrating Solar Power in California*, National Renewable Energy Laboratory Subcontract Report NREL/SR-550-39291.

42. "Oil production cost estimates by country," Reuters, July 28, 2009, http://www.reuters.com/article/idUSLS12407420090728 (accessed Jan. 31, 2010).

43. Lehr, U., J. Nitsch, M. Kratzat, C. Lutz and D. Edler, 2008, "Renewable energy and employment in Germany," *Energy Policy* 36: 108–17. The gains in employment were much more impressive if the push toward renewables was accompanied by development of export markets.

44. Preamble to the Constitution of the United States of America.

45. See Nordhaus's *Question of Balance*, p. 26, cited in note 2.

46. U.S. Environmental Protection Agency, Office of Atmospheric Programs, 2009, "Revenue recycling to reduce labor taxes," in *Supplemental EPA Analysis of the American Clean Energy and Security Act of 2009 H.R. 2454 in the 111th Congress*, p. 23, scenario 16, http://www.epa.gov/climatechange/economics/economicanalyses.html#hr2454.

47. Similarly, Nordhaus (*Question of Balance*, p. 156, cited in note 2) pointed out that the added government interference in the economy associated with a price on carbon emissions does not require more inefficiency if the money raised is used to reduce government interference elsewhere.

48. For more information, see, for example, Jorgenson, D. W. and P. J. Wilcoxen, 1993, "Reducing United States carbon-dioxide emissions—An assessment of different instruments," *Journal of Policy Modeling* 15: 491–520; Jorgenson, D. W. and P. J. Wilcoxen, 1993, "Reducing United States carbon emissions—An econometric general equilibrium assessment," *Resource and Energy Economics* 15: 7–25; Glomm, G., D. Kawaguchi and F. Sepulveda, 2008, "Green taxes and double dividends in a dynamic economy," *Journal of Policy Modeling* 30: 19–32; and Bor, Y. J. and Y. Huang, 2010, "Energy taxation and the double dividend effect in Taiwan's energy conservation policy—An empirical study using a computable general equilibrium model," *Energy Policy* 38: 2086–2100.

49. Quoting from the 2010 editorial "Progressive thinking: It is time to abandon GDP (Gross Domestic Product) as the overriding measure of social development and economic health," *Nature* 463: 849–50: "What we measure affects what we do. If we have the wrong measures, we will strive for the wrong things," says economist Joseph Stiglitz of Columbia University in New York. This was accompanied by the suggestion that indicators measuring quality of life, sustainable development, and the environment be considered with GDP (see go.nature .com/FnL7He).

50. This problem of failure to monetize some aspects of the environment can affect assessments of all competing uses of money, not just global warming.

51. See Halsnæs et al. in note 10.

52. Writers have cited the Bible, Psalm 24:1, "The earth is the Lord's and the fulness thereof, the world and those who dwell therein," and then Matthew 12:40, "You also must be ready; for the Son of man is coming at an unexpected hour" (both from the Revised Standard Version), which a believer can interpret to mean that trashing the Lord's planet before a judgment day that may be near is not a wise thing. The discussion of stewardship that follows the verse from Matthew, and the discussion in Luke 12:35-48, are often cited as well, as is Micah 6:8, "and what does the Lord require of you but to do justice." Concern for stewardship is heightened by the realization that the human-induced warming from burning all of the fossil fuels will persist longer into the future than the whole of written history to date; see, for example, Archer, D., 2008, *The Long Thaw: How Humans are Changing the Next 100,000 Years of Earth's Climate* (Princeton University Press, Princeton, NJ); Solomon, S., G.-K. Plattner, R. Knutti and P. Friedlingstein, 2009, "Irreversible climate change due to carbon dioxide emissions," *Proceedings of the National Academy of Sciences of the United States of America* 106: 1704–9.

53. Interestingly, the economists have thought about this, and may be somewhat kinder to wealthy people. Because wealth helps avoid suffering in a changing climate, making the poor people wealthier may be better for them than slowing or stopping the climate change. If CO_2 emissions are hurting poorer people in hotter places now, and a rapid reduction in CO_2 emissions would have high costs for wealthier people in cooler places, a wise response might be to avoid the rapid reduction and instead have the wealthier people send some of the money they save to the poorer people being hurt (see, e.g., Rose, A., B. Stevens, J. Edmonds and M. Wise, 1998, "International equity and differentiation in global warming policy: An application to tradeable emission permits," *Environmental and Resource Economics* 12: 25–51). We can raise the question of whether the wealthier people are really likely to do that, and whether the poorer people might be happier making their own decisions rather than having their climate changed and then charity supplied. Compensation for poor people may be required by international law (see Tol, R. S. J. and R. Verheyen, 2004, "State responsibility and compensation

for climate change damages—A legal and economic assessment," *Energy Policy* 32: 1109–30). A recent study using foreign-aid funding as a measure of our desire to avoid inequity suggests that this increases the social cost of emitting carbon, but by less than 50%; see Tol, R. S. J., 2010, "International inequity aversion and the social cost of carbon," *Climate Change Economics* 1: 21–32.

54. Or, more properly, what gives us the right to decide that the utility of consumption of future people is not as important as ours? See, for example, Broome, J., 2008, "The ethics of climate change," *Scientific American* (June): 97–102; or Broome, J., 2004, *Weighing Lives* (Oxford University Press, Oxford, UK).

55. The Rock Ethics Institute at Penn State is a great place to investigate some of the key ethical questions; http://rockethics.psu.edu/climate/ (accessed Jan. 31, 2010).

56. National Research Council, Committee on Methods for Estimating Greenhouse Gas Emissions, 2010, *Verifying Greenhouse Gas Emissions: Methods to Support International Climate Agreements* (National Academies Press, Washington, DC).

57. One nonintuitive additional argument will take a little explanation, which I give here, but if you believe that climate scientists are morons or evil liars, or that we are misguided souls trying to do an almost impossible job, you probably should join those who favor more effort now to head off climate change. First, the great majority of people on all sides of the climate-change issues, including most "skeptics," accept that burning fossil fuels releases CO_2 to the air, that CO_2 is a greenhouse gas, and that adding this greenhouse gas to the air affects climate. You can get large arguments as soon as you try to quantify these statements, with some people and even a few experts asserting that the climatic effect of CO_2 is small, beneficial, or otherwise not grounds for collective action. However, I know no scientists still in touch with reality who would qualitatively reject the existence of any human influence on climate through CO_2 from fossil-fuel burning—that requires rejecting a whole lot of physics that is very well understood and used routinely in military and civilian applications.

Next, add the common-sense results that we have built valuable stuff for the climate we have, and that the opposite of bad is not necessarily good. The opposite of "too hot" is "too cold," and either may be bad. If we are pushing the climate away from the conditions for which we have built, knowing that going too far will have bad consequences, a lot of people want confidence from climate science that we won't push the climate over the edge to a place where "there will be dragons." If you want to know how to operate a machine before you start using it at full speed, and you don't trust part of the operators' manual, then prudence might suggest that you should avoid accelerating until you can get a better manual. The optimal solutions from Nordhaus and other economists allow much climate change to occur fairly rapidly, based in part on assurances from climate scientists that we can handle it and things won't be too bad. If you don't believe this, the simplest physics showing that we are affecting the climate might lead you to think that we should slow down until we can learn more.

Not everything involving learning is simple, and under some circumstances the prospect of learning in the future may reduce the level of action recommended now; see, for example, Keller et al. in note 33.

Again, though, someone who starts from the conviction that international agreements are doomed to failure, and that governments can never get it right, probably still favors taking no actions, and the more one tends to this view, the less action seems optimal.

16: Toilets and the Smart Grid

1. Twain, M., 1916, "The McWilliamses and the Burglar Alarm," in Twain, M., 1922, *The Mysterious Stranger and Other Stories* (Harper and Brothers, New York), p. 315.
2. IPCC, 2007, "Summary for Policymakers," in Metz, B., O. R. Davidson, P. R. Bosch, R. Dave and L. A. Meyer (eds.), *Climate Change 2007: Mitigation. Contribution of Working Group III to the Fourth Assessment Report of the Intergovernmental Panel on Climate Change* (Cambridge University Press, New York).
3. The IPCC report actually was discussing stabilization levels in CO_2 equivalents. To understand this, note that we have increased methane levels as well as CO_2 levels. We can estimate how much warming the methane causes, and how much CO_2 would be required to give the warming from the methane as well as from the CO_2 we have put up. Then, rather than talking about the level of methane and the level of CO_2, we can talk about CO_2 equivalents. If you recall all of the kindergarten soccer players, each has caused some forcing in W/m^2, so add all of those together to get the total forcing in W/m^2, and then ask how much CO_2 would be required to supply that total forcing, and you have CO_2 equivalents. Because CO_2 is the biggest factor in this, the focus is on CO_2.
4. German Advisory Council on Global Change (WBGU) (Grassl, H., J. Kokott, M. Kulessa, J. Luther, F. Nuscheler, R. Sauerborn, H.-J. Schellnhuber, R. Schubert and E.-D. Schulze), 2003, *Climate Protection Strategies for the 21st Century: Kyoto and Beyond*, http://www.wbgu.de/wbgu_home_engl.html.
5. Hasselmann, K., 2009, "What to do? Does science have a role?" *European Physical Journal Special Topics* 176: 37–51. The Hasselman 2009 paper cited sources including Azar, C. and S. H. Schneider, 2002, "Are the economic costs of stabilizing the climate prohibitive?" *Ecological Economics* 42: 73–80; Hasselmann, K., M. Latif, G. Hooss, et al., 2003, "The challenge of long-term climate change," *Science* 302: 1923–25; and Stern, N., 2007, *The Economics of Climate Change: The Stern Review* (Cambridge University Press, New York), p. 692 (HM Treasury, http://www.hm-treasury.gov.uk/stern review report.htm).

6. A decade ago, a group of climate scientists, including me, led by the incomparable Wally Broecker attended a small workshop where Klaus Lackner of Columbia University addressed us about carbon capture and sequestration, which we will meet again in chapter 21. After the presentation, which included a lot of cost estimates for various steps, I asked for the total cost relative to the world economy. We promptly calculated it on the "back of an envelope," a very rough estimate, and came up with 1% of the world economy. Thinking of other solutions and paths also pointed to a cost of roughly that much. At the time, I had not spent much time reading the serious literature on this subject, but the close agreement between our back-of-the-envelope work and the more serious estimates is encouraging, suggesting that the numbers are more or less correct.

7. World Bank, http://datafinder.worldbank.org/about-world-development-indicators?cid=GPD_WDI; in 2008, world gross domestic product in U.S. dollars was $60.6 trillion.

8. See note 5.

9. International Energy Agency, Organization of the Petroleum Exporting Countries, Organisation for Economic Co-operation and Development, and the World Bank Joint Report, 2010, *Analysis of the Scope of Energy Subsidies and Suggestions for the G-20 Initiative*, prepared for submission to the G-20 Summit Meeting, Toronto, June 26–27, 2010. Note that a lot of taxes are paid on fossil fuels; the 1% subsidy level does not count those taxes. Total subsidies for renewable energy sources are much smaller than total subsidies for fossil fuels, but renewable subsidies are larger per kilowatt-hour generated.

10. Burt, E., 1876, *Burt's Letters from the North of Scotland* (William Paterson, Edinburgh), pp. 365–66.

11. Ibid.

12. See, for example, Snow, S. J., 2002, "Commentary: Sutherland, Snow and water: The transmission of cholera in the nineteenth century," *International Journal of Epidemiology* 31: 908–11; Hempel, S., 2007, *The Strange Case of the Broad Street Pump: John Snow and the Mystery of Cholera* (University of California Press, Berkeley); and Johnson, S., 2007, *The Ghost Map: The Story of London's Most Terrifying Epidemic— And How It Changed Science, Cities, and the Modern World* (Riverhead Books, New York). Also see the great range of resources at the John Snow site, http://www.ph.ucla.edu/epi/snow.html, UCLA Department of Epidemiology School of Public Health (accessed Feb. 10, 2010).

13. See Snow in note 12.

14. Just as Snow was contributing to the success of medical science, some of Snow's contemporaries were forming the threads that have been woven into our successful climate science.

15. See, for example, Goddard, N., 1996, "'A mine of wealth?' The Victorians and the agricultural value of sewage," *Journal of Historical Geography* 22: 274–90; and

Barles, S., 2007, "Urban metabolism and river systems: an historical perspective—Paris and the Seine, 1790–1970," *Hydrology and Earth System Sciences* 11: 1757–69.

16. See, for example, Kravetz, R. E., "The flush toilet?" *American Journal of Gastroenterology* 104: 522.

17. See, for example, Worthington, W., Jr., 1990, "The privy and the pump: The Matthewman & Johnson Excavating Device," *Technology and Culture* 31: 451–55.

18. See Goddard in note 15.

19. OECD (Organisation for Economic Co-operation and Development), 2002, *Social Issues in the Provision and Pricing of Water Services*, http://www.oecd.org/env/ and http://browse.oecdbookshop.org/oecd/pdfs/browseit/9703041E.pdf (accessed Feb. 9, 2010).

20. Congressional Budget Office, U.S. Congress, 2002, *Future Investment in Drinking Water and Wastewater Infrastructure*, www.cbo.gov/doc.cfm?index=3983 (accessed July 8, 2010).

21. A sewer connection costs $4400–$7100; University Area Joint Authority, http://www.uaja.com/. A water connection is in the range of $1000–$1600; Centre County Industrial Development Corporation, http://www.centrecountyidc.org/EconomicDevelopment/Demographics.aspx, under "utilities" for College Township, or the College Township Water Authority at http://www.college township.govoffice.com/. A typical home price for the Centre Region of central Pennsylvania, which includes College Township, is given as $242,574 in the College Township Planning Commission minutes, Sept. 2, 2008 (all accessed July 8, 2010). I simply took the averages of the sewer and water ranges to compare to the home price, finding 2.9%, which rounds off to 3%.

22. See Snow in note 12.

23. For some insights from people who have succeeded in business, see Greenblatt, J., 2009, "Clean Energy 2030: Google's proposal for reducing U.S. dependence on fossil fuels," http://knol.google.com/k/jeffery-greenblatt/clean-energy-2030/15x31uzlqeo5n/1# (accessed Nov. 18, 2009); or from U.S. businessman T. Boone Pickens, http://www.pickensplan.com/theplan/ (accessed Nov. 20, 2009). For thoughts of two very different Nobel laureates, see Smalley, R. E. (Nobel Prize in Chemistry, 1996), 2005, "Future global energy prosperity: The terawatt challenge," *MRS Bulletin* (Materials Research Society) 30: 412–17; and Gore, A. (Nobel Peace Prize, 2007), http://www.wecansolveit.org/pages/al_gore_a_generational_challenge_to_repower_america/ (accessed Nov. 20, 2009). I learned a lot from Jacobson, M. Z., 2009, "Review of solutions to global warming, air pollution, and energy security," *Energy & Environmental Science* 2: 148–73; and Lackner, K. S. and J. D. Sachs, 2005, "A robust strategy for sustainable energy," *Brookings Papers on Economic Activity* 2: 215–69, http://www.jstor.org/stable/3805122 (accessed Nov. 23, 2009). For additional thoughts on

energy futures, start with IPCC, 2007, "Summary for Policymakers," in Metz, B., O. R. Davidson, P. R. Bosch, R. Dave and L. A. Meyer (eds.), *Climate Change 2007: Mitigation. Contribution of Working Group III to the Fourth Assessment Report of the Intergovernmental Panel on Climate Change* (Cambridge University Press, New York). A comprehensive treatment is given by German Advisory Council on Global Change (WBGU) (Grassl, H., J. Kokott, M. Kulessa, J. Luther, F. Nuscheler, R. Sauerborn, H.-J. Schellnhuber, R. Schubert and E.-D. Schulze), 2004, *Towards Sustainable Energy Systems*, http://www.wbgu.de/wbgu_home_engl.html; together with *Climate Protection Strategies for the 21st Century: Kyoto and Beyond*, from the same authors and organization in 2003. Then consider MacKay, D. J. C., 2009, "Sustainable Energy—Without the Hot Air," www.withouthotair.com. Nathan Lewis is doing amazing things with solar cells, and his overview of the problem, written in an accessible way, is Lewis, N. S., 2007, "Powering the planet," *Engineering and Science*, no. 2., pp. 13–23, www.ccser.caltech.edu/outreach/powering.pdf (accessed July 20, 2010). There are many more good sources, with additional ones appearing quickly.

24. Pacala, S. and R. Socolow, 2004, "Stabilization wedges: Solving the climate problem for the next 50 years with current technologies," *Science* 305: 968–972; "Robert Socolow and Stephen Pacala: Strategists," *Time*, Apr. 2, 2007, http://www.time.com/time/magazine/article/0,9171,1604890,00.html (accessed Nov. 25, 2009); and "Stabilization Wedges Game," Carbon Mitigation Initiative (Princeton University, Princeton, NJ), http://cmi.princeton.edu/wedges (accessed July 9, 2010).

25. Because we have raised atmospheric CO_2 above the natural level, CO_2 is leaving the air more rapidly than naturally, primarily going into the ocean. Thus, stabilizing CO_2 requires reducing our emissions, but not stopping them completely. If warming releases CO_2 from clathrates, Arctic soils, and elsewhere, as discussed in chapter 14, then the problem becomes more difficult and more wedges are required; delay is costly in all of the analyses I have seen.

26. For example, see http://www.theclimategroup.org/our-news/interviews/2004/10/15/stephen-pacala/ (accessed July 9, 2010).

27. Full disclosure: my family's limited investments are in a balanced portfolio, not in energy futures, alternate fuels, or oil wells. If you wanted to criticize me, you could say that I don't believe in the ideas I'm promoting to others because I haven't put my money where my mouth is; if you wanted to praise me, you could say that I'm a highly moral person who is being very careful not to benefit from my "inside" knowledge. Of course, if I had invested in renewable energy, a friend could say that I have put my money where my mouth is, while a foe could accuse me of trying to make money by driving up the value of otherwise dubious investments. There is very little about this subject that cannot be "spun" to make a political point!

28. In 2004, the International Energy Agency (IEA) summarized the research expenditures of its member countries from 1974 to 2002. This is now a little out-of-date, but as the results of research take a while to reach the market, this may be the best time interval to consider. All numbers are corrected for inflation to represent true effort. The oil price shocks of the 1970s led to a rapid increase in total funding for energy research until about 1980, but this was then cut roughly in half by late in the 1980s, and remained relatively stable through 2002. International Energy Agency, 2004, *Renewable Energy: Market and Policy Trends in IEA Countries*, https://www.iea.org/textbase/nppdf/free/2004/renewable1.pdf (accessed Jan. 24, 2010). By 2008, spending for research on all renewable energy sources was about one-third of that on fission and fusion taken together across the IEA countries, and almost half as large in the United States, with funding for solar power research being about one-sixth of that for fission and fusion taken together; see Kerr, T., 2010, *IEA Report for the Clean Energy Ministerial: Global Gaps in Clean Energy RD&D: Update and Recommendations for International Collaboration*, www.iea.org/papers/2010/global_gaps.pdf (accessed July 22, 2010); and http://www.iea.org/stats/rd.asp for underlying tables (accessed July 22, 2010). A spike in funding in 2009 was linked to stimulus funds to escape the economic crisis, with indications that this was a one-time event and not a sustained level of funding, as discussed in Kerr's report.

29. Revkin, A. C., "Questions for Obama's science team," *Dot Earth*, Dec. 19, 2008, http://dotearth.blogs.nytimes.com/2008/12/19/questions-for-obamas-science-team/. Also see Revkin, A. C., "The energy challenge, budgets falling in race to fight global warming," *New York Times*, Oct. 30, 2006; and Revkin, A. C., "On the energy gap and climate crisis," *Dot Earth*, Apr. 7, 2010, http://dotearth.blogs.nytimes.com/2010/04/07/on-the-energy-gap-and-climate-crisis/.

30. See note 4.

17: Sustainable Solutions on the Wind

1. Seneca, *Morals*, Roger L'Estrange, 6th American ed., (J. B. Lippincott, 1880), p. 120.

2. See, for example, MacKay, D. J. C., 2009, "Sustainable Energy—Without the Hot Air," www.withouthotair.com.

3. As noted in chapter 4, the Energy Information Agency of the U.S. Department of Energy gave World Primary Energy Use—the part that is sold to others—as averaging about 15.7 terawatts (15.7 TW) in 2006, or 15.7×10^{12} W, or 15,700,000,000,000 W, or enough to keep one heck of a lot of lightbulbs burning.

4. U.S. energy expenditures were $3881/person, or $1,157,910,000,000 for the country in 2006; http://www.eia.doe.gov/emeu/aer/txt/ptb0105.html (accessed Nov. 20, 2009).

5.   On a U.S. football field, the region in play between the goal lines is 160 feet (49 m) by 300 feet (91 m), so 4459 m². I didn't use a soccer pitch because the size isn't specified exactly, being required to have a length of 100–130 yards (90–120 m) and width of 50–100 yards (45–90 m), but not to be square; http://www.fifa.com/worldfootball/lawsofthegame.html (accessed July 15, 2010).

6.   The energy supply from the sun is what is available at the top of the atmosphere, but to get it, you would need to use space-based solar cells. About 30% is reflected, leaving about 122,000 TW to heat the planet. When spread around the world, this averages 240 W/m². Some of this warms the air or clouds, while much reaches the surface. A good textbook, such as Peixoto, J. P. and A. H. Oort, 1992, *Physics of Climate* (American Institute of Physics, New York), is the best source. The simple figure at http://asd-www.larc.nasa.gov/ceres/brochure/rad_bal.gif (accessed Nov. 19, 2009) is a helpful shortcut.

7.   About 1% of the sun's energy absorbed by Earth drives the wind. If the sun stopped shining, much of the wind would stop soon afterward because of "friction"—the wind makes big tornadoes and little whirlpools, which make smaller and smaller disturbances until the energy is spread out into random motion of molecules, which is heat. This heat is added to the air, or to surfaces beneath. The approximately 2.4 W/m² generated by the wind is spread out through the atmosphere as well as at the surface, and so isn't nearly enough for you to feel—the wind cools you by evaporating perspiration much more than it warms you by friction. Lu, X., M. B. McElroy and J. Kiviluoma, 2009, "Global potential for wind-generated electricity," *Proceedings of the National Academy of Sciences of the United States of America* 106: 10,933–38; Lorenz, E. N., 1967, *The Nature and Theory of the General Circulation of the Atmosphere* (World Meteorological Organization, Geneva); and Peixoto and Oort in note 6.

8.   This is net primary production, with 90 TW on land and the rest in the ocean, from Field, C. B., M. J. Behrenfeld, J. T. Randerson and P. Falkowski, 1998, "Primary production of the biosphere: Integrating terrestrial and oceanic components," *Science* 281: 237–40, who give production of 104.9×10^{15} g C/year, with 56.4×10^{15} g C/year on land and 48.5×10^{15} g C/year in the oceans. To convert to terawatts, note that from http://bioenergy.ornl.gov/papers/misc/energy_conv.html (accessed Nov. 19, 2009), the energy of dry wood is 20 gigajoules/ton, and taking this as 40% carbon gives 50 gigajoules/ton of carbon.

9.   Some of the wind's energy is turned to waves and currents in the ocean, which eventually lose their energy as heat. Globally, this amounts to about 65 TW, of which roughly 2 TW eventually ends up in waves breaking on the coast. Ferrari, R. and C. Wunsch, 2009, "Ocean circulation kinetic energy: Reservoirs, sources and sinks," *Annual Reviews of Fluid Mechanics* 41: 253–82.

10. Pollack, H. N., S. J. Hurter and J. R. Johnson, 1993, "Heat-flow from the Earth's interior: Analysis of the global data set," *Reviews of Geophysics* 31: 267–80.

11. See note 3.

12. Munk, W., 1997, "Once again: Once again—Tidal friction," *Progress in Oceanography* 40: 7–35. Note that the tides get their energy from the rotation of the planet, left over from formation of the solar system. If we used a lot of tidal energy, and if this usage increased dissipation of that energy as opposed to transferring it from natural heat generation, then we would be doing something unsustainable. However, we are very unlikely to be able to extract enough tidal power to notably deplete the stored energy over mere millions of years, so I left this in the sustainable category. Whenever we try to make a "simple" division, such as sustainable/unsustainable, we find that the world isn't designed to make such simple divisions easy.

13. The sun evaporates water, which falls as rain or snow. Most of the rain that falls on land eventually evaporates from plants, but some flows downhill to the ocean. If we could get the energy from all of the falling rain over the entire Earth from the moment it leaves the cloud until it reaches sea level, it would be about 500 TW. The average rainfall is 1 cubic meter of water per square meter of surface per year, and with an assumed cloud height of 3000 m, water density of 1000 kg/m^3, acceleration of gravity of 10 m/s^2 (for simplicity), and 3.16 $\times$ 10^7 s in a year, the potential energy is converted to heat at the rate 1*3000*1000*10/3.16 $\times$ 10^7 = 1 W/m^2. For a surface area of the planet of 5.1 $\times$ 10^{14} m^2, this gives 510 TW. There is no way to capture all of this energy, but the size may be of interest. The part that might be available is the energy of the rainfall on land that does not evaporate. A crude estimate can be made as follows. From Fekete, M. and C. J. Vörösmarty, 2007, "The current status of global river discharge monitoring and potential new technologies complementing traditional discharge measurements," *International Association of Hydrological Sciences Publication* 309: 129–36, the global river flow is about 4 $\times$ 10^{12} m^3/year, for a land area of 1.49 $\times$ 10^{14} m^2. Assuming a mean vertical drop of 600 m, this yields 0.005 W/m^2, or 0.8 TW globally. Higher mountains tend to be wetter; allowing for this and doing the calculation better almost triples the total available energy (Jacobson, M. Z. and M. A. Delucchi, 2009, "A path to sustainable energy by 2030," *Scientific American* (Nov.): 58–65; also see Jacobson, M. Z., 2009, "Review of solutions to global warming, air pollution, and energy security," *Energy & Environmental Science* 2: 148–73).

14. See the Jacobson and Delucchi, and Jacobson, references in note 13; and U.S. Department of Energy, Energy Information Agency. Projected energy use for the globe for the year 2030 was 16.9 TW, but this was reduced to 11.5 TW if all energy were generated sustainably by wind, water, and solar, with no fossil fuels or biomass burning.

15. Jacobson, M. Z., 2009, "Review of solutions to global warming, air pollution, and energy security," *Energy & Environmental Science* 2: 148–73. See table 3 and

the accompanying discussion in section 4, with 1.6–4.3 months energy payback given for the cases considered.

16. Nemet, G. F., 2009, "Net radiative forcing from widespread deployment of photovoltaics," *Environmental Science and Technology* 43: 2173–78. For a lighter, but highly substantive, version of some of this material, see Pierrehumbert, R., "An open letter to Steve Levitt," *Realclimate*, Oct. 29, 2009, http://www.realclimate .org/index.php/archives/2009/10/an-open-letter-to-steve-levitt/.

17. Keith, D. W., J. F. DeCarolis, D. C. Denkenberger, et al., 2004, "The influence of large-scale wind power on global climate," *Proceedings of the National Academy of Sciences of the United States of America* 101: 16,115–20; and Brostrom, G., 2008, "On the influence of large wind farms on the upper ocean circulation," *Journal of Marine Systems* 74: 585–91.

18. Haxel, G. B., J. B. Hedrick and G. J. Orris, 2002, *Rare Earth Elements—Critical Resources for High Technology*, USGS Fact Sheet 087-02.

19. Morcos, T., 2009, "Harvesting wind power with (or without) permanent magnets," *Magnetics and Business Technology* (Summer): 26, www.magneticsmaga zine.com/images/PDFs/Online%20Issues/2009/Magnetics_Summer09 .pdf (accessed July 21, 2010); and U.S. Geological Survey (USGS), 2010, "Rare Earths," in *Mineral Commodity Summaries*, http://minerals.usgs.gov/minerals/ pubs/commodity/rare_earths/mcs-2010-raree.pdf (accessed July 21, 2010).

20. Archer, C. L. and K. Caldeira, 2009, "Global assessment of high-altitude wind power," *Energies* 2: 307–319, doi:10.3390/en20200307.

21. *Discoveries and Inventions: A Lecture by Abraham Lincoln Delivered in 1860* (John Howell, San Francisco, 1915).

22. Some dispute exists as to who first harnessed the wind with a windmill, probably for grinding grain or lifting water. Heron of Alexandria designed a wind-powered organ almost two millennia ago (Drachmann, A. G., 1961, "Heron's windmill," *Centaurus* 7: 145–51). Persians were using windmills more than a millennium ago (Hassan, A. Y. and D. R. Hill, 1986, *Islamic Technology: An Illustrated History* [Cambridge University Press, New York], p. 54), and the Chinese used windmills as early as the twelfth century and perhaps before (Zhang, B. C., 2009, "Ancient Chinese windmills," in Yan, H. S. and Ceccarelli, M. [eds.], *International Symposium on History of Machines and Mechanisms, Proceedings of HMM 2008*, pp. 203–14). Windmills were used widely in Europe, including for their key role of pumping water out of the polders in the Netherlands.

23. See Baker, T. L., 1985, *A Field Guide to American Windmills* (University of Oklahoma Press, Norman); also, Baker, T. L., 1980, "Turbine-type windmills of the Great Plains and Midwest," *Agricultural History* 54: 38–51, which includes much information on the many manufacturers and the rapid rate of production and installation. The estimate of nearly 100,000 windmills/year at the peak is from Davenport, J., 2007, "Quirky old-style contraptions make water from wind on

the mesas of West Texas," *San Antonio Express-News*, http://www.mysanantonio .com/news/MYSA092407_01A_State_windmills_3430a27_html (accessed July 14, 2010). Note that decades later, pioneering efforts in Denmark to harness wind power led to major export markets that helped the economy; see Hansen, J. D., C. Jensen and E. S. Madsen, 2003, "The establishment of the Danish wind-mill industry—Was it worthwhile?" *Review of World Economics* 139: 324–47.

24. "Harnessing icy polar gales for power," *Popular Science Monthly* 128, no. 3 (March 1936): 25. For more on Debenham, see Walsh, G. P., 1993, "Debenham, Frank (1883–1965)," in the *Australian Dictionary of Biography*, vol. 13 (Melbourne University Press), pp. 602–3; and http://about.nsw.gov.au/collections/doc/frank-debenham/ (accessed Nov. 21, 2009).

25. See Rejcek, P., 2010, "Winding up: New wind farm to help power U.S., New Zealand research stations in Antarctica," *Antarctic Sun*, Jan. 19, http://antarctic sun.usap.gov/features/contenthandler.cfm?id=2014 (accessed July 15, 2010). Also, http://www.nsf.gov/news/news_summ.jsp?cntn_id=116281&org=OPP &from=home.

26. For example, Wiser, R. and M. Bolinger, 2007, *Annual Report on U.S. Wind Power Installation, Cost, and Performance Trends: 2006* (Energy Efficiency and Renewable Energy, U.S. Department of Energy, Washington, DC).

27. See Lu, X., M. B. McElroy and J. Kiviluoma, 2009, "Global potential for wind-generated electricity," *Proceedings of the National Academy of Sciences of the United States of America* 106: 10,933–38.

28. The slightly more than 10,000 W of usage per person in the United States should be reduced by about one-third for the extra efficiency of electricity, following note 14.

29. The power at 330 feet height compared to 33 feet is approximately $P_{330}/P_{33} = (330/33)^{3/7} = 2.7$, or at heights z_1 and z_2, $P_{z2}/P_{z1} = (z_2/z_1)^{3/7}$; see Peterson, E. W. and J. P. Hennessey, 1978, "Use of power laws for estimates of wind power potential," *Journal of Applied Meteorology* 17: 390–94. The power also depends on the air density, which decreases upward, but not enough to matter much going from 33 to 330 feet.

30. See note 28.

31. Available offshore wind power might be 39 TW, well over human use; Capps, S. B. and C. S. Zender, 2010, "Estimated global ocean wind power potential from QuikSCAT observations, accounting for turbine characteristics and siting," *Journal of Geophysical Research* 115: D09101.

32. See note 26.

33. In 2006, the price of installed wind power in the United States was averaging $1500/kW, with larger projects costing a bit less and some projects coming in at about $1000/kW, although with some signs of rising costs. See ibid.

34. World Bank, http://datafinder.worldbank.org/about-world-development-indicators?cid=GPD_WDI; in 2008, world gross domestic product in U.S. dollars was $60.6 trillion.

35. Thoreau, H. D., *The Writings of Henry David Thoreau: Journal*, vol. 9, *August 16, 1856–August 7, 1857*, ed. Bradford Torrey (Houghton Mifflin, Boston, 1906), p. 245.

36. It is also possible to use pricing to shift demand to follow supply, and to use accurate forecasting of potential supply and demand to guide rapid wise decisions on responses. See Jacobson in note 13.

37. See note 27.

38. Mills, A., R. Wiser and K. Porter, 2009, *The Cost of Transmission for Wind Energy: A Review of Transmission Planning Studies*, Lawrence-Berkeley National Laboratory LBNL-1471E, http://eetd.lbl.gov/EA/EMP/reports/lbnl-1471e.pdf (accessed July 16, 2010); and Energy Efficiency and Renewable Energy, U.S. Department of Energy, 2008, *20% Wind Energy by 2030: Increasing Wind Energy's Contribution to U.S. Electricity Supply*, DOE/GO-102008-2567, www.eere.energy.gov/windandhydro; http://www.nrel.gov/docs/fy08osti/41869.pdf (accessed July 16, 2010).

39. See note 27.

40. See, for example, Jacobson, and Jacobson and Delucchi, in note 13; or Lu, X., M. B. McElroy and J. Kiviluoma, 2009, "Global potential for wind-generated electricity," *Proceedings of the National Academy of Sciences of the United States of America* 106: 10,933–38.

41. Fridleifsson, I. B., 2001, "Geothermal energy for the benefit of the people," *Renewable and Sustainable Energy Reviews* 5: 299–312.

42. Ekman, C. K. and S. H. Jensen, 2010, "Prospects for large scale electricity storage in Denmark," *Energy Conversion and Management* 51: 1140–47.

43. The ability for individuals to plug their hybrid or all-electric vehicles into the grid to buy and sell electricity is now being called "vehicle-to-grid," or "V2G," by many people. Information sources (with or without that acronym) include Jacobson and Delucchi in note 13; and chapter 4 in World Bank, 2009, *World Development Report 2010* (World Bank, Washington, DC). Also see Lund, H. and W. Kempton, 2008, "Integration of renewable energy into the transport and electricity sectors through V2G," *Energy Policy* 36: 3578–87. However, plug-in hybrids are not a panacea; see National Research Council, 2009, *Transitions to Alternative Transportation Technologies: Plug-In Hybrid Electric Vehicles* (National Academies Press, Washington, DC); and Sovacool, B. K. and R. F. Hirsh, 2009, "Beyond batteries: An examination of the benefits and barriers to plug-in hybrid electric vehicles (PHEVs) and a vehicle-to-grid (V2G) transition," *Energy Policy* 37: 1095–1103.

44. Kempton, W. and J. Tomic, 2005, "Vehicle-to-grid power implementation: From stabilizing the grid to supporting large-scale renewable energy," *Journal of Power Sources* 144: 280–94.

45. The wind project was facilitated in part by the Island Institute, http://www .islandinstitute.org/. For full disclosure, its amazing president, Philip Conkling, is also a good friend of mine.

46. Curry, A., 2009, "Deadly flights," *Science* 325: 386–87; Sovacool, B. K., 2009, "Contextualizing avian mortality: A preliminary appraisal of bird and bat fatalities from wind, fossil-fuel, and nuclear electricity," *Energy Policy* 37: 2241–48.

47. Philip Conkling, president of the Island Institute, quoted in Waterman, M., "Vinalhaven celebrates Fox Islands wind turbines," *Free Press*, Nov. 19, 2009, http://freepressonline.com/main.asp?SectionID=52&SubSectionID=78& ArticleID=3980.

48. See, for example, Burnett, J., 2007, "Winds of change blow into Roscoe, Texas," National Public Radio, Nov. 27, 2007, http://www.npr.org/templates/story/ story.php?storyId=16658695. For an interesting discussion of similar conflicts playing out over transmission lines in Colorado, and in particular, lines for electricity made at solar farms, see Jaffe, M., "Battle lines drawn over San Luis Valley electric-transmission plans," *Denver Post*, Dec. 13, 2009, http://www.denverpost .com/business/ci_13982743#ixzz0dkNls3Op (accessed Jan. 26, 2010).

49. Rubinkam, M., "Pa. residents sue gas driller over polluted wells," Associated Press; published in many places including the *Atlanta Journal-Constitution*, http:// www.ajc.com/business/pa-residents-sue-gas-207374.html (accessed Nov. 22, 2009).

50. For example, Kirby, C. S., B. McInerney and M. D. Turner, 2008, "Groundtruthing and potential for predicting acid deposition impacts in headwater streams using bedrock geology, GIS, angling, and stream chemistry," *Science of the Total Environment* 393: 249–61; and Kahl, J. S., J. L. Stoddard, R. Haeuber, et al., 2004, "Have US surface waters responded to the 1990 Clean Air Act Amendments?" *Environmental Science and Technology* 38: 484A–490A.

51. For example, White, E. M., G. J. Keeler and M. S. Landis, 2009, "Spatial variability of mercury wet deposition in eastern Ohio: Summertime meteorological case study analysis of local source influences," *Environmental Science and Technology* 43: 4946–53; and Hammerschmidt, C. R. and W. F. Fitzgerald, 2006, "Methylmercury in freshwater fish linked to atmospheric mercury deposition," *Environmental Science and Technology* 40: 7764–70.

52. The damage from the sulfur dioxide, nitrous oxides, and particulate matter from coal-fired power plants is estimated as about $0.03/kWh (National Research Council, 2010, *Hidden Costs of Energy: Unpriced Consequences of Energy Production and Use* [National Academies Press, Washington, DC]). That study estimated a similar cost from the global warming from the emitted CO_2; as discussed in

chapter 15, Nordhaus obtained about one-third of that. From data from the Energy Information Agency of the U.S. Department of Energy, in its *Monthly Energy Review* (June 2010, www.eia.gov/mer) and *Coal Prices and Outlook*, http://www.eia.doe.gov/energyexplained/index.cfm?page=coal_prices, the cost of the coal used in energy generation in the United States, at the power plant, is roughly $0.02/kWh, much less than the "unpriced" damages. The average cost of electricity to industrial customers in the United States (who pay lower prices than residential, commercial, or transportation customers) was $0.068/kWh in 2008, with customers in states having much coal-fired generation often paying on the low side of that, and with the average price for all users in West Virginia, which has much coal-fired generation, being $0.056/kWh. Some of this cost, such as billing, advertising, profit, and distribution, would apply to electricity from any source. Especially if mercury deposition, acid mine drainage, and other costs were included, the societal cost of coal-fired electricity thus seems to be safely more than twice as much as the price paid to generate the electricity from coal.

53. Hoen, B., R. Wiser, P. Cappers, M. Thayer and G. Sethi, 2009, *The Impact of Wind Power Projects on Residential Property Values in the United States: A Multi-site Hedonic Analysis*, Lawrence Berkeley National Laboratory LBNL 2829E, http://eetd.lbl.gov/EA/EMP/reports/lbnl-2829e.pdf. In the abstract: "Although the analysis cannot dismiss the possibility that individual homes or small numbers of homes have been or could be negatively impacted, it finds that if these impacts do exist, they are either too small and/or too infrequent to result in any widespread, statistically observable impact."

54. Pryor, S. C. and R. J. Barthelmie, 2010, "Climate change impacts on wind energy: A review," *Renewable and Sustainable Energy Reviews* 14: 430–37.

55. See note 18.

18: Sun and Water

1. Entry from Aug. 18, 1830, in Emerson, R. W., *Journals of Ralph Waldo Emerson, 1820–1872*, vol. 2, ed. Edward Waldo Emerson and Waldo Emerson Forbes (Houghton Mifflin, Boston, 1909), p. 304.

2. See, for example, Kryza, F. T., 2003, *The Power of Light: The Epic Story of Man's Quest to Harness the Sun* (McGraw-Hill, New York); Velasquez-Manoff, M., "Sunrise for solar heat power," *Christian Science Monitor*, Aug. 18, 2009, http://www.csmonitor.com/Innovation/Energy/2009/0818/sunrise-for-solar-heat-power (accessed Jan. 24, 2010); Belessiotis, V. and E. Delyannis, 2000, "The history of renewable energies for water desalination," *Desalination* 128: 147–59; Sempler, K., "Nothing new under the sun," reprinted in *New Science Journalism*, July 23, 2009, http://www.newsciencejournalism.net/index.php?/news_articles/view/nothing_new_under_the_sun/ (accessed Jan. 24, 2010), originally published in

Swedish in *NyTeknik*, Feb. 18, 2009, http://www.nyteknik.se/popular_teknik/kaianders/article518080.ece (also accessed Jan. 24, 2010); Shuman, F., 1911, "Solar power," *Scientific American* 71: 78; and "Method to harness the sun is found," *New York Times*, Dec. 3, 1911.

3.　For example, see "Shower of glass in a school. A falling wall breaks a skylight, injuring four persons," *New York Times*, Jan. 25, 1884, p. 8, http://query.nytimes.com/gst/abstract.html?res=9903E6D61538E033A25756C2A9679C94659FD7CF (accessed July 17, 2010).

4.　See Shuman, p. 78, in note 2.

5.　Ackermann, A. S. E., 1912, "The Shuman sun-heat absorber," *Nature* 89: 122–23, doi:10.1038/089122a0.

6.　Clemensen, B., 1979, *Historic Resource Study, Cape Cod National Seashore, Massachusetts* (Denver Service Center, National Park Service, U.S. Department of the Interior, Denver), FX-79. See p. 18 for the original extent of woodlands, p. 19 for loss of those woodlands, and pp. 23–24 for the history of salt-making.

7.　MacKay, D. J. C., 2009, "Sustainable Energy—Without the Hot Air," www.withouthotair.com. Note that this is the average over a year, not the peak in midsummer at noon.

8.　Fsthenakis, V., J. E. Mason and K. Zweibel, 2009, "The technical, geographical, and economic feasibility for solar energy to supply the energy needs of the US," *Energy Policy* 37: 387–99.

9.　See Schnatbaum, L., 2009, "Solar thermal power plants," *European Physical Journal Special Topics* 176: 127–40. As reported by the BBC, the Desertec Consortium, including such companies as Deutsche Bank and Siemens, launched a $400 billion project in November 2009, building a serious version of Shuman's solar-thermal plant in the Sahara to make electricity for Europe and Africa, with a goal of supplying energy continuously through high-voltage, low-loss transmission lines. "Sahara sun 'to help power Europe,'" BBC, Nov. 2, 2009, http://news.bbc.co.uk/2/hi/africa/8337735.stm (accessed Jan. 24, 2010).

10.　Service, R. F., 2009, "Sunlight in your tank," *Science* 326: 1472–75. Also see Licht, S., B. Wang, S. Ghosh, et al., 2010, "A new solar carbon capture process: Solar thermal electrochemical photo (STEP) carbon capture," *Journal of Physical Chemistry Letters* 11: 2363–68.

11.　Each silicon atom brings four electrons that are looking for action (valence electrons), each of these electrons forms a shared-electron bond with an electron from a neighboring silicon atom, and the electrons then are tied up in their bonds and not free to do much that is interesting unless they are kicked really hard. But, put a little phosphorus in the silicon, with each phosphorus atom bringing five valence electrons, and you end up with a solid material in which a lot of electrons are mobile and ready to move around because they don't have a bond of their own. You could call this an "n-doped material"—adding the phosphorus is called "doping," and the phosphorus frees up *n*egatively charged electrons. If instead you added boron atoms, which bring three valence electrons each, you would have a shortage of the mobile electrons. You can think of a missing electron as a *p*ositive (p) hole; a neighboring electron can jump into this "hole," and wherever that electron came from is now short one electron, so you can think of the positive hole moving around. Now, lay a p-doped against an n-doped layer. Because concentrations of things tend to spread out by the usual random wanderings that we call "diffusion," some of the extra mobile electrons will diffuse from the n-layer, where they are initially common, to the p-layer, where they are initially rare. Meanwhile, extra holes will wander the other way. This adds negative charge to the p-doped layer right near the interface, and positive charge to the n-doped layer, creating an electric field across the boundary. This electric-field charge difference pulls back on the electrons and holes, eventually reaching a balance so things quit changing. At this point, there is an electrical field across the boundary, or junction, between the two layers. If a sufficiently energetic photon now hits near the junction in the region with the electric field, knocking an electron loose from its atom to leave a hole, the electrical field will push the electrons in one direction (toward the n-doped layer) and the holes the other way. If you attach a wire that goes out, then through something such as a lightbulb or a heart-lung machine, and then back, the pushed electrons will follow that path and do something useful for you on the way. This is a solar cell, supplying electricity. You can find a rigorous treatment of how solar cells work, at some level of friendliness, from a good physics text. The web site "How Stuff Works" has a reasonably friendly treatment, at http://science.howstuffworks .com/solar-cell1.htm (accessed Jan. 24, 2010).

12. Steinhagen, C., M. G. Panthani, V. Akhavan, et al., 2009, "Synthesis of $Cu_2Zn SnS_4$ nanocrystals for use in low-cost photovoltaics," *Journal of the American Chemical Society* 131 (Communication): 12,554–55; also see Li, C. and S. Mitra, 2007, "Processing of fullerene-single wall carbon nanotube complex for bulk heterojunction photovoltaic cells," *Applied Physics Letters* 91: paper 253112.

13. Dedeo, M. T., K. E. Duderstadt, J. M. Berger and M. B. Francis, 2010, "Nanoscale protein assemblies from a circular permutant of the tobacco mosaic," *Nano Letters* 10: 181–86.

14. Bahners, T., U. Schlosser, R. Gutmann and E. Schollmeyer, 2008, "Textiles for solar light collectors based on models for polar bear hair," *Solar Energy Materials and Solar Cells* 92: 1661–67; Stegmaier, T., M. Linke and H. Planck, 2009, "Bionics in textiles: Flexible and translucent thermal insulations for solar thermal applications," *Philosophical Transactions of the Royal Society of London A: Mathematical, Physical and Engineering Sciences* 367: 1749–58; also see Kelzenberg, M. D., S. W. Boettcher,

J. A. Petykiewicz, et al., 2010, "Enhanced absorption and carrier collection in Si wire arrays for photovoltaic applications," *Nature Materials* (advanced online publication Feb. 14): doi:10.1038/nmat2635.

Plants have been photosynthesizing for a lot of billions of years, and they must split hydrogen from water and combine with rearranged CO_2 to make plant material while releasing oxygen. The efficiency is not especially impressive—maybe 1% (see note 7), but some of the inefficiency at least is related to the requirement that the plant keep itself alive, defend itself from competitors or things that want to eat it, and make more of itself. If we could genetically engineer plants to concentrate on energy, or make fake plants to concentrate on energy, or use catalysts and sunlight to split water or CO_2 completely outside of plants, maybe we could increase the efficiency of solar collection enough to really help us; see, for example, Service, R. F., 2009, "New trick for splitting water with sunlight," *Science* 325: 1200–1201; Angamuthu, R., P. Byers, M. Lutz, A. L. Spek and E. Bouwman, 2010, "Electrocatalytic CO_2 conversion to oxalate by a copper complex," *Science* 327: 313–15; and Service, R. F., 2010, "Catalyst offers new hope for capturing CO_2 on the cheap," *Science* 327: 257.

15. International Energy Agency, Organization of the Petroleum Exporting Countries, Organisation for Economic Co-operation and Development, and the World Bank Joint Report, 2010, *Analysis of the Scope of Energy Subsidies and Suggestions for the G-20 Initiative*, prepared for submission to the G-20 Summit Meeting, Toronto, June 26–27, 2010. Recall that a lot of taxes are paid on fossil fuels; the 1% subsidy level does not count those taxes.

16. See note 8.

17. Fourteen cents per kilowatt-hour for solar-thermal electricity, versus twelve cents for elecricity from gas and six cents for that from coal; see Velasquez-Manoff, M., "Sunrise for solar heat power," *Christian Science Monitor*, Aug. 18, 2009, http://www.csmonitor.com/Innovation/Energy/2009/0818/sunrise-for-solar-heat-power (accessed Jan. 24, 2010).

18. Lackner, K. S. and J. D. Sachs, 2005, "A robust strategy for sustainable energy," *Brookings Papers on Economic Activity* 2: 215–69, http://www.jstor.org/stable/3805122 (accessed Nov. 23, 2009).

19. Refer back to notes 8 and 18; also see Jacobson, M. Z. and M. A. Delucchi, 2009, "A path to sustainable energy by 2030," *Scientific American* (Nov.): 58–65. An instructive analysis for Germany especially is Bhandari, R. and I. Stadler, 2009, "Grid parity analysis of solar photovoltaic systems in Germany using experience curves," *Solar Energy* 83: 1634–44. Also see Yang, C. J., 2010, "Reconsidering solar grid parity," *Energy Policy* 38: 3270–73; Wynn, G., "Solar power edges towards boom time," Reuters, Oct. 19, 2007, http://www.reuters.com/article/idUSL1878986220071019 (accessed Jan. 24, 2010); and Harrabin, R., "Solar panel

costs 'set to fall,'" BBC, Nov. 30, 2009, http://news.bbc.co.uk/2/hi/science/nature/8386460.stm (accessed Jan. 24, 2010).

20. International Energy Agency, 2004, *Renewable Energy: Market and Policy Trends in IEA Countries*, https://www.iea.org/textbase/nppdf/free/2004/renewable1.pdf (accessed Jan. 24, 2010). By 2008, funding for research on all renewable energy sources was about one-third of that for research on fission and fusion taken together across the IEA countries, and was almost half as much in the United States, with research on solar energy accounting for up to about one-sixth of the funding for research on fission and fusion taken together; see Kerr, T., 2010, *IEA Report for the Clean Energy Ministerial: Global Gaps in Clean Energy RD&D: Update and Recommendations for International Collaboration*, www.iea.org/papers/2010/global_gaps.pdf (accessed July 22, 2010), and http://www.iea.org/stats/rd.asp for underlying tables (accessed July 22, 2010).

21. Heating water accounted for 9.1% of household electricity use in the United States, for 2001, from the U.S. Department of Energy, Energy Information Agency, http://www.eia.doe.gov/emeu/reps/enduse/er01_us_tab1.html (accessed Jan. 22, 2010).

22. Feist, W. and J. Schnieders, 2009, "Energy efficiency—A key to sustainable housing," *European Physical Journal Special Topics* 176: 141–53.

23. See note 19.

24. Ramanathan, V. and K. Balakrishnan, 2007, "Reduction of air pollution and global warming by cooking with renewable sources: A controlled and practical experiment in rural India," white paper, http://www-ramanathan.ucsd.edu/ProjectSurya.html (accessed July 21, 2010).

25. Anandakrishnan, S., D. E. Voigt, P. G. Burkett, B. Long and R. Henry, 2000, "Deployment of a broadband seismic network in west Antarctica," *Geophysical Research Letters* 27: 2053–56.

26. Ezra Cornell, co-founder of Cornell University, as quoted in "The Ezra Cornell Bicentennial Exhibition," http://rmc.library.cornell.edu/ezra/exhibition/ithaca/index.html. Ezra Cornell designed and built a new mill to grind grain for flour, a tunnel to take water from a reservoir to the mill, and a dam to fill the reservoir. Cornell Papers, Memorandum Book, II, Feb. 1, 1861, Cornell University, Ithaca, NY.

27. The ability to quickly follow demand for power is often called "peaking and gap-filling power." See U.S. Department of the Interior, Bureau of Reclamation Power Resources Office, 2005, *Reclamation: Managing Water in the West: Hydroelectric Power*, http://www.usbr.gov/power/edu/pamphlet.pdf.

28. Naiman, R. J., C. A. Johnston and J. C. Kelley, "Alteration of North American streams by beaver," *BioScience* 38: 753–62; they quote an older source giving 60–400 million.

29. See, for example, Chatanantavet, P. and G. Parker, 2009, "Physically based modeling of bedrock incision by abrasion, plucking, and macroabrasion," *Journal of Geophysical Research* 114: F04018, doi:10.1029/2008JF001044.

30. Helms, S. W., 1981, *Jawa: Lost City of the Black Desert* (Cornell University Press, Ithaca, NY); and Whitehead, P. G., S. J. Smith, A. J. Wade, et al., 2008, "Modelling of hydrology and potential population levels at Bronze Age Jawa, Northern Jordan: A Monte Carlo approach to cope with uncertainty," *Journal of Archaeological Science* 35: 517–29. Use of water mills for grinding grain appears to date back at least to Sumerian times (Vowles, H. P., 1931, "Pliny's water-mill," *Nature* 127: 889).

31. Walter, R. C. and D. J. Merritts, 2008, "Natural streams and the legacy of water-powered mills," *Science* 319: 299–304.

32. Ibid.

33. Ibid.

34. See, for example, Jacobson, M. Z., 2009, "Review of solutions to global warming, air pollution, and energy security," *Energy & Environmental Science* 2: 148–73.

35. ISI Web of Knowledge produced 190 papers on "Three Gorges Dam," Google search of the web produced "about 711,000 results," and Bing produced 3,380,000 results (all accessed July 22, 2010).

36. Kelner, D. E. and B. E. Sietman, 2000, "Relic populations of the ebony shell, *Fusconaia ebena* (Bivalvia : Unionidae), in the upper Mississippi River drainage," *Journal of Freshwater Ecology* 15: 371–77. For a friendlier introduction, see "America's Mussels: Silent Sentinels," U.S. Fish and Wildlife Service, http://www.fws .gov/midwest/endangered/clams/mussels.html (accessed July 22, 2010).

37. A wealth of material on the impacts of the Glen Canyon dam, and the attempts to limit these impacts through actions, including human-released floods, is available through sources, including the U.S. Geological Survey. Start at http://walrus.wr .usgs.gov/grandcan/flood.html (accessed on Aug. 11, 2010).

38. One carefully studied case in the United States involves two dams built in the early 1900s on the Elwha River, which flows north from Olympic National Park in Washington State to an arm of the Pacific Ocean (the Strait of Juan de Fuca). The dams reduced annual salmon runs from roughly 300,000 to roughly 3000 fish—fish ladders were not built, and a fish hatchery was quickly abandoned. Sand and gravel bars below the dams were washed away by the sediment-free water released, contributing to loss of fish. Then, the lack of additional sediment supply to the coast led to loss of beaches along the Strait, interfering with traditional shellfishing of the native people, while engineering was required to protect the nearby harbor of Port Angeles after it lost its natural sediment-fed bars. An ambitious plan to remove the dams is moving forward slowly. See, for example, the U.S. National Park Service resources on the Elwha Ecosystem

Restoration, at http://www.nps.gov/olym/naturescience/elwha-ecosystem-restoration.htm (accessed Nov. 25, 2009). Also see Morley, S. A., J. J. Duda, H. J. Coe, K. K. Kloehn and M. L. McHenry, 2008, "Benthic invertebrates and periphyton in the Elwha River basin: Current conditions and predicted response to dam removal," *Northwest Science* 82: 179–96.

39. Steinmann, P., J. Keiser, R. Bos, M. Tanner and J. Utzinger, 2006, "Schistoso-miasis and water resources development: Systematic review, meta-analysis, and estimates of people at risk," *Lancet Infectious Diseases* 6: 411–25.

40. Nilsson, C., C. A. Reidy, M. Dynesius and C. Revenga, 2005, "Fragmentation and flow regulation of the world's large river systems," *Science* 308: 405–8.

41. Bagla, P., 2010, "News Focus: Along the Indus River, saber rattling over water security," *Science* 328: 1226–27.

42. Bawa, K. S., L. P. Koh, T. M. Lee, et al., 2010, "Policy Forum: China, India and the environment," *Science* 327: 1457–58.

43. Lejon, A. G. C., B. M. Renofalt and C. Nilsson, 2009, "Conflicts associated with dam removal in Sweden," *Ecology and Society* 14: article 4 2009; also see note 36.

44. See note 42.

45. Part 2 in Food and Agriculture Organization of the United Nations, 2009, *The State of Food and Agriculture 2009*, http://www.fao.org/docrep/012/i0680e/i0680e00.htm (accessed Aug. 11, 2010), p. 104.

46. Wisser, D., S. Frolking, E. M. Douglas, et al., 2010, "The significance of local water resources captured in small reservoirs for crop production—A global-scale analysis," *Journal of Hydrology* 384: 264–75.

47. Mote, P. W., A. F. Hamlet, M. P. Clark and D. P. Lettenmaier, 2005, "Declining mountain snowpack in western North America," *Bulletin of the American Meteorological Society* 86: 39–49.

48. Immerzeel, W. W., L. P. H. van Beek and M. F. P. Bierkens, 2010, "Climate change will affect the Asian water towers," *Science* 328: 1382–85.

49. See note 40.

19: Down by the Sea, Where the Water Power Grows

1. "For thogh we slepe, or wake, or rome, or ryde, Ay fleeth the tyme; it nyl no man abyde," from Chaucer, G., "The Clerk's Tale," in *The Ellesmere Manuscript of Chaucer's Canterbury Tales*, part 1 (Trubner, London, 1868), p. 407.

2. Sharples, J., C. M. Moore, A. E. Hickman, et al., 2009, "Internal tidal mixing as a control on continental margin ecosystems," *Geophysical Research Letters* 36: L23603.

3. Start with Munk, W., 1997, "Once again: Once again—Tidal friction," *Progress in Oceanography* 40: 7–35; you might also enjoy Munk, W. and C. Wunsch, 1998,

"Abyssal recipes II: Energetics of tidal and wind mixing," *Deep Sea Research Part I: Oceanographic Research Papers* 45: 1977–2010. Winds as well as tides are important in mixing the ocean.

4. Ibid.

5. Tidal heating can be much larger on moons close to gigantic Jupiter and Saturn, with their strong gravity; for example, Meyer, J. and J. Wisdom, 2007, "Tidal heating in Enceladus," *Icarus* 188: 535–39.

6. Dallison, I., "A Timeline History of Yarmouth, Isle of Wight," http://freespace .virgin.net/iw.history/yarmouth/history.htm (accessed Nov. 29, 2009); and Milton, S., "The winds of time," *Cape Cod Times*, Apr. 9, 2009; for additional information on tide mills, see http://www.tidemillinstitute.org/ (accessed Nov. 29, 2009).

7. For example, New York State Department of Environmental Protection, "World's first free-flow tidal-power turbines activated in New York," *Environment DEC*, July 2007, http://www.dec.ny.gov/environmentdec/35380.html (accessed Nov. 29, 2009).

8. Shanahan, G., 2009, "Tidal range technologies," in Brito-Melo, A. and G. Bhuyan (eds.), *2008 Annual Report: International Energy Agency Implementing Agreement on Ocean Energy Systems (IEA-OES)*.

9. Jacobson, M. Z., 2009, "Review of solutions to global warming, air pollution, and energy security," *Energy & Environmental Science* 2: 148–73. Also see the European Marine Energy Centre, http://www.emec.org.uk/ (accessed Nov. 29, 2009). A recent cost estimate is given by Denny, E., 2009, "The economics of tidal energy," *Energy Policy* 37: 1914–24.

10. "Geology Field Notes: Acadia National Park, Maine," http://www.nature.nps .gov/Geology/parks/acad/index.cfm (accessed Nov. 26, 2009).

11. Ferrari, R. and C. Wunsch, 2009, "Ocean circulation kinetic energy: Reservoirs, sources and sinks," *Annual Reviews of Fluid Mechanics* 41: 253–82.

12. See note 9.

13. Falcão, A. F. de O., 2009, "The development of wave energy utilisation," in Brito-Melo, A. and G. Bhuyan (eds.), *2008 Annual Report, International Energy Agency Implementing Agreement on Ocean Energy Systems (IEA-OES)*, pp. 30–37.

14. Ibid.

15. Siegel, S., T. Jeans and T. McLaughlin, 2009, "Deep ocean wave cancellation using a cycloidal turbine (abstract)," 62nd Annual Meeting of the American Physical Society Division of Fluid Dynamics, Minneapolis, MN.

16. Franklin, B., 1786, "A Letter from Dr. Benjamin Franklin, to Mr. Alphonsus le Roy, Member of Several Academies, at Paris. Containing Sundry Maritime Observations," *Transactions of the American Philosophical Society* 2: 294–329. Also see the wonderful book Dolin, E. J., 2007, *Leviathan: The History of Whaling in America* (W. W. Norton, New York). Also see Richardson, P. L., 1980, "Benjamin

Franklin and Timothy Folger's first printed chart of the Gulf Stream," *Science* 207: 643–45.

17. Franklin, B., *The Works of Benjamin Franklin*, vol. 6 (C. Tappan, Louisville, 1844), p. 497.

18. See Franklin in note 16.

19. "Technology White Paper on Ocean Current Energy Potential on the U.S. Outer Continental Shelf, Minerals Management Service, Renewable Energy and Alternate Use Program," U.S. Department of the Interior, http://ocsenergy.anl.gov. The origin of the number provided there—total worldwide power in ocean currents of 5 TW—is not given clearly. But there can be no doubt that the currents carry a lot more water than the world's rivers, and there is enough hydroelectric to be important, so there is surely useful power in the big currents.

20. Gorlov, A. M., 1998, "Helical turbines for the gulf stream: Conceptual approach to design of a large-scale floating power farm," *Marine Technology and SNAME News* 35: 175–82; and Finkl, C. W. and R. Charlier, 2009, "Electrical power generation from ocean currents in the Straits of Florida: Some environmental considerations," *Renewable and Sustainable Energy Reviews* 13: 2597–2604.

21. Nihous, G. C., 2007, "An estimate of Atlantic Ocean thermal energy conversion (OTEC) resources," *Ocean Engineering* 34: 2210–21; d'Arsonval, A., 1881, "Utilisation des forces naturelles. Avenir de l'é lectricité," *Revue Scientifique* 17: 370–72; and Claude, G., 1930, "Power from the tropical seas," *Mechanical Engineering* 52: 1039–44.

20: Power from the Land

1. The full quote refers to the value of wood, both for fuel and as a building material: "It is remarkable what a value is still put upon wood even in this age and in this new country, a value more permanent and universal than that of gold. After all our discoveries and inventions no man will go by a pile of wood. It is as precious to us as it was to our Saxon and Norman ancestors." Thoreau, H. D., 1854, "House-warming," in *Walden*, vol. 2 (Houghton Mifflin, New York, 1897), pp. 388–89.

2. A useful review is given by Kullander, S., 2009, "Energy from biomass," *European Physical Journal Special Topics* 176: 115–25. Also see Dornburg, V., D. van Vuuren, G. van de Ven, et al., 2010, "Bioenergy revisited: Key factors in global potentials of bioenergy," *Energy & Environmental Science* 3: 258–67.

3. You may need to average over a century or longer to see the balance between uptake and release of CO_2 in one forest, because the trees grow slowly and then burn in a great conflagration. Or you can average over many places and shorter times, because trees will be growing in most places and burning in only a few of them at any moment. However you look, a broad enough view would show that

naturally the world's trees and other plants would not be changing the amount of CO_2 in the air. As I discussed in chapter 12, the trees do exchange a lot of CO_2 with the air, taking up a lot more than they release in the spring, and releasing that much more than they take up in the fall. Thus, a given molecule of CO_2 stays in the air only a few years before it cycles through a tree, or swaps places with a molecule in a plant or the soil or the ocean. But these "trades" of CO_2 between air and trees don't affect the average concentration of CO_2 in the air for a year; if we raise that concentration, it will remain high for centuries, millennia, and longer.

4.   Prentice, I. C., G. D. Farquhar, M. J. R. Fasham, M. L. Goulden, M. Heimann, V. J. Jaramillo, H. S. Kheshgi, C. Le Quéré, R. J. Scholes and D. W. R. Wallace, 2001, "The Carbon Cycle and Atmospheric Carbon Dioxide," in Houghton, J. T., Y. Ding, D. J. Griggs, M. Noguer, P. J. van der Linden, X. Dai, K. Maskell and C. A. Johnson (eds.), *Climate Change 2001: The Scientific Basis. Contribution of Working Group I to the Third Assessment Report of the Intergovernmental Panel on Climate Change* (Cambridge University Press, New York), http://www.grida.no/publications/other/ipcc_tar/.

5.   Roughly 50 years is a starting point; Fargione, J., J. Hill, D. Tilman, S. Polasky and P. Hawthorne, "Land clearing and the biofuel carbon debt," *Science* 319: 1235–38; and Field, C. B., J. E. Campbell and D. B. Lobell, 2008, "Biomass energy: The scale of the potential resource," *Trends in Ecology and Evolution* 23: 65–72.

6.   Crutzen, P. J., A. R. Mosier, K. A. Smith and W. Winiwarter, 2008, "N_2O release from agro-biofuel production negates global warming reduction by replacing fossil fuels," *Atmospheric Chemistry and Physics* 8: 389–95. For alcohol made from sugar cane grown on abandoned farmland and replacing fossil fuels, 50–90% of the cooling from reduced CO_2 emissions is estimated to be offset by increased levels of nitrous oxides, and for alcohol made from grain corn, 90–150% of the cooling is offset by increased levels, with anything above 100% indicating net warming from switching to the biofuel.

7.   Some variability exists in estimates, and there is a fair amount of discord about the number. The ones here are from Field et al. in note 5; Hill, J., E. Nelson, D. Tilman, S. Polasky and D. Tiffany, 2006, "Environmental, economic, and energetic costs and benefits of biodiesel and ethanol biofuels," *Proceedings of the National Academy of Sciences of the United States of America* 103: 11,206–10; and International Energy Agency, 2004, *Biofuels for Transport: An International Perspective* (International Energy Agency). Also see Hertel, T. W., A. A. Golub, A. D. Jones, et al., 2010, "Effects of US maize ethanol on global land use and greenhouse gas emissions: Estimating market-mediated responses," *Bioscience* 60: 223–31.

8.   See, for example, the Letters to the Editor in *Science* magazine from June 13, 2008, vol. 320, pp. 1419–22, and July 11, 2008, vol. 321, pp. 199–201. Also see "New studies portray unbalanced perspective on biofuels, DOE committed to environmentally sound biofuels development," Office of Biomass Program; Argonne

National Lab, National Renewable Energy Lab, Oak Ridge National Lab, Pacific Northwest National Lab; USDA, www.colorado.gov/energy/in/uploaded_pdf/ BiofuelsScienceResponse.pdf (accessed Nov. 27, 2009).

9. Field, C. B., J. E. Campbell and D .B. Lobell, 2008, "Biomass energy: The scale of the potential resource," *Trends in Ecology and Evolution* 23: 65–72.

10. National Renewable Energy Laboratory, "Biomass Research," http://www.nrel .gov/biomass/ (accessed July 26, 2010).

11. ISI Web of Science (accessed Nov. 27, 2009).

12. Tyner, W. E., 1981, "Food versus fuel: The moral issue in using corn for ethanol," *Technology Review* 837: 76–77.

13. http://en.wikipedia.org/wiki/Food_vs._fuel (accessed Nov. 27, 2009).

14. Organisation for Economic Co-operation and Development (OECD), 2008, *Biofuel Support Policies: An Economic Assessment.*

15. Saunders, C., W. Kaye-Blake, L. Marshall, S. Greenhalgh and M. D. Pereira, 2009, "Impacts of a United States' biofuel policy on New Zealand's agricultural sector," *Energy Policy* 37: 3448–54. Also see Dornburg et al. in note 2, and Hertel et al. in note 7.

16. Kerr, R. A., 2009, "From burning dung to global warming and back again," *Science* 326: 362–63.

17. Letter printed in *Philadelphia Gazette and Daily Advertiser*, Sept. 27, 1827, and reprinted in *Yellowstone Nature Notes* (Yellowstone Park, WY): 21 (Sept. 1947), 50–56; also see Mattes, M. J., 1949, "Behind the legend of Colter's Hell: The early exploration of Yellowstone National Park," *Mississippi Valley Historical Review* 36: 251–82.

18. See, for example, Hutchinson, R. A., J. A. Westphal and S. W. Kieffer, 1997, "In situ observations of Old Faithful geyser," *Geology* 25: 875–78; and references therein.

19. New Zealand's Rotorua, with its geysers and hot springs, is part of a long zone of volcanoes that includes "Mount Doom" from the *Lord of the Rings* movies (which was, at different times in the movies, either Mount Ngauruhoe or Mount Ruapehu, both of which are in the Taupo Volcanic Zone that runs through Rotorua). These volcanoes on New Zealand's North Island have been, over the last couple of million years, the planet's most frequently erupting silica-rich volcanoes (rhyolitic, hence especially good for geysers), and the largest eruption in the Southern Hemisphere for at least the last few hundred thousand years occurred here (Froggatt, P. C., C. S. Nelson, L. Carter, G. Griggs and K. P. Black, 1986, "An exceptionally large late Quaternary eruption from New Zealand," *Nature* 319: 578–82). The volcanoes of Taupo and Rotorua put out about as much heat, and magma, and have been active for about as long as Yellowstone's silica-rich volcano, in Wyoming and adjacent Montana and Idaho in the United States. However, in that time New Zealand had many small, medium, and large erup-

tions, while Yellowstone saved up and then produced three immense ones, each about 1000 times the size of the famous 1980 eruption of Mt. St. Helens in Washington State. For comparison of Yellowstone and the Taupo Volcanic Zone, see Houghton, B. F., C. J. N. Wilson, M. O. McWilliams, et al., 1995, "Chronology and dynamics of a large silicic magmatic system: Central Taupo Volcanic Zone, New Zealand," *Geology* 23: 13–16; for more on Yellowstone's volcano, see Lowenstern, J. B. and S. Hurwitz, 2008, "Monitoring a supervolcano in repose: Heat and volatile flux at the Yellowstone caldera," *Elements* 4: 35–40.

20. Barrick, K. A., 2007, "Geyser decline and extinction in New Zealand: Energy development impacts and implications for environmental management," *Environmental Management* 39: 783–805; and White, P. A. and T. A. Hunt, "Simple modeling of the effects of exploitation of hot springs, Geyser Valley, Wairakei, New Zealand," *Geothermics* 34: 783–805.

21. Person, M., A. Banerjee, A. Hofstra, D. Sweetkind and Y. L. Gao, 2008, "Hydrologic models of modern and fossil geothermal systems in the Great Basin: Genetic implications for epithermal Au-Ag and Carlin-type gold deposits," *Geosphere* 4: 888–917.

22. Barrick, K. A., 2010, "Protecting the geyser basins of Yellowstone National Park: Toward a new national policy for a vulnerable environmental resource," *Environmental Management* 45: 192–202.

23. Soeder, D. J. and Kappel, W. M., 2009, *Water Resources and Natural Gas Production from the Marcellus Shale*, U.S. Geological Survey Fact Sheet 2009–3032.

24. Although most of the water in geysers and hot springs and geothermal plants percolated down from the surface, a little bit may be coming out of the magma beneath. The magma brings a lot of the other gases too. The circulating waters also dissolve minerals from the surrounding rocks, and concentrate them in veins and pipes above the magma. Gold, silver, copper, and other valuable ores have been formed in this way, but sulfur and other things are also concentrated by the hot flowing waters.

25. See, for example, Rebhan, E., 2009, "Challenges for future energy usage," *European Physical Journal Special Topics* 176: 53–80; and Jacobson, M. Z., 2009, "Review of solutions to global warming, air pollution, and energy security," *Energy & Environmental Science* 2: 148–73.

26. See, for example, "Energy in Iceland: Historical Perspective, Present Status, Future Outlook," 2004, http://www.os.is/page/english/ (accessed Dec. 22, 2009); also see Orkustofnun, the National Energy Authority of Iceland, http://www.os.is/page/energystatistics (accessed Feb. 22, 2010).

27. See Jacobson in note 25.

28. See, for example, Bixley, P. F., A. W. Clotworthy and W. Mannington, 2009, "Evolution of the Wairakei geothermal reservoir during 50 years of production," *Geothermics* 38; 145–54.

29. Massachusetts Institute of Technology, 2006, *The Future of Geothermal Energy: Impact of Enhanced Geothermal Systems (EGS) on the United States in the 21st Century*, assessment by an MIT-led interdisciplinary panel.

30. Healy, J. H., W. W. Rubey, D. T. Griggs and C. B. Raleigh, 1968, "The Denver earthquakes," *Science* 161: 1301–10.

31. I am indebted to Terry Engelder of Penn State for this. To learn more about joints and faults, see, among other sources, Engelder, T., M. P. Fischer and M. R. Gross, 1993, *Geological Aspects of Fracture Mechanics* (Geological Society of America Short Course Notes, Boulder, CO); Engelder, T., B. F. Haith and A. Younes, "Horizontal slip along Alleghanian joints of the Appalachian plateau: Evidence showing that mild penetrative strain does little to change the pristine appearance of early joints," *Tectonophysics* 336: 31–41; and Engelder, T., K. Schulmann and O. Lexa, 2004, "Indentation pits: A product of incipient slip on joints with a mesotopography," in Cosgrove, J. W. and T. Engelder (eds.), "The initiation, propagation and arrest of joints and other fractures," *Geological Society of London Special Publication* 231: 315–24.

32. Kraft, T., P. M. Mai, S. Wiemer, et al., 2009, "Enhanced geothermal systems: Mitigating risk in urban areas," *Eos* 90: 273–74; the quote is from p. 273.

33. Section 5.7, pp. 5–8, of the Massachusetts Institute of Technology report cited in note 29.

21: Put It Where the Sun Doesn't Shine

1. Coleridge, S. T., 1816, *Kubla Khan; or, A Vision in a Dream: A Fragment*. See, for example, Coleridge, S. T., 1899, *Coleridge's Ancient Mariner, Kubla Khan and Christabel* (Macmillan, New York).

2. Burt, Edward, 1876, *Burt's Letters from the North of Scotland* (William Paterson, Edinburgh), pp. 365–366.

3. In turn, these options require public trust. And as someone who has seen how some members of the public can lose confidence in well-founded science, I suspect that it will be easier to promote solutions that are better understood by the public.

4. The nuclear force arises from the strong nuclear force between the still-smaller quarks that comprise the neutrons and protons. A good introductory physics textbook will supply the key information on these topics.

5. Most commonly, small nuclei spit out an electron (beta decay) or a helium nucleus (two protons with two neutrons, called alpha decay), but other types of radioactivity happen.

6. A breeder reactor uses some of its radiation to convert uranium-238 or thorium-232 into material that fissions better, such as making uranium-233 from thorium-232.

7. See, for example, Coogen, L. A. and J. T. Cullen, 2009, "Did natural reactors form as a consequence of the emergence of oxygenic photosynthesis during the Archean?" *GSA Today* 19, no. 10: 4–10; also Gauthier-Lafaye, F., P. Holliger and P.-L. Blanc, 1996, "Natural fission reactors in the Franceville basin, Gabon: A review and results of a 'critical event' in a geologic system," *Geochimica et Cosmochimica Acta* 60: 4831–52.

8. See, for example, Spirakis, C. S., 1996, "The roles of organic matter in the formation of uranium deposits in sedimentary rocks," *Ore Geology Reviews* 11: 53–69. Other uranium deposits started as "placers," local concentrations of dense minerals at the bottom of sediment piles deposited by moving water, much like the gold that was panned by prospectors. On early Earth, with its low oxygen content, uranium-bearing minerals did not break down as easily as they do in our high-oxygen environment, and because those minerals were anomalously dense compared to most surrounding material, they tended to end up together. Such uranium deposits must have remained sufficiently isolated from oxygen-rich groundwaters since formation, or they would have broken down and been carried away. Sufficiently concentrated deposits to have made natural reactors seem to be quite rare, and natural reactors have become harder to produce over time as Earth's uranium supply has slowly decayed radioactively.

9. See note 7.

10. Rebhan, E., 2009, "Challenges for future energy usage," *European Physical Journal Special Topics* 176: 53–80. France obtained 79% of its electricity from nuclear—the highest fraction for any nation. The United States generated more total electricity from nuclear than France, but a smaller fraction of the much-larger U.S. energy supply.

11. Jacobson, M. Z., 2009, "Review of solutions to global warming, air pollution, and energy security," *Energy & Environmental Science* 2: 148–73.

12. Proven reserves are 145 years for coal, 43 years for conventional oil, and 62 years for conventional gas; these exclude oil shales, tar sands, and other unconventional resources; see Rebhan in note 10; also Mueller, A., 2009, "Prospects for transmutation of nuclear waste and associated proton accelerator technology," *European Physical Journal Special Topics* 176: 179–91.

13. MacKay, D. J. C., 2009, "Sustainable Energy—Without the Hot Air," www.withouthotair.com.

14. U.S. Environmental Protection Agency, Office of Air and Radiation, Office of Radiation and Indoor Air, 2007, *Radiation Risks and Realities*, EPA-402-K-07-006.

15. McBride, J. P., R. E. Moore, J. P. Witherspoon and R. E. Blanco, 1978, "Radiological impact of airborne effluents of coal and nuclear-plants," *Science* 202: 1045–50; and Papastefanou, C., 2010, "Escaping radioactivity from coal-fired power plants

(CPPs) due to coal burning and the associated hazards: A review," *Journal of Environmental Radioactivity* 101: 191–200.

16. Zeevaert, T., L. Sweeck and H. Vanmarcke, 2006, "The radiological impact from airborne routine discharges of a modern coal-fired power plant," *Journal of Environmental Radioactivity* 85: 1–22.

17. See Weightman, M., 2007, *Report of the Investigation into the Leak of Dissolver Product Liquor at the Thermal Oxide Reprocessing Plant (THORP), Sellafield, Notified to HSE on 20 April 2005*, www.hse.gov.uk/nuclear/thorpreport.pdf; and MacKay in note 13.

18. *Chernobyl's Legacy: Health, Environmental and Socio-Economic Impacts and Recommendations to the Governments of Belarus, the Russian Federation and Ukraine*, Chernobyl Forum: 2003–2005, second revised version, http://www.iaea.org/Publications/Booklets/Chernobyl/chernobyl.pdf, under the aegis of International Atomic Energy Agency (IAEA), World Health Organization (WHO), United Nations Development Programme (UNDP), Food and Agriculture Organization (FAO), United Nations Environment Programme (UNEP), United Nations Office for the Coordination of Humanitarian Affairs (UN-OCHA), United Nations Scientific Committee on the Effects of Atomic Radiation (UNSCEAR), the World Bank, and the governments of Belarus, the Russian Federation, and Ukraine.

19. Ibid.

20. See MacKay in note 13, pp. 168, 175.

21. See Rebhan in note 10 and Mueller in note 12.

22. See note 13.

23. See "Long Term Safety," at National Cooperative for the Disposal of Radioactive Waste (NAGRA), Switzerland, http://www.nagra.ch/g3.cms/s_page/84410/s_name/ourmandate (accessed July 28, 2010). Also see Crowley, K. D., 1997, "Nuclear waste disposal: The technical challenges," *Physics Today* (June): 32–39.

24. Gauthier-Lafaye, F., P. Holliger and P.-L. Blanc, 1996, "Natural fission reactors in the Franceville basin, Gabon: A review and results of a 'critical event' in a geologic system," *Geochimica et Cosmochimica Acta* 60: 4831–52.

25. Ibid.

26. For concerns about volcanic eruptions through a preferred site for a U.S. repository, see Parsons, T., G. A. Thompson and A. H. Cogbill, 2006, "Earthquake and volcano clustering via stress transfer at Yucca Mountain, Nevada," *Geology* 34: 785–88; or Whitney, J. W. and W. R. Keefer (eds.), 2000, *Geologic and Geophysical Characterization Studies of Yucca Mountain, Nevada, A Potential High-Level Radioactive-Waste Repository*, U.S. Geological Survey Digital Data Series 058. For glacier erosion, we have probably headed off the next ice age or two (Archer, D. and A. Ganopolski, 2005, "A movable trigger: Fossil fuel CO_2 and the onset of the next glaciation," *Geochemistry Geophysics Geosystems* 6: article Q05003), but others

after that are possible if we don't retain the ability to stop them. Glacier erosion can be very fast (Hallet, B., L. Hunter and J. Bogen, 1996, "Rates of erosion and sediment evacuation by glaciers: A review of field data and their implications," *Global and Planetary Change* 12: 213–35), and so might possibly reach even a quite deep repository. I have been part of a team that suggested that worries might not be huge (Alley, R. B., D. E. Lawson, G. J. Larson, E. B. Evenson and G. S. Baker, 2003, "Stabilizing feedbacks in glacier-bed erosion," *Nature* 424: 758–60), but I remain nervous that our understanding of glacial erosion is very sketchy, causing any projections we might make to have very large uncertainties.

27. See note 10.

28. International Energy Agency, 2004, *Renewable Energy: Market and Policy Trends in IEA Countries*, https://www.iea.org/textbase/nppdf/free/2004/renewable1 .pdf (accessed Jan. 24, 2010); and Kerr, T., 2010, *IEA Report for the Clean Energy Ministeria: Global Gaps in Clean Energy RD&D: Update and Recommendations for International Collaboration*, www.iea.org/papers/2010/global_gaps.pdf (accessed July 22, 2010), and http://www.iea.org/stats/rd.asp for underlying tables (accessed July 22, 2010). A spike in funding in 2009 was linked to stimulus funds to escape the economic crisis, with indications that this was a one-time event and not a sustained level of funding, as discussed in this Kerr report. During the era of relatively stable expenditures for energy research from 1987 to 2002, funding for nuclear fission was five times the funding for all renewables combined. Since then, renewables have gained a little relative to nuclear, and accounting has changed a little; by 2008, funding for the International Energy Agency members for all renewables had risen to one-third the level for nuclear fission and fusion combined; in the United States, funding for research on renewables had reached half that of fission and fusion combined, and two-thirds that of fission alone. Still, that means that sun, wind, tides, hydropower, ocean thermal energy, biomass, and other renewables were outspent relative to nuclear fission alone.

29. For example, green malachite and blue azurite are copper carbonates, containing CO_2.

30. For an accessible introduction, see the following group of articles, all published in 2009 in *Science* 325: Rochelle, G. T., "Amine scrubbing for CO_2 capture," 1652–54; Keith, D. W., "Why capture CO_2 from the atmosphere," 1654–55; Orr, F. M., Jr., 2009, "Onshore geologic storage of CO_2," 1656–58; and Schrag, D. P., "Storage of carbon dioxide in offshore sediments," 1658–59. Also see Shafer, G., 2010, "Long-term effectiveness and consequences of carbon dioxide sequestration," *Nature Geoscience*, published online June 27, 2010, doi:10.1038/NGEO896.

31. See, for example, Mater, J. M. and P. B. Kelemen, 2009, "Permanent storage of carbon dioxide in geological reservoirs by mineral carbonation," *Nature Geoscience* 2: 837–41.

32. Rau, G. H., K. G. Nau, W. H. Langer and K. Caldeira, 2007, "Reducing energy-

related CO_2 emissions using accelerated weathering of limestone," *Energy* 32: 1471–77.

33. Lackner, K., 2009, "Capture of carbon dioxide from ambient air," *European Physical Journal Special Topics* 176: 93–106.

34. For example, see Service, R. F., 2009, "Catalyst offers new hope for capturing CO_2 on the cheap," *Science* 327: 257; and Angamuthu, R., P. Byers, M. Lutz, A. L. Spek and E. Bouwman, 2010, "Electrocatalytic CO_2 conversion of oxalate by a copper complex," *Science* 327: 313–15.

35. See note 30.

36. Lackner, K. S. and J. D. Sachs, 2005, "A robust strategy for sustainable energy," *Brookings Papers on Economic Activity* 2: 215–69, http://www.jstor.org/stable/ 3805122 (accessed Nov. 23, 2009).

37. Bickle, M. J., 2009, "Geological carbon storage," *Nature Geoscience* 2: 815–18.

38. See, for example, Sigvaldason, G. E., 1989, "International Conference on Lake Nyos Disaster, Yaounde, Cameroon, 16–20 March 1987—Conclusions and recommendations," *Journal of Volcanology and Geothermal Research* 39: 97–107. We do not expect CO_2 in the free atmosphere to come close to the approximately 10% level where people die; see the *Materials Safety Data Sheet for CO_2*; or National Institute for Occupational Safety and Health (NIOSH), 1976, *Criteria for a Recommended Standard, Occupational Exposure to Carbon Dioxide*; and, NIOSH, 1996, *Documentation for Immediately Dangerous to Life or Health Concentrations (IDLHs) for Carbon Dioxide*, www.cdc.gov.

39. Chiodini, G., D. Granieri, R. Avino, S. Caliro, A. Costa, C. Minopoli and G. Vilardo, 2010, "Non-volcanic CO_2 Earth degassing: Case of Mefite d'Ansanto (southern Apennines), Italy," *Geophysical Research Letters* 37: L11303, doi:10.1029/2010GL042858.

40. Van Noorden, R., 2010, "Buried trouble," *Nature* 463: 871–73.

41. Chisholm, S. W., P. G. Falkowski and J. J. Cullen, 2001, "Discrediting ocean fertilization," *Science* 294: 309–10; and Trick, C. G., B. D. Bill, W. P. Cochlan, et al., 2010, "Iron enrichment stimulates toxic diatom production in high-nitrate, low-chlorophyll areas," *Proceedings of the National Academy of Sciences of the United States of America* 107: 5887–92.

42. Schrag, D., 2009, "Coal as a low-carbon fuel?" *Nature Geoscience* 2: 818–20; Schrag, D. P., 2009, "Storage of carbon dioxide in offshore sediments," *Science* 325: 1658–59; House, K. Z., D. P. Schrag, C. F. Harvey and K. S. Lackner, 2006, "Permanent carbon storage in deep-sea sediments," *Proceedings of the National Academy of Sciences of the United States of America* 103: 12,291–95.

22: Conservation

1.   Theodore Roosevelt, Apr. 15, 1907, "Arbor Day—A Message to the School-Children of the United States."

2.   For a fascinating, authoritative introduction, see Meyer-Rochow, V. B., 2007, "Glowworms: A review of *Arachnocampa* spp. and kin," *Luminescence* 22: 251–65. Also check out Viviani, V. R., J. W. Hastings and T. Wilson, 2002, "Two biolumi-nescent Diptera: The North American *Orfelia fultoni* and the Australian *Arachno-campa flava*. Similar niche, different bioluminescence systems," *Photochemistry and Photobiology* 75: 22–27.

3.   Why? Good question, and it isn't clear which of the many hypotheses is/are correct.

4.   See Lewis, S. M. and C. K. Cratsley, 2008, "Flash signal evolution, mate choice, and predation in fireflies," *Annual Review of Entomology* 53: 293–321.

5.   Eisner, T., M. A. Goetz, D. E. Hill, S. R. Smedley and J. Meinwald, 1997, "Fire-fly 'femme fatales' acquire defensive steroids (lucibufagins) from their firefly prey," *Proceedings of the National Academy of Sciences of the United States of America* 94: 9723–28.

6.   See, for example, Widder, E. A. and S. Johnsen, 2000, "3D spatial point patterns of bioluminescent plankton: A map of the 'minefield,'" *Journal of Plankton Research* 22: 409–20; also Widder, E. A., 2010, "Bioluminescence in the ocean: Origins of biological, chemical, and ecological diversity," *Science* 328: 704–8.

7.   The increase in resting metabolism is 37%; see Woods, W. A., H. Hendrickson, J. Mason and S. M. Lewis, 2007, "Energy and predation costs of firefly court-ship signals," *American Naturalist* 170: 702–8; also see p. 299 in Lewis, S. M. and C. K. Cratsley, 2008, "Flash signal evolution, mate choice, and predation in fire-flies," *Annual Review of Entomology* 53: 293–321.

8.   Azevedo, I. L., M. G. Morgan and F. Morgan, 2009, "The transition to solid-state lighting," *Proceedings of the IEEE* 97: 481–510.

9.   From ibid., in lumens of light per watt, candles are in the 0.05–0.09 range, oil lamps are in the range of 0.2–0.3, mantle gas lantern around 1, incandescent typically 15 but as good as 18, and compact fluorescents are typically four or five times better than incandescents, with their peak approaching 100 lumens per watt; light-emitting diodes (LEDs) are comparable to compact fluorescents, with an expectation of doubling this. Considering everything (how much light is going into the shade rather than the room?), LEDs may be ten times better than incandescents.

10.   Capelli, R., S. Toffanin, G. Generali, et al., 2010, "Organic light-emitting transis-tors with an efficiency that outperforms the equivalent light-emitting diodes," *Nature Materials*, published online May 2, 2010, doi:10.1038/NMAT2751.

11. U.S. Department of Energy, Energy Efficiency and Renewable Energy, "Energy Savers: Benefits of Geothermal Heat Pump Systems," http://www.energysavers .gov/your_home/space_heating_cooling/index.cfm/mytopic=12660 (accessed Jan. 23, 2010).

12. U.S. Environmental Protection Agency, Office of Atmospheric Programs, 2009, *Economic Impacts of S. 1733: The Clean Energy Jobs and American Power Act of 2009*, www.epa.gov/climatechange/economics/pdfs/EPA_S1733_Analysis.pdf (accessed Jan. 23, 2010).

13. On December 29, 2008, the average of gasoline prices in the United States was $1.613 per gallon, and by December 28, 2009, this had risen to $2.607, an increase of $0.994/gallon, which is pretty close to a dollar. U.S. Department of Energy, Energy Information Agency, http://www.eia.doe.gov/oil_gas/petroleum/ data_publications/wrgp/mogas_history.html (accessed Jan. 23, 2010).

14. Dietz, T., G. T. Gardner, J. Gilligan, P. C. Stern and M. P. Vandenberghe, 2010, "Household actions can provide a behavioral wedge to rapidly reduce U.S. carbon emissions," *Proceedings of the National Academy of Sciences of the United States of America* 106: 18,452–56.

23: Game-Changers?

1. Quoted in Arnott, N., 1856, *Elements of Physics* (Blanchard and Lea, Philadelphia), p. 98.

2. Hoffert, M. I., K. Caldeira, G. Benford, et al., 2002, "Advanced technology paths to global climate stability: Energy for a greenhouse planet," *Science* 298: 981–87; and Feichter, J. and T. Leisner, 2009, "Climate engineering: A critical review of approaches to modify the global energy balance," *European Physical Journal Special Topics* 176: 81–92.

3. See, for example, Cicerone, R. J., 2006, "Geoengineering: Encouraging research and overseeing implementation," *Climatic Change* 77: 221–26 (Cicerone was president of the U.S. National Academy of Sciences at the time); Crutzen, P. J., 2006, "Albedo enhancement by stratospheric sulfur injections: A contribution to resolve a policy dilemma?" *Climatic Change* 77: 211–20 (Crutzen was a Nobel laureate in chemistry); Schneider, S. H., 1996, "Geoengineering: Could or should we do it?" *Climatic Change* 33: 291–302; and Hoffert et al. in note 2.

4. ISI Web of Science (accessed Feb. 5, 2010). The subject "geoengineering" had no more than four published papers in any year through 2005, then 10 in 2006, rising to 45 in 2009; the number of indexed citations did not exceed 10/year until 2000, but reached 282 in 2009.

5. Penner, S. S., 1993, "A low-cost no-regrets view of greenhouse-gas emissions and global warming," *Journal of Clean Technology and Environmental Sciences* 3: 255–59.

6. Penner, S. S., A. M. Schneider and E. M. Kennedy, 1984, "Active measures for

reducing the global climatic impacts of escalating CO_2 concentrations," *Acta Astronautica* 11: 345–48; cited twice in the indexed literature.

7. Budyko, M. I., 1982, *The Earth's Climate: Past and Future* (Academic Press, New York); see chapter 28 in National Research Council, 1992, *Policy Implications of Greenhouse Warming: Mitigation, Adaptation, and the Science Base* (National Academies Press, Washington, DC).

8. See note 4.

9. Goodell, J., "Can Dr. Evil save the world?" *Rolling Stone,* Nov. 3, 2006, http://www.rollingstone.com/news/story/12343892/can_dr_evil_save_the_world (accessed Feb. 5, 2010).

10. Mooney, C., "Can a million tons of sulfur dioxide combat climate change?" *Wired*, June 23, 2008, http://www.wired.com/print/science/planetearth/magazine/16-07/ff_geoengineering.

11. For example, Levitt, S. D. and S. J. Dubner, 2009, *SuperFreakonomics: Global Cooling, Patriotic Prostitutes and Why Suicide Bombers Should Buy Life Insurance* (William Morrow, New York), pp. 193–94.

12. See, for example, Pierrehumbert, R., "An open letter to Steve Levitt," *RealClimate*, Oct. 29, 2009, http://www.realclimate.org/index.php/archives/2009/10/an-open-letter-to-steve-levitt/#more-1488.

13. Fuller, T., 1732, *Gnomologia* (Barker, London), p. 237.

14. See, for example, Marley, M., D. J. Hazenbronck and M. T. Walsh, 1992, "The application of in situ air sparging as an innovative soils and groundwater remediation technology," *Ground Water Monitoring Review* 12: 137–45; Martin, L. M., R. J. Sarnelli and M. T. Walsh, 1992, "Pilot-scale evaluation of groundwater air sparging: Site-specific advantages and limitations," in *Proceedings of R&D 92—National Research and Development Conference on the Control of Hazardous Materials, Hazardous Materials Control Research Institute, Greenbelt, MD*; U.S. Environmental Protection Agency, 1992, *A Technology Assessment of Soil Vapor Extraction and Air Sparging* (Office of Research and Development, Washington, DC), EPA/600/R-92/173; and U.S. Environmental Protection Agency, "Underground Storage Tanks: Air Sparging," http://www.epa.gov/oust/cat/airsparg.htm (accessed Feb. 5, 2010).

15. See U.S. Environmental Protection Agency in note 14.

16. See, for example, Feichter, J. and T. Leisner, 2009, "Climate engineering: A critical review of approaches to modify the global energy balance," *European Physical Journal Special Topics* 176: 81–92.

17. Hegerl, G. C. and S. Solomon, 2009, "Risks of climate engineering," *Science* 325: 955–56.

18. Caldeira, K. and L. Wood, 2008, "Global and Arctic climate engineering: Numerical model studies," *Philosophical Transactions of the Royal Society A: Mathematical, Physical and Engineering Sciences* 366: 4039–56.

19. Robock, A., L. Oman and G. L. Stenchikov, "Regional climate responses to

geoengineering with tropical and Arctic SO$_2$ injections," *Journal of Geophysical Research—Atmospheres* 113(D): D16101; and Lunt, D. J., A. Ridgwell, P. J. Valdes and A. Seale, 2008, "'Sunshade World': A fully coupled GCM evaluation of the climatic impacts of geoengineering," *Geophysical Research Letters* 35: L12710.

20. Trenberth, K. E. and A. Dai, 2007, "Effects of Mount Pinatubo volcanic eruption on the hydrological cycle as an analog of geoengineering," *Geophysical Research Letters* 34: L15702.

21. Tilmes, S., R. Muller and R. Salawitch, 2008, "The sensitivity of polar ozone depletion to proposed geoengineering schemes," *Science* 320: 1201–4; Heckendorn, P., D. Weisenstein, S. Fueglistaler, et al., 2009, "The impact of geoengineering aerosols on stratospheric temperature and ozone," *Environmental Research Letters* 4: 045108.

22. Irvine, P. J., D. J. Lunt, E. J. Stone and A. Ridgwell, 2009, "The fate of the Greenland Ice Sheet in a geoengineered, high CO$_2$ world," *Environmental Research Letters* 4: 045109.

23. See Feichter and Leisner in note 16, p. 91.

24. Brovkin, V., V. Petoukhov, M. Claussen, et al., 2009, "Geoengineering climate by stratospheric sulfur injections: Earth system vulnerability to technological failure," *Climatic Change* 92: 243–59; and Robock, A., A. Marquardt, B. Kravitz and G. Stenchikov, 2009, "Benefits, risks, and costs of stratospheric geoengineering," *Geophysical Research Letters* 36: L19703.

25. Nordhaus, W., 2008, *A Question of Balance: Weighing the Options on Global Warming Policies* (Yale University Press, New Haven, CT).

26. Cockcroft, J., 1958, *New Scientist*, Jan. 30, p. 14.

27. Goodstein, D., 2005, "The End of the Age of Oil," http://pr.caltech.edu/periodicals/caltechnews/articles/v38/oil.html (accessed Feb. 5, 2010).

28. Smith, C. L., 2009, "The path to fusion power," *European Physical Journal Special Topics* 176: 167–78.

29. Cook, I., N. Taylor, D. Ward, L. Baker and T. Hender, 2005, "Accelerated development of fusion power," UKAEA FUS 521, available at http://www.fusion.org.uk/, the Culham Centre for Fusion Energy, the United Kingdom's national fusion research laboratory (accessed Aug. 11, 2010).

30. U.S. Energy Research and Development Administration, 1976, *Fusion Power by Magnetic Confinement Program Plan*, ERdA-78/110; Rowberg, R. E., 1999, "Congress and the fusion energy sciences program: A historical analysis," *Journal of Fusion Energy* 18: 29–46; and Smith in note 28.

31. International Energy Agency member countries from 1987 to 2002 devoted 11% of their energy research funding to fusion, almost equal to support for fossil fuels (12%) and well above that for all renewables taken together (just under 8%), although far less than the 40% for fission. By 2008, though, U.S. fusion funding had fallen behind funding for all renewables taken together, although fusion

research funding still exceeded funding for any individual renewable source such as solar energy. International Energy Agency, 2004, *Renewable Energy: Market and Policy Trends in IEA Countries*, https://www.iea.org/textbase/nppdf/free/20016/renewable1.pdf (accessed Jan. 24, 2010); and Kerr, T., 2010, *IEA Report for the Clean Energy Ministeria: Global Gaps in Clean Energy RD&D: Update and Recommendations for International Collaboration*, www.iea.org/papers/2010/global_gaps.pdf (accessed July 22, 2010), and http://www.iea.org/stats/rd.asp for underlying tables (accessed July 22, 2010).

32. See Smith in note 28.

33. Ibid.

34. Moses, E. I., R. N. Boyd, B. A. Remington, C. J. Keane and R. Al-Ayat, 2009, "The National Ignition Facility: Ushering in a new age for high energy density science," *Physics of Plasmas* 16: 041006; also see the National Ignition Facility web site, https://lasers.llnl.gov/ (accessed Feb. 5, 2010).

35. If the unexpected event occurs a few decades in the future, we may have needed to delay the warming until then, so depending on what you're imagining, the slowdown either was or wasn't economically justified.

36. Caption, figure 5-9, in Nordhaus's *Question of Balance*, cited in note 25.

37. The story about salting the fields of Carthage is probably not accurate, but there is historical record of the salting of the conquered Italian city of Palestrina directed by Pope Boniface VIII in 1299; see Warmington, B. H., 1988, "The destruction of Carthage: A retractatio," *Classical Philology* 83: 308–10.

38. It also required the courage of Lincoln and a lot of other people . . .

24: Ten Billion and Smiling

1. Thanks to good friend and world-leading scientist Bob Bindschadler for reminding us of this proverb. See, for example, slide 17 at Information for Financial Aid Professionals of the U.S. Government, www.ifap.ed.gov/qadocs/Tools forSchools/QAProgramPreConf.ppt, (accessed Aug. 11, 2010).

2. For example, see Fleming, J. R., 2004, "Sverre Petterssen and the contentious (and momentous) weather forecasts for D-Day," *Endeavour* 28: 59–63.

3. See, for example, U.S. National Oceanographic and Atmospheric Administration Economics, http://www.economics.noaa.gov/?goal=climate&file=users/business/energy (accessed July 29, 2010); or U.S. National Oceanographic and Atmospheric Administration, 2007, *Research and Networks for Decision Support in the NOAA Sectoral Applications Research Program*; National Research Council, 2008, *Understanding and Responding to Climate Change: Highlights of National Academies Reports* (National Academies Press, Washington, DC); and Hamlet, A. F., D. Huppert and D. P. Lettenmaier, 2002, "Economic value of long-lead stream-

flow forecasts for Columbia River hydropower," *Journal of Water Resources Planning and Management* 128: 91–101.

4. Doyle, A. C., 1904, "The Adventure of the Abbey Grange," reprinted in *The Return of Sherlock Holmes* (A. Wessels, New York, 1907), pp. 319–48; quote is on p. 319.

5. Doyle, A. C., 1912, "The Adventure of the Bruce-Partington Plans," reprinted in *His Last Bow: A Remininscence of Sherlock Holmes* (Review of Reviews, New York, 1917), pp. 130–78; quote is on p. 130.

6. Urbinato, D., 1994, "London's historic 'pea-soupers,'" *U.S. Environmental Protection Agency EPA Journal*, http://www.epa.gov/history/topics/perspect/london.htm (accessed Feb. 7, 2010).

7. Different sources give different dates for the ban on coal. I have quoted from p. 4 of the *Report of the Commissioners Appointed to Inquire into the Several Matters Relating to Coal in the United Kingdom, Report of Committee E, Presented to Both Houses of Parliament by Command of Her Majesty*, 1871. However, the ban was credited to Edward II in 1316 on p. 113 of Pepper, J. H., 1866, *The Playbook of Metals* (George Routledge, London), which also notes subsequent bans on coal in the reign of Elizabeth and again in 1648. The reference in note 6 places the first ban on coal under Edward I, but in 1272, and raises Pepper's penalty (a fine) to death.

For a fictional view of burning for light rather than for heat, see Isaac Asimov's *Nightfall*, from 1941, voted the best science-fiction short story prior to 1965 by the Science Fiction Writers of America.

8. See note 6.

9. McConnell, J. R., R. Edwards, G. L. Kok, et al., 2007, "20th-century industrial black carbon emissions altered Arctic climate forcing," *Science* 317: 1381–84. Also see Alley, R. B., 2007, "Geochemistry: 'C' ing Arctic climate with black ice (Perspective)," *Science* 317: 1333–34. Note that other regions today have high and often rising pollution from burning, including fossil fuels but also biomass. See, for example, Carmichael, G. R., B. Adhikary, S. Kulkarni, et al., 2009, "Asian aerosols: Current and year 2030 distributions and implications to human health and regional climate change," *Environmental Science and Technology* 43: 5811–17.

10. Hunter, P., 2008, "A toxic brew we cannot live without," *European Molecular Biology Organization (EMBO) Reports* 9: 15–18; translated from the German.

11. Arrhenius, S., 1896, "On the influence of carbonic acid in the air upon the temperature of the ground," *Philosophical Magazine* 41: 237–76. Also see Chamberlin, T., 1899, "An attempt to frame a working hypothesis of the cause of glacial periods on an atmospheric basis," *Journal of Geology* 7: 545–84, 667–85, 751–87 (the paper was sufficiently long that it was split into three blocks).

12. See Weart, S., "The Discovery of Global Warming" (American Institute of Physics), http://www.aip.org/history/climate/co2.htm.

13. National Academy of Sciences, 1979, *Carbon Dioxide and Climate: A Scientific Assessment* (National Academies Press, Washington, DC); this is the "Charney" report, chaired by Jule G. Charney. As given in note 15 of chapter 5, the authors, and the people they thank for advice and unpublished results, are a remarkable "who's who" of a large part of climate science.

14. See chapter 10; also, Alley, R. B., 2009, "The Biggest Control Knob: Carbon Dioxide in Earth's Climate History" (presented at A23A, Bjerknes Lecture, AGU, San Francisco, December), http://www.agu.org/meetings/fm09/lectures/ videos.php. Also Young, S. A., M. R. Saltzman, K. A. Foland, J. S. Linder and L. R. Kump, 2009, "A major drop in seawater 87Sr/86Sr during the Middle Ordovician (Darriwilian): Links to volcanism and climate?" *Geology* 37: 951–54.

15. My thanks to Mr. Hoeffler, Mr. Mansfield, Ms. Pyron, Mr. Routson, and Mrs. Saylor, who are not to blame for any shortcomings of this book. Also, thanks to a lot of great teachers who taught me other subjects!

16. See, for example, the U.S. Congress, *Congressional Record*, 2003, Proceedings and debates of the 108th Congress, 1st session, vol. 149, no. 152, Washington, DC, where Senator James Inhofe refers to "this hoax called global warming." Also see the 2009 quote from Senator Inhofe in which he quotes from his 2003 speech: "With all of the hysteria, all of the fear, all of the phony science, could it be that manmade global warming is the greatest hoax ever perpetrated on the American people? It sure sounds like it." http://epw.senate.gov/public/index .cfm?FuseAction=Minority.Speeches&ContentRecord_id=08d7b2d2-802a-23ad-41d8-332a1ef4715e (accessed Feb. 8, 2010).

17. For those who see politics in physics, I am happy to confirm the public record, that I vote consistently, register as a Republican, and I am a loyal member of St. Paul's United Methodist Church in State College, Pennsylvania, affiliated with a "mainline" Christian denomination. Who I vote for in any given election, and what I say in my prayers, are between me and my God, and you can take the matter up with him if you're interested.

18. UN Department of Economic and Social Affairs, Population Division, 2009, *World Population Prospects: The 2008 Division*, ST/ESA/SER.A/287/ES, see http:// esa.un.org/unpd/wpp2008/peps_documents.htm (accessed Aug. 11, 2010).

19. Penn State University, where I am proud to work, is home to the largest student-run philanthropy in the world; see http://www.thon.org/ (accessed Feb. 15, 2010).

20. The UN identified fossil-fuel use, and agriculture and food consumption, as being especially unsustainable aspects of our lives, but a lot of the agriculture is linked to fossil fuels; Hertwich, E., E. van der Voet, S. Suh, et al., UNEP (United Nations Environment Programme), 2010, *Assessing the Environmental Impacts of Consumption and Production: Priority Products and Materials*, a report of the Working Group on the Environmental Impacts of Products and Materials to the International Panel for Sustainable Resource Management.

ACKNOWLEDGMENTS

A project like this requires a team; I've been fortunate to work with the best, and I thank them. This book is related to a larger project by Geoffrey Haines-Stiles and Erna Akuginow, funded by the U.S. National Science Foundation (under grant 0917564), and I am grateful to the NSF for funding that project, and to Geoff and Erna for including me, and for their good humor, bright ideas, guidance, and much more. The filming trips were amazing, and I thank the incomparable Art Howard and Andy Quinn for their insights and friendship, and Anna Belle Peevey for her discoveries. I also thank a lot of good people who helped us along the way, including Hugh Barnard and Davie Robinson on the Franz Josef Glacier in New Zealand, Geoff Hargreaves, Eric Cravens and colleagues at the National Ice Core Lab, and Jim McMillan and Joe Verrengia at the National Renewable Energy Lab. Aaron and Dave Putnam showed me great things in the Wind Rivers.

President Ralph Cicerone, and archivists Dan Barbiero and Janice Goldblum, of the National Academy of Sciences helped us get that important story straight, and in other ways; thanks also to Elaine Engst and the reference staff at the Cornell University libraries. I leaned on many colleagues who provided background information, filled my knowledge gaps, and reviewed the results, including Klaus Keller, Jim Kasting, Kate Freeman, Lee Kump, Tim Bralower, Michael Mann, Jeff Severinghaus, David Battisti, Tony Broccoli, Bill Nordhaus, and Rich Mullins; Terry Engelder, Peter Wilf, Andrew Mackintosh, and many other colleagues also shared their knowledge freely. Early versions of the book benefited from the suggestions of Bill Ross, Joan Fitzpatrick, and Patrick Applegate, all of whom also encouraged me when I needed it. Despite this great help, the blame for the unavoidable errors, imprecisions, or ambiguities rests entirely with me.

I am fortunate to conduct research in a wonderful community, working on CReSIS, WAIS, WAISCORES, and other projects, usually funded by the National

Science Foundation and with additional funding from NASA, and I am grateful for this vote of confidence and support. I especially thank Kurt Cuffey for his ideas on book writing, and Wally Broecker, George Denton, and the Comer family for believing in me and taking me along for a great ride.

The ice researchers at Penn State are without parallel, and I thank these special colleagues and students, most notably Sridhar Anandakrishnan, the heart of "our" group, and Byron Parizek. The administration of Penn State is enlightened, supportive, and forward-looking, and I thank them for their support, especially Tim Bralower and Bill Easterling. Thanks also to the good folk at WPSU. I started on this project while a visitor at the Woods Hole Oceanographic Institution, and I am indebted to many friends there, and especially Bill and Ruth Curry and Delia Oppo.

Thanks to editor Jack Repcheck, who has guided me and put up with me through two books now with rare good humor, as well as steering us to the best toilet story ever.

I have benefited most from my family, including elder daughter Janet, a teacher, nature-enthusiast, and occasional park ranger, younger daughter Karen, who helped with the filming and bungy-jumping and much more, and the love of my life Cindy, who made figures, guided travel, offered wisdom on writing, and put up with an amazing amount of disruption from this project and everything else in my life. Thanks!

FIGURE CREDITS AND NOTES

Figure 1.1: Photograph by Cindy Alley.

Figure 1.2: Photograph by Richard Alley.

Figure 1.3: Concentrations are in pg/g. Data from McConnell, J. R., and R. Edwards, 2008, "Greenland Ice Core ACT2 Toxic Heavy Metals Data," IGBP PAGES / World Data Center for Paleoclimatology Data Contribution Series # 2008-079, NOAA/NCDC Paleoclimatology Program, Boulder, CO. Graph prepared by Richard Alley.

Figure 2.1: Photograph by Geoffrey Haines-Stiles.

Figure 2.2: Photograph by Richard Alley.

Figure 3.1: Photograph by Cindy Alley; Janet and Richard Alley for scale.

Figure 3.2: Photograph by Richard Alley.

Figure 3.3: Historical photograph from the U.S. National Archives, Photo ARC 2129418 / Local ID 419979, Historic Photographs, compiled ca. 1880–ca. 1970. U.S. Department of Agriculture Forest Service Region 9.

Figure 3.4: Photograph by Geoffrey Haines-Stiles.

Figure 3.5: Frontispiece of William Scoresby's 1920 account of Arctic whaling. From the National Oceanic and Atmospheric Administration Photo Library, http://www.photolib.noaa.gov/bigs/libr0415.jpg. Scanned from Scoresby, W., Jr., 1920, *An Account of the Arctic Regions, with a History and Description of the Northern Whale-Fishery*, vol. 2 (Archibald Constable and Co., Edinburgh).

Figure 3.6: From *Vanity Fair*, April 1861, p. 186. Courtesy of the New Bedford Whaling Museum, New Bedford, Massachusetts. This figure was displayed at the New Bedford Whaling National Historical Park in their whaling timeline at http://www.nps.gov/nebe/historyculture/upload/timeline.pdf.

Figure 3.7: Photograph by Richard Alley, in State College, Pennsylvania.

Figure 4.1: Photograph by Richard Alley.

Figure 4.2: Jean Lafitte National Historical Park and Preserve, Louisiana. Photograph by Richard Alley.

Figure 4.3: Diagram by Cindy Alley.

Figure 4.4: Photograph by Richard Alley.

Figure 5.1 March 9, 1862. Calvert Lithographing Company, Detroit, Michigan, published by the McCormick Harvesting Machine Company, March 2, 1891. The whole lithograph includes two insets illustrating McCormick farm equipment, and can be seen at the Library of Congress Prints and Photographs Division, Washington, DC, and online at http://www.loc.gov/pictures/item/2003666423.

Figure 5.2: Photograph by Geoffrey Haines-Stiles; compass courtesy of Lannan Galleries, Boston.

Figure 5.3: The patent refers both to "bouying" on the graphics page and to "buoying" on the text page. Graphic enhancement by Crazybridge Studios for the *Earth—The Operators' Manual* TV series, courtesy of Geoffrey Haines-Stiles.

Figure 5.4: Courtesy of the U.S. National Academy of Sciences.

Figure 6.1: These data, from NASA's Goddard Space Flight Center, were collected over the central part of the tropical Pacific Ocean by the Incorporated Research Institutions for Seismology (IRIS). The smooth curve shows what would be expected for the Earth's average temperature if we had no air and emitted as a "black body," while the thinner, jagged curve shows actual satellite data. At some wavelengths the data are above the expectation, mostly because the greenhouse effect has raised the Earth's temperature to emit more (in addition, the tropical position of this observation means that it should be a little warmer than for the global average). At wavelengths where CO_2 and ozone are active, and at other wavelengths not shown where these and other gases (and especially water vapor) are active, less energy leaves than in the no-air case. The "valley" in the plot from CO_2 doesn't go down to zero even in the center and certainly not near the valley edges, so adding more CO_2 will block more outgoing radiation and warm the planet more. Satellite observations of this type have measured the expected effects of changing atmospheric composition. Redrafted by Richard Alley from http://climate.gsfc.nasa.gov/static/cahalan/Radiation/. The original data were plotted against wavenumber (the inverse of wavelength), which is why the horizontal scale may look odd.

Figure 6.2: Still image from a U.S. Department of Defense film, 342-USAF-18700, courtesy of the U.S. National Archives and Records Administration, College Park, MD.

Figure 6.3: Still image from a U.S. Department of Defense film, 342-USAF-49391, courtesy of the U.S. National Archives and Records Administration, College Park, MD.

Figure 7.1: Photograph by Geoffrey Haines-Stiles.

Figure 7.2: Shown courtesy of the University of Alaska Museum of the North,

Fairbanks, AK. Photograph courtesy of POLAR-PALOOZA, Passport to Knowledge.

Figure 7.3: Photograph by Richard Alley.

Figure 8.1: Temperature is local, and includes some uncertainty related to the transfer from isotopic ratio. CO_2 is a global signal. The midsummer sunshine is calculated from features of Earth's orbit. Full-year sunshine also matters, but as explained in the text, neither the midsummer sunshine nor the full-year sunshine explains the timing and size of the temperature changes if the CO_2 curve is ignored. The figure was modified by Richard Alley from the figure in Alley, R. B., 2004, "Abrupt climate changes: Oceans, ice and us," *Oceanography* 17, 194–206. In turn, all the data are held at the World Data Center for Paleoclimatology and National Oceanic and Atmospheric Administration Paleoclimatology Program, Boulder, CO, http://www.ncdc.noaa.gov/paleo/. See Petit, J. R., J. Jouzel, D. Raynaud, N. I. Barkov, J. M. Barnola, I. Basile, M. Bender, J. Chappellaz, J. Davis, G. Delaygue, M. Delmotte, V. M. Kotlyakov, M. Legrand, V. Lipenkov, C. Lorius, L. Pipin, C. Ritz, E. Saltzman, and M. Stievenard, 1999, "Climate and atmospheric history of the past 420,000 years from the Vostok ice core, Antarctica," *Nature* 399, 429–36. The estimated South Pole sunshine is from Berger, A., and M. F. Loutre, 1991, "Insolation values for the climate of the last 10 million years," *Quaternary Science Reviews* 10, 297–317, also at the World Data Center.

Figure 8.2 For projections of future CO_2, see Solomon, S., D. Qin, M. Manning, Z. Chen, M. Marquis, K. B. Averyt, M. Tignor, and H. L. Miller (eds.), *Climate Change 2007: The Physical Science Basis. Contribution of Working Group I to the Fourth Assessment Report of the Intergovernmental Panel on Climate Change* (Cambridge University Press, Cambridge, UK, and New York).

Figure 9.1: Photograph by Richard Alley. In especially cold regions, trees may be more sensitive to temperature and provide paleothermometers. See, for example, Jansen, E., J. Overpeck, K. R. Briffa, J.-C. Duplessy, F. Joos, V. Masson-Delmotte, D. Olago, B. Otto-Bliesner, W. R. Peltier, S. Rahmstorf, R. Ramesh, D. Raynaud, D. Rind, O. Solomina, R. Villalba, and D. Zhang, 2007, "Palaeoclimate," in Solomon, S., D. Qin, M. Manning, Z. Chen, M. Marquis, K. B. Averyt, M. Tignor, and H. L. Miller (eds.), *Climate Change 2007: The Physical Science Basis. Contribution of Working Group I to the Fourth Assessment Report of the Intergovernmental Panel on Climate Change* (Cambridge University Press, Cambridge, UK, and New York).

Figure 9.2: Photograph by Richard Alley.

Figure 9.3: In the right-hand picture, the "zebra striping" between two crystals is caused by the boundary between the crystals being almost horizontal rather than almost vertical as viewed here. Photograph by Joan Fitzpatrick. Sample from the WAIS Divide core.

Figure 10.1: The ice actually flowed up this bedrock bump, moving from lower right to upper left in the picture. Photograph by Richard Alley.

Figure 10.2: On the left is 400 million years ago; today is on the right. The results from several different ways of estimating past CO_2 levels in the air generally agree, as shown along the bottom. The bars hanging down from above show how extensive ice was on the planet. Since this figure was made, new evidence points to a drop in CO_2 at the time of the oldest of the glacier advances shown here. The figure is modified from Alley, R. B., J. Brigham-Grette, G. H. Miller, L. Polyak, and J. W. C. White, 2009, *Past Climate Variability and Change in the Arctic and at High Latitude*, a report by the U.S. Climate Change Program and Subcommittee on Global Change Research U.S. Geological Survey, Reston, VA, 461 pp.; product Lead, U.S. Geological Survey, Joan J. Fitzpatrick.

Figure 10.3: Morrison Fossil Area National Natural Landmark, Morrison, Colorado. Photograph by Richard Alley.

Figure 10.4: Photograph by Richard Alley.

Figure 10.5: The fossil is from the early-middle Jurassic, so it is slightly less than 200 million years old. This sample was collected by the author for a National Science Foundation–sponsored expedition led by David Elliot, Ohio State University. Photograph by Richard Alley.

Figure 11.1: "Tilt" is also called "obliquity." The figure was modified from Alley, R. B., J. Brigham-Grette, G. H. Miller, L. Polyak, and J. W. C. White, 2009, *Past Climate Variability and Change in the Arctic and at High Latitude*, a report by the U.S. Climate Change Program and Subcommittee on Global Change Research, U.S. Geological Survey, Reston, VA, 461 pp.; product lead, U.S. Geological Survey, Joan J. Fitzpatrick.

Figure 11.2 Ice also grew during the ice age in places getting more total sunshine. The effects of lower CO_2 during the ice age explain most of this behavior, which has not been explained successfully when ignoring the CO_2. Photograph by Cindy Alley.

Figure 11.3: Image by Art Howard, from the *Earth—The Operators' Manual* TV series.

Figure 12.1: Where data are available, they show that glaciers in Greenland advanced from medieval times to reach their Little Ice Age size before retreating recently. Photograph by Richard Alley.

Figure 12.2: Photograph by Richard Alley.

Figure 12.3: Figure modified from United States Global Change Research Program, 2009, *Global Climate Change Impacts in the United States*, http://www.globalchange .gov/publications/reports/scientific-assessments/us-impacts, and http://www .globalchange.gov/resources/gallery?func=viewcategory&catid=2. Also published in Karl, Thomas R., Jerry M. Melillo, and Thomas C. Peterson (eds.), 2009, *Global Climate Change Impacts in the United States* (Cambridge University Press, New York), p. 16, which was modified from IPCC, 2007, "Summary for Policymakers," in Solomon, S., D. Qin, M. Manning, Z. Chen, M. Marquis, K. B. Averyt, M. Tignor and H. L. Miller (eds.), *Climate Change 2007: The Physical Science Basis.*

Contribution of Working Group I to the Fourth Assessment Report of the Intergovernmental Panel on Climate Change (Cambridge University Press, Cambridge, UK, and New York), figure SPM-2.

Figure 13.1: The data are from NASA's Goddard Institute for Space Studies; plot made by Richard Alley.

Figure 13.2 Regression lines are shown from 1880–1900, 1900–1920, and so on, and then each year beginning with 1980–2000, 1981–2001, and on to 1989–2009. Those since 1980 give high confidence that warming is occurring, and all of them give higher confidence than any of the older regression lines shown. The data are from NASA's Goddard Institute for Space Studies; plot made by Richard Alley.

Figure 13.3: Plot made by Richard Alley, by tracing the projections and uncertainty bars of Figure SPM.5 of the IPCC Fourth Assessment Report, Working Group I, onto figure 13.2 in this chapter. Following closely from the caption in SPM.5, the curves are "multi-model global averages of surface warming (relative to 1980–1999) for the scenarios A2, A1B and B1," and the dashed line is "the experiment where concentrations were held constant at year 2000 values." The "bars at right indicate the best estimate" (gray band) "and the likely range assessed for the . . . SRES marker scenarios. The assessment of the best estimate and likely ranges in the . . . bars includes the AOGCMs in the left part of the figure, as well as results from a hierarchy of independent models and observational constraints." The IPCC showed these projections with their estimate of the twentieth-century temperature trend, which is similar but not identical to the data shown here because the IPCC considered other temperature reconstructions, but I have used one for simplicity. See Solomon, S., D. Qin, M. Manning, Z. Chen, M. Marquis, K. B. Averyt, M. Tignor, and H. L. Miller (eds.), *Climate Change 2007: The Physical Science Basis. Contribution of Working Group I to the Fourth Assessment Report of the Intergovernmental Panel on Climate Change* (Cambridge University Press, Cambridge, UK, and New York).

Figure 13.4: If you really want to know which curve is which, you can see a color version of this as Figure 3.35 in CCSP (Alley, R. B., J. Brigham-Grette, G. H. Miller, L. Polyak, and J. W. C. White [coordinating lead authors]), 2009, *Past Climate Variability and Change in the Arctic and at High Latitude*, a report by the U.S. Climate Change Program and Subcommittee on Global Change Research, U.S. Geological Survey, Reston, VA, 461 pp; product lead, U.S. Geological Survey, Joan J. Fitzpatrick. In turn, this is modified from Mann, M. E., A. Z. Zhang, M. K. Hughes, et al., 2008, "Proxy-based reconstructions of hemispheric and global surface temperature variations over the past two millennia," *Proceedings of the National Academy of Sciences of the United States of America* 105: 13,252–57.

Figure 14.1: Photograph by Richard Alley.

Figure 14.2: Nauset Light Beach, Cape Cod National Seashore, Massachusetts. Photograph by Richard Alley.

Figure 14.3: Photograph by Richard Alley, at University Park, Pennsylvania.

Figure 14.4: Photograph by Richard Alley, at University Park, Pennsylvania.

Figure 15.1: Rear Admiral David Titley: ". . . we are operating in nature's casino; I intend to count the cards." Freeman, B., 2009, "Navy task force assesses changing climate," Special to American Forces Press Service, Washington, DC, July 31, 2009, http://www.defense.gov/news/newsarticle.aspx?id=55327. Image by Art Howard, from the *Earth—The Operators' Manual* TV series.

Figure 15.2: These are on the Daugaard-Jensen Glacier in East Greenland. Photograph by Richard Alley.

Figure 15.3: Upstream B camp, Antarctica. Photograph by Richard Alley.

Figure 16.1: Photograph by Richard Alley.

Figure 16.2: Courtesy of Mary King's Close, Edinburgh. Image by Art Howard, from the *Earth—The Operators' Manual* TV series.

Figure 16.3: Courtesy of Mary King's Close, Edinburgh. Image by Art Howard, from the *Earth—The Operators' Manual* TV series.

Figure 17.1: The "sails," or arms of the windmill, were covered with sailcloth to catch the wind's power and to grind grain. This windmill was originally built in Plymouth and later moved to Eastham. Photograph by Rochelle Pearson.

Figure 17.2: Photograph from the U.S. National Archives, from the U.S. Department of the Interior, Bureau of Indian Affairs, Phoenix Area Office, Forestry and Grazing Division, File Unit 339.3, Maps for the Forestry and Grazing Annual Report, Ft. Apache, 1939, ARC ID 295182.

Figure 17.3: Photograph by Anna Belle Peevey.

Figure 18.1: Abengoa Solar's Solúcar Platform. Parts of the project are in operation, with more under construction. Photograph by Erna Akuginow.

Figure 18.2: Photograph by Richard Alley.

Figure 18.3: Photograph by Richard Alley.

Figure 18.4: Photograph from the U.S. National Archives, Photo ARC 532725; photographed in 1933 by Lewis Hine for the Tennessee Valley Authority.

Figure 19.1: Photograph by Karen Alley.

Figure 19.2: Photograph by Karen Alley.

Figure 19.3: Franklin, Benjamin, 1786, "A Letter from Dr. Benjamin Franklin, to Mr. Alphonsus le Roy, Member of Several Academies at Paris. Containing Sundry Maritime Observations. At Sea, on board the London Packet, Capt. Truxton, August 1785," *Transactions of the American Philosophical Society, held at Philadelphia, for Promoting Useful Knowledge* 2: 294–329. Includes chart and diagrams. Held by the National Oceanic and Atmospheric Administration Central Library, Silver Spring, MD, http://oceanexplorer.noaa.gov/library/readings/gulf/gulf.html.

Figure 19.4: From the National Oceanic and Atmospheric Administration Photo Library, http://www.photolib.noaa.gov/bigs/figb0187b.jpg. Scanned from

Goode, G. B., 1887, *The Fisheries and Fishery Industries of the United States, Section V, History and Methods of the Fisheries* (Government Printing Office, Washington, DC).

Figure 20.1: Photograph by Richard Alley.

Figure 20.2: Photograph by Richard Alley.

Figure 20.3: Kawerau Geothermal Plant, New Zealand. Photograph by Karen Alley.

Figure 21.1: Photograph by Richard Alley.

Figure 21.2: Nile River Cave, Paparoa National Park, New Zealand. Photograph by Karen Alley.

Figure 21.3: Photograph by Geoffrey Haines-Stiles, at Shidongkou Second Power Plant, Huaneng Power International, Inc.

Figure 22.1: Photograph by Cindy Alley.

Figure 22.2: Image by Art Howard, from the *Earth—The Operators' Manual* TV series.

Figure 23.1: U.S. Geological Survey photograph taken on May 18, 1980, by Austin Post, http://vulcan.wr.usgs.gov/Volcanoes/MSH/SlideSet/ljt_slideset.html.

Figure 23.2: Photograph by Richard Alley.

Figure 23.3 At this depth, the pressure has squeezed the bubbles to make solid clathrates, but the air is still in the core. Good records of eruptions are available from ice cores, and the material in the ice cores can often be tied to the volcano that did the deed. Photograph by Matthew Spencer.

Figure 24.1: Image by Andy Quinn, from the *Earth—The Operators' Manual* TV series, Shanghai, China.

Figure 24.2: Photograph by Richard Alley.

Figure 24.3: Photograph by Richard Alley.

Figure 24.4: Photograph by Richard Alley.

INDEX

Page numbers in *italics* refer to illustrations.